INFORMATION AND COMMUNICATION THEORY

IEEE Press
445 Hoes Lane
Piscataway, NJ 08854

INFORMATION AND COMMUNICATION THEORY

STEFAN HÖST
Lund University, Sweden

IEEE SERIES ON DIGITAL & MOBILE COMMUNICATION

IEEE PRESS

WILEY

Published by John Wiley & Sons, Inc., Hoboken, New Jersey.
Published simultaneously in Canada.

Library of Congress Cataloging-in-Publication Data is available.

ISBN 978-1-119-43378-1

Printed in the United States of America.

V10008794_031219

CONTENTS

PREFACE

Information theory started as a topic in 1948 when Claude E. Shannon published the paper "A mathematical theory of communication." As the name reveals, it is a theory about what communication and information are in a mathematical sense. Shannon built a theory first to quantify and measure the information of a source. Then, by viewing communication as reproduction of information, the theory of communication came into view. So, without information, or choices, there cannot be any communication. But these two parts go hand in hand. The information measure is based on the amount of data needed for reconstruction, and communication is based on the amount of data needed to be transmitted to transfer a certain amount of information. In a mathematical sense, choices mean probabilities, and the theory is based on probability theory.

Humans have always strived to simplify the process of communication and spreading knowledge. In the 1450s when Gutenberg managed to get his printing press in operation, it simplified the spreading of the written words. Since then, we have seen many different technologies that have spread the information—from the telegraph, via the telephone and television, to the Internet as we know it today. What will come next we can only speculate, but all these existing and forthcoming technologies must comply with the theories that Shannon stated. Therefore, information theory as a topic is the base for everyone working with communication systems, both past, present, and forthcoming.

This book is intended to be used in a first course in information theory for communication engineering students, typically in higher undergraduate or lower graduate level. Since the theory is based on mathematics and probability theory, a certain level of maturity in these subjects is expected. This means that the students should have a couple of mathematics courses in their trunk, such as basic calculus and probability theory. It is also recommended, but not required, that they have some understanding of digital communication as a concept. This together will give a solid ground for understanding the ideas of information theory. The first requirement of a level of maturity in mathematics comes from information theory that sets up a mathematical model of information in very general terms, in the sense that it is valid for all kinds of communication. On some occasions, the theory can be seen as pretty abstract by students the first time they engage with it. Then, to have some understanding of communication on a physical layer beforehand might help the understanding.

The work with this book, as well as the content, has grown while lecturing the course on information and communication. By the time it started, I used one of the most recommended books in this field. But it lacked the engineering part of the course, such as the intuitive understanding and a description of what it means in reality. The theory sets up bounds for how much a source can be compressed and how much data can be transmitted over a noisy channel. This is the pure information theory. The understanding of what this means when designing a communication system is one of the most important results from the theory.

Information theory has grown since its origin in 1948, particularly in the subjects of compressing a source and getting a high data rate on the communication link. Therefore, I wanted to include the basics of data compression and error-correcting codes in the course that I lectured. Since I did not find a book on these topics, I started writing about data compression and the Lempel–Ziv algorithms, as a complementary material to the students. Since not all students have the up-to-date knowledge of probability theory, I also wrote a couple of pages about this. With this at hand, and the slides I used during the lectures, I started working on a somewhat wider scope with the material. After some time, while working with the material as lecture notes, I realized that I had almost all the material I needed for a course. Then I decided to use this as the main course literature, and the text continued to develop during the years. I feel now that the text is mature enough to take the next step and worth publishing to a broader audience.

The work with this book has taken about 10 years, and over the years I have had a lot of help and discussions to learn and understand the subject. Especially, I would like to thank all the students who have passed the course and patiently endured my attempts to find intuitive explanations for the theory. It has not always been the best at the first attempt. It is interesting that even after working with the material for long time, I still get new inputs and questions from the students that challenge my understanding and help with the description. Second, I would like to thank all my colleagues and co-workers those who have enlightened me during discussions and explanations over the years, to name a few: Rolf Johannesson, Viktor Zyablov, Michael Lentmaier, John B Andersson, Fredrik Rusek, Per-Erik Eriksson, Miguel Berg, Boris Dortchy, and Per Ola Börjesson. A special thank goes to John B. Andersson, without his encouragement this book would have never been completed. I am also grateful to reviewers for their many relevant comments on the manuscript. I would also like to thank the teaching assistants associated with the course over the years, who have detected and corrected many mistakes: Adnan Prlja, Eduardo Medeiros, Yezi Huang, and Umar Farooq.

Finally, I want to thank my wife and children, Camilla, Karla, and Filip, for all their support and understanding throughout the years of working on the manuscript.

STEFAN HÖST
Lund, Sweden

CHAPTER 1

INTRODUCTION

At some point in the scientific history, new revolutionary ideas serve as starting points of new topics. Information theory started in 1948 when Claude Shannon's paper "A mathematical theory of information" was published in the *Bell System Technical Journal* [1]. The objectives in the article in many ways were pioneering, but the first thoughts in this direction started more than 20 years earlier.

In 1924 and 1928, Nyquist [2, 3] showed that a signal with a bandwidth W Hz and a durability of T seconds could not contain more than $2WT$ distinguishable pulses. This was later reformulated into what is today known as the sampling theorem. This is an important concept of all systems converting between discrete and analog representations of signals and is widely used in signal processing and communication theory. In 1924, Hartley was first to propose a measure of information in his paper "Transmission of information" [4]. In this study, he realized that the limiting factor in communication is the noise. Without noise there would not be any problem. And then, 20 years later, Claude Shannon further developed the concepts and concluded that a measure of information must be based on probability theory.

The idea is that without choices there is no information. If a variable can have only one outcome, it does not give any information to an observer. If there are multiple outcomes, their presence is determined by their probabilities, and the information measure is therefore derived from the distribution on which the symbol is generated. That means the developed information measures are purely probabilistic functions. At first glance, it might be surprising that the information measure developed, and on which communication systems still rely, does not have anything to do with the content of a sequence. The measure can instead be interpreted as the amount of information required to determine, or reproduce, the sequence. In that perspective, a completely random sequence also contains a lot of information, even though it does not have a meaning to us. The strength in this view is that the theory is valid for all types of information sources and thus all types of communications.

Information theory deals with two closely related terms: information and communication. As stated above, the information measure will depend on the amount of information needed to reproduce the source sequence. Then, if the aim is to transport this sequence or message from a source to a destination, it is only this amount of information that is needed to be transported. In his paper, Shannon expressed that the "fundamental problem of communication is that of reproducing at one point exactly

Information and Communication Theory, First Edition. Stefan Höst.

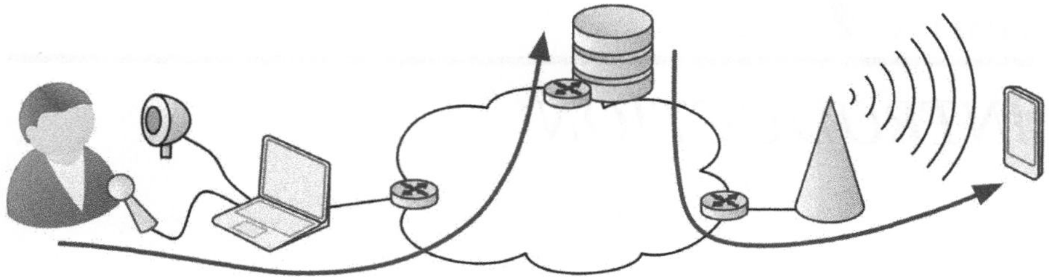

Figure 1.1 A communication system.

or approximately a message selected at another point." In this rather general statement, a *point* can be referred to as *place* and *time*, meaning that the message selected at on place and one time should be reproduced at another place and another time.

In Figure 1.1, an example of communication is shown. It shows someone recording a video with a computer and then uploading it to a server, e.g., Youtube. Then, at later time someone else downloads the video and displays it on a screen. In this case, the stated *another point* refers to the time and place where the video is displayed. The video that is uploaded is typically compressed using a video compression format, e.g., H.264 or MPEG. Videos are typically compressed quite hard, using a lossy compression. When the video is downloaded and decompressed, it differs quite a lot from the raw video recorded at the source. But for the human eye and perception ability, it should *look* approximately the same. If the message transmitted was a text or a computer program, the same principle can be used, the text is saved on a server and downloaded at a later time. But in case of a program, the reconstruction needs to be an exact copy of the uploaded file. This depicts the difference between the statement *exactly* or *approximately* in the cited case above.

In the world of information theory and communication theory, Figure 1.1, and most other communication situations, is better represented by Figure 1.2. There the source is the recording of the (raw) video that gives the sequence X. To save disk space as well as uploading and downloading time, the source sequence X should be compressed through a source encoder. The aim of the source encoder is to extract the pure information, and in the case of lossy compression a representation with less

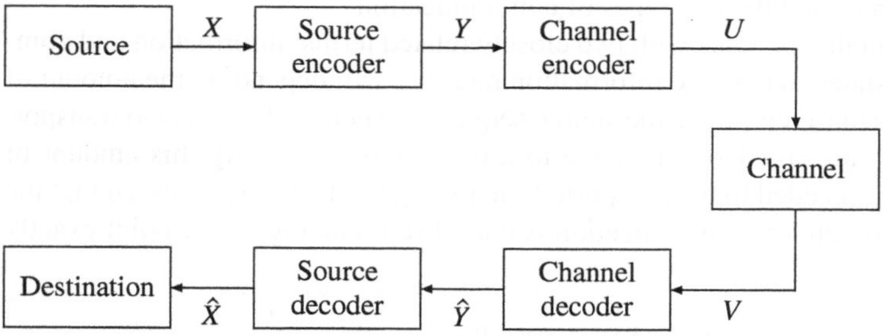

Figure 1.2 Shannon's model for a communication system.

information gives approximately the same experience as the raw video. Since pure information is very sensitive to errors in transmission and storage, it is protected by a channel code. The purpose of this code is to introduce redundancy such that the channel decoder can correct errors. In the case of the video uploading to a data center, there are several occasions of encoding and decoding. First when the file is uploaded, there is encoding and decoding in the transmission protocols. At the data center, there are often error-correcting codes, protecting from hard drive failure. At downloading, there is again the transmission protocol using error-correcting codes. Finally at the destination, the compressed sequence is decompressed and sent to the screen.

The aim of this text is to give an introduction to the information theory and communication theory. It should give an understanding of the information measures and the bounds given by them, e.g., bounds on source coding and channel coding. It also describes source coding and channel coding on the basis of some well-known algorithms. Chapter 2 gives a short review of probability theory, where most of the basic theory needed in the text is presented. The intention is to present a quick and handy lookup of the theory regarding probabilities. Even though there are attempts to explain intuitively the theory, most of the mathematical proofs are omitted. For a more thorough treatment of the subject, the reader should refer to the literature, e.g., [5, 6]. The chapter is complemented by an appendix where some of the most common distributions are shown, both discrete and continuous.

In Chapter 3, the information measures used in the theory are defined. It also presents their relations, and how they can be interpreted. In Chapter 4, the measures are used to set up bounds on source coding. For the case of vectors with identical and independently distributed (i.i.d.) variables, a simple, yet optimal, code construction due to Huffman is given. Even though Huffman's code construction gives optimal codes for the i.i.d. case, in reality sources for compression are seldom independent variables and the distribution is seldom known. In Chapter 5, adaptive compression methods are described. These include the widely used the Lempel–Ziv algorithm, that is used in, for example, zip, png, and other well-known standards.

In Chapter 6, the concept of asymptotic equipartition property is presented, which is a consequence of the law of large numbers. It will give the possibility to show two of the most important theorems on information theory, i.e., the source-coding theorem and the channel-coding theorem. In this chapter, the channel capacity will also be introduced as the possible amount of information carried in a transmission over a noisy channel. In Chapter 7, the concepts of error-correcting codes are discussed, introducing both block codes and convolutional codes.

In Chapter 8, the information measures defined in Chapter 3 for discrete variables are extended to the continuous case. The relation between the interpretations for the measures in the discrete and continuous cases is discussed. Depending on the logical level of the channel model, it can be discrete or continuous. The discrete channels were treated in Chapter 6, whereas in Chapter 9 the continuous counterpart is discussed, with a special focus on the case for Gaussian noise. To reach the channel capacity for this case, it turns out that the transmitted variable should be Gaussian distributed, whereas in reality it is often both discrete and uniformly distributed. This case is treated specially in Chapter 10, where a closer relationship is presented with practically used modulation schemes like pulse amplitude modulation and quadrature amplitude modulation.

In Chapter 11, the concept of distortion is introduced. In reality, it is often acceptable with a certain amount of distortion, e.g., in image and video coding the human eye will tolerate distortions and in a communication scenario there might be tolerance for few errors in the transmission. This is incorporated in the previous theory by using the rate distortion functions. The chapter is concluded with a description of lossless compression and an introduction to transform decoding, used in, e.g., jpeg.

Over the years there has been several books written in the subject of information theory, e.g., [11, 13, 67, 71, 80, 81, 84, 86, 92]. The aim of this book is to present the theory for both discrete and continuous variables. It also aims at applying the theory towards communication theory, which is especially seen in Chapter 10 where discrete input Gaussian channels are considered.

CHAPTER 2

PROBABILITY THEORY

ONE OF THE MOST important insights when setting up a measure for information is that the observed quantity must have multiple choices to contain information [4]. These multiple choices are best described by means of probability theory. This chapter will recapitulate the parts of probability theory that are needed throughout the rest of the text. It is not, in any way, a complete course, for that purpose the reader may refer to standard textbooks, e.g., [5, 6].

2.1 PROBABILITIES

In short, a probability is a measure of how likely an event may occur. It is represented by a number between zero and one, where zero means it will not happen and one that it is certain to happen. The sum of the probabilities for all possible events is one, since it is certain that one of them will happen.

It was the Russian mathematician Kolmogorov who in 1933 proposed the axioms for the theory as it is known today. The *sample space* Ω is the set of all possible outcomes of the random experiment. For a discrete source, the sample space is denoted as

$$\Omega = \{\omega_1, \omega_2, \dots, \omega_n\}$$

where ω_i are the outcomes.

An *event* is a subset of the sample space, $\mathcal{A} \subseteq \Omega$, and the event is said to occur if the outcome from the random experiment is found in the event. Examples of specific events are

- The certain event Ω (the full subset) and
- The impossible event $\emptyset$ (the empty subset of Ω).

Each of the outcomes in the sample space, $\omega_1, \omega_2, \dots, \omega_n$, are also called elementary events or, sometimes, atomic events. To each event, there is assigned a

Information and Communication Theory, First Edition. Stefan Höst.

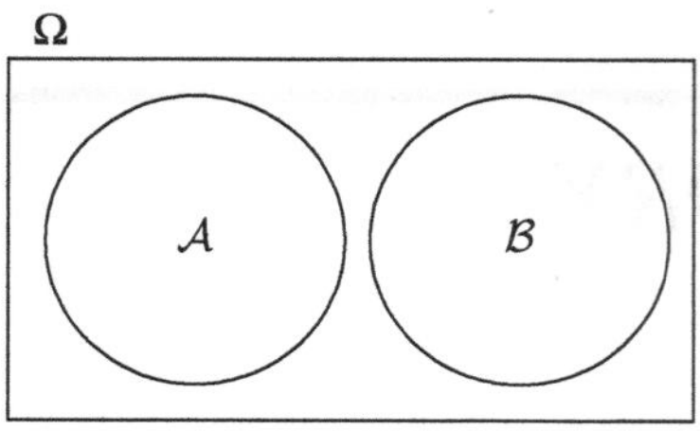

Figure 2.1 Two disjoint sets $\mathcal{A}$ and $\mathcal{B}$, where $\mathcal{A} \cap \mathcal{B} = \emptyset$.

probability measure P, where $0 \leq P \leq 1$, which is a measure of how probable the event is to happen. The probabilities of the certain event and the impossible event are

$$P(\Omega) = 1$$
$$P(\emptyset) = 0$$

If $\mathcal{A}$ and $\mathcal{B}$ are two disjoint events (see Figure 2.1), then the probability for the union is

$$P(\mathcal{A} \cup \mathcal{B}) = P(\mathcal{A}) + P(\mathcal{B}), \text{ where } \mathcal{A} \cap \mathcal{B} = \emptyset \tag{2.1}$$

This implies that

$$P(\mathcal{A}) = \sum_{\omega_i \in \mathcal{A}} P(\omega_i) \tag{2.2}$$

If the sets $\mathcal{A}$ and $\mathcal{B}$ are not disjoint, i.e., $\mathcal{A} \cap \mathcal{B} \neq \emptyset$ (see Figure 2.2), the probability for the union is

$$P(\mathcal{A} \cup \mathcal{B}) = P(\mathcal{A}) + P(\mathcal{B}) - P(\mathcal{A} \cap \mathcal{B}) \tag{2.3}$$

where the probability for the intersection $\mathcal{A} \cap \mathcal{B}$ is represented in both $P(\mathcal{A})$ and $P(\mathcal{B})$ and has to be subtracted.

An important concept in probability theory is how two events are related. This is described by the *conditional probability* and is related to the probability of event $\mathcal{A}$, when it is known that event $\mathcal{B}$ has occurred. Then, the atomic events of interest are those in both $\mathcal{A}$ and $\mathcal{B}$, i.e., $\mathcal{A} \cap \mathcal{B}$. Since the certain event now is $\mathcal{B}$, which should give probability 1, the conditional probability should be normalized with $P(\mathcal{B})$. The definition becomes

$$P(\mathcal{A}|\mathcal{B}) = \frac{P(\mathcal{A} \cap \mathcal{B})}{P(\mathcal{B})} \tag{2.4}$$

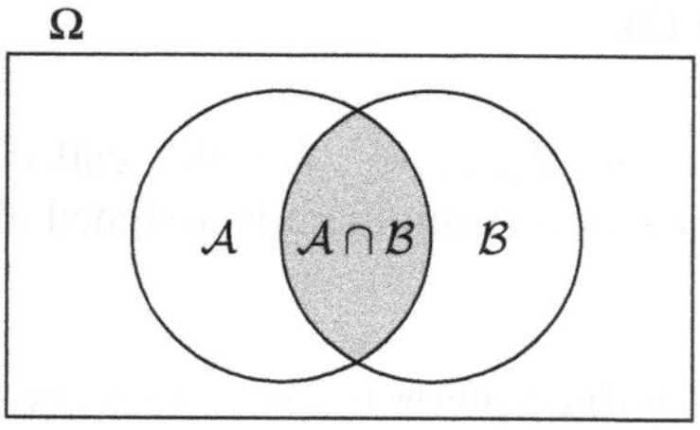

Figure 2.2 Two sets $\mathcal{A}$ and $\mathcal{B}$, where $\mathcal{A} \cap \mathcal{B} \neq \emptyset$.

If the two events $\mathcal{A}$ and $\mathcal{B}$ are independent, the probability for $\mathcal{A}$ should not change if it is conditioned on $\mathcal{B}$ or not. Hence,

$$P(\mathcal{A}|\mathcal{B}) = P(\mathcal{A}) \tag{2.5}$$

Applying this to the definition of conditional probabilities (2.4), it is concluded that $\mathcal{A}$ and $\mathcal{B}$ are independent if and only if the probability of their intersection is equal to the product of the individual probabilities:

$$P(\mathcal{A} \cap \mathcal{B}) = P(\mathcal{A}) \cdot P(\mathcal{B}) \tag{2.6}$$

2.2 RANDOM VARIABLE

A *random variable*, or a stochastic variable, X is a function from the sample space onto a specified set, most often the real numbers,

$$X : \Omega \rightarrow \mathcal{X} \tag{2.7}$$

where $\mathcal{X} = \{x_1, x_2, \dots, x_k\}$, $k \leq n$, is the set of possible values. To simplify notations, the event $\{\omega | X(\omega) = x_i\}$ is denoted by $\{X = x_i\}$.

To describe the probability distribution of the random variable, for the discrete case the *probability function* is used, $p_X(x) = P(X = x)$, and for the continuous case the *density function* $f_X(x)$ is used. The *distribution function* for the variable is defined as

$$F_X(x) = P(X \leq x) \tag{2.8}$$

which can be derived as[1]

$$F_X(x) = \sum_{k \leq x} p_X(k), \qquad \text{discrete case} \tag{2.9}$$

$$F_X(x) = \int_{-\infty}^{x} f_X(\nu) d\nu, \qquad \text{continuous case} \tag{2.10}$$

On many occasions later in the text, the index $(\cdot)_X$ will be omitted, if the intended variable is definite from the content. In Appendix A, the probability or density functions of some common probability distributions are listed together with some of their important properties.

To consider how two or more random variables are jointly distributed, view a vector of random variables. This vector will represent a multidimensional random variable and can be treated the same way. So, in the two-dimensional case, consider (X, Y), where X and Y are random variables with possible outcomes $\mathcal{X} = \{x_1, x_2, \dots, x_M\}$ and $\mathcal{Y} = \{y_1, y_2, \dots, y_N\}$, respectively. Then the set of joint outcomes is

$$\mathcal{X} \times \mathcal{Y} = \Big\{(x_1, y_1), (x_1, y_2), \dots, (x_M, y_N)\Big\} \tag{2.11}$$

[1] Often in this chapter, the result is given for both discrete and continuous random variables, as done here.

Normally, for the discrete case, the *joint probability function* $p_{X,Y}(x,y)$ means the event $P(X = x$ and $Y = y)$. In the continuous case, the *joint density function* is denoted by $f_{X,Y}(x,y)$.

The *marginal distribution* is derived from the joint distribution as

$$p_X(x) = \sum_y p_{X,Y}(x,y), \qquad \text{discrete case} \tag{2.12}$$

$$f_X(x) = \int_{\mathbb{R}} f_{X,Y}(x,y)dy, \qquad \text{continuous case} \tag{2.13}$$

for the discrete and continuous case, respectively. Analogous to (2.4), the *conditional probability distribution* is defined as

$$p_{X|Y}(x|y) = \frac{p_{X,Y}(x,y)}{p_Y(y)}, \qquad \text{discrete case} \tag{2.14}$$

$$f_{X|Y}(x|y) = \frac{f_{X,Y}(x,y)}{f_Y(y)}, \qquad \text{continuous case} \tag{2.15}$$

This gives a probability distribution for the variable X when the outcome for the variable Y is known. Repeating this iteratively concludes the *chain rule* for the probability of an n-dimensional random variable as

$$\begin{aligned} P(X_1 \ldots X_{n-1}X_n) &= P(X_n|X_1 \ldots X_{n-1})P(X_1 \ldots X_{n-1}) \\ &= P(X_1)P(X_2|X_1) \cdots P(X_n|X_1 \ldots X_{n-1}) \\ &= \prod_{i=1}^{n} P(X_i|X_1 \ldots X_{i-1}) \end{aligned} \tag{2.16}$$

Combining the above results, two random variables X and Y are *statistically independent* if and only if

$$p_{X,Y}(x,y) = p_X(x)p_Y(y), \qquad \text{discrete case} \tag{2.17}$$

$$f_{X,Y}(x,y) = f_X(x)f_Y(y), \qquad \text{continuous case} \tag{2.18}$$

for all x and y.

If X and Y are two random variables and $Z = X + Y$ their sum, then the distribution of Z is given by the convolution

$$p_Z(z) = (p_X * p_Y)(z) = \sum_k p_X(k-z)p_Y(k), \qquad \text{discrete case} \tag{2.19}$$

$$f_Z(z) = (f_X * f_Y)(z) = \int_{\mathbb{R}} f_X(t-z)f_Y(t)dt, \qquad \text{continuous case} \tag{2.20}$$

for the discrete and continuous case, respectively.

Example 2.1 Consider two exponentially distributed random variables, $X \sim \text{Exp}(\lambda)$ and $Y \sim \text{Exp}(\phi)$. Their density functions are

$$f_X(x) = \lambda e^{-\lambda x} \tag{2.21}$$
$$f_Y(y) = \phi e^{-\phi y} \tag{2.22}$$

Let $Z = X + Y$, then the density function of Z is

$$\begin{aligned} f_Z(z) &= \int_{\mathbb{R}} f_X(z-t) f_Y(t) dt \\ &= \int_0^z \lambda e^{-\lambda(z-t)} \phi e^{-\phi t} dt \\ &= \lambda\phi e^{-\lambda z} \int_0^z e^{(\lambda-\phi)t} dt \\ &= \lambda\phi e^{-\lambda z} \frac{e^{(\lambda-\phi)z} - 1}{\lambda - \phi} = \frac{\lambda\phi}{\lambda - \phi}\left(e^{-\phi z} - e^{-\lambda z}\right) \end{aligned} \tag{2.23}$$

If X and Y are equally distributed, $X \sim \text{Exp}(\lambda)$ and $Y \sim \text{Exp}(\lambda)$, the density function of $Z = X + Y$ can be derived from (2.23) by using the limit value, given by l'Hospital's rule,

$$\lim_{\alpha \to 0} \frac{e^{\alpha z} - 1}{\alpha} = \lim_{\alpha \to 0} z e^{\alpha z} = z \tag{2.24}$$

as

$$f_Z(z) = \lambda^2 z e^{-\lambda z} \tag{2.25}$$

which is the density function for an Erlang distributed variable, indicating that $Z \sim \text{Er}(2, \lambda)$.

2.3 EXPECTATION AND VARIANCE

Consider an example where a random variable X is the outcome of a roll with a fair die, i.e. the probabilities for each outcome are

$$p_X(x) = \frac{1}{6}, \quad x = 1, 2, \dots, 6 \tag{2.26}$$

A counterfeit version of the die can be obtained by inserting a small weight close to one side. Let Y be the a random variable, describing the outcome for this case. Simplifying the model, assume that the number one will never occur and number six will occur twice as often as before. That is, $p_Y(1) = 0$, $p_Y(i) = \frac{1}{6}$, $i = 2, 3, 4, 5$, and $p_Y(6) = \frac{1}{3}$. The arithmetic mean of the outcomes for both cases becomes

$$\overline{X} = \overline{Y} = \frac{1}{6} \sum_x x = 3.5 \tag{2.27}$$

However, intuitively the average value of several consecutive rolls should be larger for the counterfeit die than the fair. This is not reflected by the arithmetic mean above. Therefore, introducing the *expected value*, which is a weighted mean where the probabilities are used as weights,

$$E[X] = \sum_{x} x p_X(x) \tag{2.28}$$

In the case with the fair die, the expected value is the same as the arithmetic mean, but for the manipulated die it becomes

$$E[Y] = 0 \cdot 1 + \tfrac{1}{6} \cdot 2 + \tfrac{1}{6} \cdot 3 + \tfrac{1}{6} \cdot 4 + \tfrac{1}{6} \cdot 5 + \tfrac{1}{3} \cdot 6 \approx 4.33 \tag{2.29}$$

Later, in Section 2.4, the law of large numbers is introduced, stating that the arithmetic mean of a series of outcomes from a random variables will approach the expected value as the length grows. In this sense, the expected value is a most natural definition of the mean for the outcome of a random variable.

A slightly more general definition of the expected value is as follows:

Definition 2.1 Let $g(X)$ be a real-valued function of a random variable X, then the *expected value* of $g(X)$ is

$$E\big[g(X)\big] = \sum_{x \in \mathcal{X}} g(x) p_X(x), \qquad \text{discrete case} \tag{2.30}$$

$$E\big[g(X)\big] = \int_{\mathbb{R}} g(v) f_X(v) dv, \qquad \text{continuous case} \tag{2.31}$$

□

Returning to the true and counterfeit die, the function $g(x) = x^2$ can be used. This will lead to the so-called second-order moment for the variable X. In the case for the true die

$$E[X^2] = \sum_{i=1}^{6} \frac{1}{6} i^2 \approx 15.2 \tag{2.32}$$

and for the counterfeit

$$E[Y^2] = \sum_{i=2}^{5} \frac{1}{6} i^2 + \frac{1}{3} 6^2 = 21 \tag{2.33}$$

Also other functions can be useful, for example, the Laplace transform of the density function can be expressed as

$$\mathcal{L}\big(f_X(x)\big) = E\big[e^{-sX}\big] = \int_{\mathbb{R}} f_X(x) e^{-sx} dx \tag{2.34}$$

In (2.20), it is stated that the density function for $Z = X + Y$ is the convolution of the density functions for X and Y. Often it is easier to perform a convolution as a multiplication in the transform plane,

$$E\big[e^{-sZ}\big] = E\big[e^{-sX}\big] E\big[e^{-sY}\big] \tag{2.35}$$

For the discrete case, the $\mathcal{Z}$-transform can be used in a similar way as

$$E\left[z^{-Z}\right] = E\left[z^{-X}\right]E\left[z^{-Y}\right] \tag{2.36}$$

As an alternative to the derivations in Example 2.1, in the next example the Laplace transform is used to derive the density function of the sum.

Example 2.2 Let $X \sim \text{Exp}(\lambda)$ and $Y \sim \text{Exp}(\phi)$. The Laplace transforms for the density functions can be derived as

$$E\left[e^{-sX}\right] = \lambda \int_0^\infty e^{-(s+\lambda)x} dx = \frac{\lambda}{s+\lambda} \tag{2.37}$$

$$E\left[e^{-sY}\right] = \phi \int_0^\infty e^{-(s+\phi)x} dx = \frac{\phi}{s+\phi} \tag{2.38}$$

Thus, the transform of $f_Z(z)$ of the sum $Z = X + Y$, is

$$E\left[e^{-sZ}\right] = \frac{\lambda}{s+\lambda} \cdot \frac{\phi}{s+\phi} = \frac{\lambda\phi}{\lambda-\phi}\left(\frac{1}{s+\phi} - \frac{1}{s+\lambda}\right) \tag{2.39}$$

where the inverse transform is given by

$$f_Z(z) = \frac{\lambda\phi}{\lambda-\phi}\left(e^{-\phi z} - e^{-\lambda z}\right) \tag{2.40}$$

If the variables are equally distributed, $X \sim \text{Exp}(\lambda)$ and $Y \sim \text{Exp}(\lambda)$, the transform of $f_Z(z)$ becomes

$$E\left[e^{-sZ}\right] = \left(\frac{\lambda}{s+\lambda}\right)^2 \tag{2.41}$$

which gives the inverse transform[2]

$$f_Z(z) = \lambda^2 z e^{-\lambda z}$$

The expected value for the multidimensional variable is derived similar to the one-dimensional case, using the joint distribution. In the case of a two-dimensional vector (X, Y), it becomes

$$E\left[g(X,Y)\right] = \sum_{x,y} g(x,y) p_{X,Y}(x,y), \qquad \text{discrete case} \tag{2.42}$$

$$E\left[g(X,Y)\right] = \int_{\mathbb{R}^2} g(\nu,\mu) f_X(\nu,\mu) d\nu d\mu, \qquad \text{continuous case} \tag{2.43}$$

[2] From a Laplace transform pair lookup table, e.g., [7],

$$e^{-t\alpha}u(t) \longleftrightarrow \frac{1}{s+\alpha}$$

$$te^{-t\alpha}u(t) \longleftrightarrow \frac{1}{(s+\alpha)^2},$$

where $u(t)$ is the unit step function.

From the definition of the expected value, it is easy to verify that the expected value is a linear mapping, i.e.

$$\begin{aligned} E[aX+bY] &= \sum_{x,y}(ax+by)p_{XY}(x,y) \\ &= \sum_{x,y} axp_{XY}(x,y) + \sum_{x,y} byp_{XY}(x,y) \\ &= a\sum_x x \sum_y p_{XY}(x,y) + b\sum_y y \sum_x p_{XY}(x,y) \\ &= a\sum_x xp_X(x) + b\sum_y yp_Y(y) \\ &= aE[X] + bE[Y] \end{aligned} \tag{2.44}$$

In Theorem 2.1, a more general version of this result is stated.

For the case when X and Y are independent, the expectation of their product equals the product of their expectations,

$$\begin{aligned} E[XY] &= \sum_{x,y} xyp_{X,Y}(x,y) \\ &= \sum_{x,y} xyp_X(x)p_Y(y) \\ &= \sum_x xp_X(x)\sum_y xp_Y(y) = E[X]E[Y] \end{aligned} \tag{2.45}$$

While the expected value is a measure of the weighted mean for the outcome of a random variable, there is also a need for a measure showing how the outcome varies. This is the *variance*, and it is defined as the expected value of the squared distance to the mean.

Definition 2.2 Let X be a random variable with an expected value $E[X]$, then the *variance* of X is

$$V[X] = E\big[(X-E[X])^2\big] \tag{2.46}$$

□

The variance is often derived from

$$\begin{aligned} V[X] &= E\big[(X-E[X])^2\big] \\ &= E\big[X^2 - 2XE[X] + E[X]^2\big] \\ &= E[X^2] - 2E[X]E[X] + E[X]^2 = E[X^2] - E[X]^2 \end{aligned} \tag{2.47}$$

where $E[X^2]$ is the second-order moment of X. In many descriptions, the expected value is denoted by m, but here the notation $E[X]$ is presented as a case of clarity. Still, it should be regarded as a constant in the derivations.

The variance is a measure of the squared deviation to the mean value. To get a measure in the same scale as the mean, the *standard deviation* is defined as the square root of the variance:

$$\sigma_X = \sqrt{V[X]}. \tag{2.48}$$

For the true and counterfeit die described earlier in this section, the variances become

$$V[X] = E[X^2] - E[X]^2 \approx 2.9 \tag{2.49}$$
$$V[Y] = E[Y^2] - E[Y]^2 \approx 2.2 \tag{2.50}$$

and the standard deviations

$$\sigma_X = \sqrt{V[X]} \approx 1.7 \tag{2.51}$$
$$\sigma_Y = \sqrt{V[Y]} \approx 1.5 \tag{2.52}$$

To understand more about the variance, first one of its relatives, the covariance function, is considered. This can be seen as a measure of the dependencies between two random variables.

Definition 2.3 Let X and Y be two random variables with expected values $E[X]$ and $E[Y]$. The *covariance* between the variables is

$$\mathrm{Cov}(X, Y) = E\big[(X - E[X])(Y - E[Y])\big] \tag{2.53}$$

□

Similar to (2.47), the covariance can be derived as

$$\begin{aligned}
\mathrm{Cov}(X, Y) &= E\big[(X - E[X])(Y - E[Y])\big] \\
&= E\big[XY - XE[Y] - E[X]Y + E[X]E[Y]\big] \\
&= E[XY] - E[X]E[Y]
\end{aligned} \tag{2.54}$$

Hence, for independent X and Y, the covariance is zero, $\mathrm{Cov}(X, Y) = 0$.

The variance for a linear combination $aX + bY$ can now be derived as

$$\begin{aligned}
V[aX + bY] &= E\big[(aX + bY)^2\big] - E[aX + bY]^2 \\
&= E\big[a^2X^2 + b^2Y^2 + 2abXY\big] - \big(aE[X] + bE[Y]\big)^2 \\
&= a^2E[X^2] + b^2E[Y^2] + 2abE[XY] \\
&\quad - a^2E[X]^2 - b^2E[Y]^2 - 2abE[X]E[Y] \\
&= a^2V[X] + b^2V[Y] + 2ab\mathrm{Cov}(X, Y)
\end{aligned} \tag{2.55}$$

For independent X and Y, the covariance disappears, and the result is

$$V[aX + bY] = a^2V[X] + b^2V[Y] \tag{2.56}$$

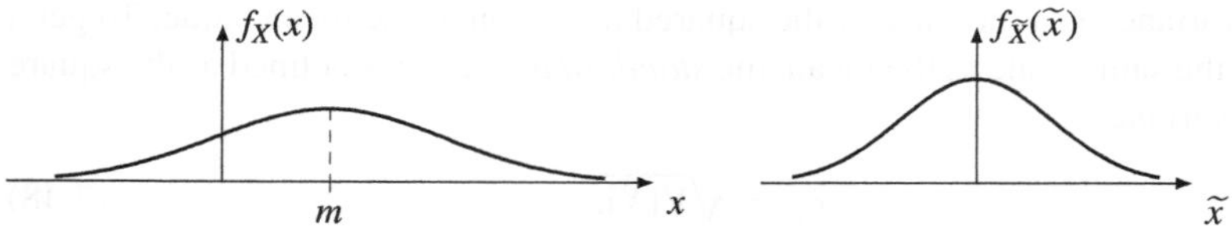

Figure 2.3 Normalization of $f_X(x)$ by $\widetilde{X} = (X - m)/\sigma$ to achieve zero mean and unit variance.

Example 2.3 Sometimes it is suitable to consider a normalized random variable with zero mean and unit variance. This can be obtained from the function

$$\widetilde{X} = \frac{X - m}{\sigma} \tag{2.57}$$

where $E[X] = m$ and $V[X] = \sigma^2$. Then the expectation and variance become

$$E[\widetilde{X}] = \frac{1}{\sigma}\big(E[X] - m\big) = 0 \tag{2.58}$$

$$V[\widetilde{X}] = \frac{1}{\sigma^2}E[(X - m)^2] = 1 \tag{2.59}$$

The procedure is illustrated in Figure 2.3.

The results in (2.44), (2.55), and (2.56) can be generalized using similar derivations as explained above to hold for N variables, which are stated in the next theorem.

Theorem 2.1 Given a set of N random variables X_n and scalar constants α_n, $n = 1, 2, \dots, N$, the sum

$$Y = \sum_{n=1}^{N} \alpha_n X_n \tag{2.60}$$

is a new random variable with the expected value

$$E[Y] = \sum_{n=1}^{N} \alpha_n E[X_n] \tag{2.61}$$

and variance

$$V[Y] = \sum_{n=1}^{N} \alpha_n^2 V[X_n] + 2\sum_{m<n} \alpha_n \alpha_m \mathrm{Cov}(X_n, X_m) \tag{2.62}$$

If the random variables X_n are all independent, then

$$V[Y] = \sum_{n=1}^{N} \alpha_n^2 V[X_n] \tag{2.63}$$

□

In the next two examples, the geometric distribution and the Gaussian distributions will be studied. Both are important in mathematics as well as engineering disciplines.

Example 2.4 [Geometric distribution] The *geometric distribution* is a discrete distribution that can be explained from a series of coin flips. Assume that the probability for head and tail is given by $P(\text{head}) = \alpha$ and $P(\text{tail}) = 1 - \alpha$. Let K be a random variable describing the number of flips until a tail shows up. The probability for this number to be k is

$$p_K(k) = \alpha^{k-1}(1-\alpha), \quad k = 1, 2, \ldots \tag{2.64}$$

and the distribution function is

$$\begin{aligned} F_K(k) = P(K \leq k) &= \sum_{n=1}^{k} p_K(n) \\ &= \sum_{n=1}^{k} \alpha^{n-1}(1-\alpha) \\ &= (1-\alpha)\frac{1-\alpha^k}{1-\alpha} = 1 - \alpha^k \end{aligned} \tag{2.65}$$

To see that this is indeed a probability distribution, let $k \to \infty$ to get $F_K(k) \to 1$. Alternatively, the same result is obtained by summing up all probabilities,

$$\sum_{n=1}^{\infty} \alpha^{n-1}(1-\alpha) = (1-\alpha)\frac{1}{1-\alpha} = 1. \tag{2.66}$$

The expected value and the second-order moment can be found as

$$\begin{aligned} E[K] &= \sum_{k=1}^{\infty} k\alpha^{k-1}(1-\alpha) = \frac{1-\alpha}{\alpha}\sum_{k=1}^{\infty} k\alpha^k \\ &= \frac{1-\alpha}{\alpha} \cdot \frac{\alpha}{(1-\alpha)^2} = \frac{1}{1-\alpha} \end{aligned} \tag{2.67}$$

$$\begin{aligned} E[K^2] &= \sum_{k=1}^{\infty} k^2\alpha^{k-1}(1-\alpha) = \frac{1-\alpha}{\alpha}\sum_{k=1}^{\infty} k^2\alpha^k \\ &= \frac{1-\alpha}{\alpha} \cdot \frac{\alpha(1+\alpha)}{(1-\alpha)^3} = \frac{1+\alpha}{(1-\alpha)^2} \end{aligned} \tag{2.68}$$

Hence, the variance is

$$V[K] = E[K^2] - E[K]^2 = \frac{1+\alpha}{(1-\alpha)^2} - \left(\frac{1}{1-\alpha}\right)^2 = \frac{\alpha}{(1-\alpha)^2}. \tag{2.69}$$

For $\alpha = \frac{1}{2}$, i.e. a fair coin, $p_k(k) = (\frac{1}{2})^k$, which gives

$$E[K] = 2 \qquad E[K^2] = 6 \qquad V[K] = 2 \qquad \sigma_K = \sqrt{2}. \tag{2.70}$$

In the derivations given above, the following well-known sums are used:

$$\sum_{n=0}^{\infty} \alpha^n = \frac{1}{1-\alpha}, \quad \sum_{n=0}^{\infty} n\alpha^n = \frac{\alpha}{(1-\alpha)^2}, \quad \text{and} \quad \sum_{n=0}^{\infty} n^2\alpha^n = \frac{\alpha(1+\alpha)}{(1-\alpha)^3} \tag{2.71}$$

for $|\alpha| < 1$.

The next example considers the Gaussian distribution.

Example 2.5 [Gaussian distribution] The Gaussian, or normal, distribution is a very central distribution in probability theory. It is also a very common distribution to use for channel modeling in communication theory. The distribution is denoted $X \sim \mathrm{N}(m, \sigma)$, and the density function

$$f_X(x) = \frac{1}{\sqrt{2\pi\sigma^2}} e^{-\frac{(x-m)^2}{2\sigma^2}} \tag{2.72}$$

To show that this is actually a density function, and to derive the expectation and variance, start with a version centered around 0, i.e. $Y \sim \mathrm{N}(0, \sigma)$, with density function

$$f_Y(y) = \frac{1}{\sqrt{2\pi\sigma^2}} e^{-\frac{y^2}{2\sigma^2}} \tag{2.73}$$

Then, consider the squared integral of the function,

$$\begin{aligned}
\left(\int_{\mathbb{R}} \frac{1}{\sqrt{2\pi\sigma^2}} e^{-\frac{y^2}{2\sigma^2}} dy\right)^2 &= \frac{1}{2\pi\sigma^2} \int_{\mathbb{R}} e^{-\frac{y^2}{2\sigma^2}} dy \int_{\mathbb{R}} e^{-\frac{z^2}{2\sigma^2}} dz \\
&= \frac{1}{2\pi\sigma^2} \int_{\mathbb{R}^2} e^{-\frac{y^2+z^2}{2\sigma^2}} dydz \\
&= \frac{1}{2\pi\sigma^2} \int_0^{2\pi} \int_0^{\infty} re^{-\frac{r^2\cos^2\phi + r^2\sin^2\phi}{2\sigma^2}} drd\phi \\
&= \frac{1}{2\pi} \int_0^{2\pi} \int_0^{\infty} \frac{r}{\sigma^2} e^{-\frac{r^2}{2\sigma^2}} drd\phi \\
&= \frac{1}{2\pi} \int_0^{2\pi} \left[-e^{-\frac{r^2}{2\sigma^2}}\right]_0^{\infty} d\phi \\
&= \frac{1}{2\pi} \int_0^{2\pi} d\phi = 1
\end{aligned} \tag{2.74}$$

where a variable change to polar coordinates is utilized. Since $f_Y(y)$ is strictly positive, it can be concluded that $\int_{\mathbb{R}} \frac{1}{\sqrt{2\pi\sigma^2}} e^{-\frac{y^2}{2\sigma^2}} dy = 1$, proving that it is indeed a distribution.

The expectation and the second-order moment can be derived as

$$E[Y] = \int_{\mathbb{R}} y \frac{1}{\sqrt{2\pi\sigma^2}} e^{-\frac{y^2}{2\sigma^2}} dy = \frac{\sigma^2}{\sqrt{2\pi\sigma^2}} \int_{\mathbb{R}} \frac{y}{\sigma^2} e^{-\frac{y^2}{2\sigma^2}} dy$$
$$= \frac{\sigma^2}{\sqrt{2\pi\sigma^2}} \left[-e^{-\frac{y^2}{2\sigma^2}} \right]_{-\infty}^{\infty} = 0. \tag{2.75}$$

$$E[Y^2] = \int_{\mathbb{R}} y^2 \frac{1}{\sqrt{2\pi\sigma^2}} e^{-\frac{y^2}{2\sigma^2}} dy = \frac{\sigma^2}{\sqrt{2\pi\sigma^2}} \int_{\mathbb{R}} y \frac{y}{\sigma^2} e^{-\frac{y^2}{2\sigma^2}} dy$$
$$= \frac{\sigma^2}{\sqrt{2\pi\sigma^2}} \left(\left[-ye^{-\frac{y^2}{2\sigma^2}} \right]_{-\infty}^{\infty} - \int_{\mathbb{R}} -e^{-\frac{y^2}{2\sigma^2}} dy \right)$$
$$= \sigma^2 \int_{\mathbb{R}} \frac{1}{\sqrt{2\pi\sigma^2}} e^{-\frac{y^2}{2\sigma^2}} dy = \sigma^2 \tag{2.76}$$

To make the same derivations in the more general case, $X \sim N(m, \sigma)$, utilize the variable change $y = x - m$ (implying $dy = dx$ and $x = y + m$), where $Y \sim N(0, \sigma)$, and

$$\int_{\mathbb{R}} \frac{1}{\sqrt{2\pi\sigma^2}} e^{-\frac{(x-m)^2}{2\sigma^2}} dx = \int_{\mathbb{R}} \frac{1}{\sqrt{2\pi\sigma^2}} e^{-\frac{y^2}{2\sigma^2}} dy = 1 \tag{2.77}$$

Hence, the expectation and second-order moment is

$$E\big[X\big] = E\big[Y + m\big] = m \tag{2.78}$$
$$E\big[X^2\big] = E\big[(Y + m)^2\big] = E[Y^2] + 2mE[Y] + m^2 = \sigma^2 + m^2 \tag{2.79}$$

and the variance is

$$V[X] = E[X^2] - E[X]^2 = \sigma^2 \tag{2.80}$$

2.4 THE LAW OF LARGE NUMBERS

The weak law of large numbers will play a central role when deriving both the source-coding theorem that bounds the compression ratio and the channel-coding theorem, a limit on the capability of reliable communication when transmitting over a noisy channel. Both of them are very important, resulting in information theory and communication theory.

The basics of the law of large numbers begins with two famous bounds in probability theory, Markov's inequality and Chebyshev's inequality. The first is stated in the following theorem. In this description, only the discrete case is considered, but the result also holds for continuous variables.

Theorem 2.2 (Markov's inequality) Let X be a nonnegative random variable with a finite expected value. Then, for any positive integer a

$$P(X > a) \leq \frac{E[X]}{a} \tag{2.81}$$

□

To show the theorem, begin with the expected value of X,

$$\begin{aligned} E[X] &= \sum_{k=0}^{\infty} kp(k) \geq \sum_{k=a+1}^{\infty} kp(k) \\ &\geq \sum_{k=a+1}^{\infty} ap(k) = a \sum_{k=a+1}^{\infty} p(k) = aP(X > a) \end{aligned} \tag{2.82}$$

which presents the result.

If instead of the expected value, the variance is considered, and the positive constant a is replaced with ε^2, where ε is a positive number,

$$P\left(\left(X - E[X]\right)^2 > \varepsilon^2\right) \leq \frac{E[(X - E[X])^2]}{\varepsilon^2} = \frac{V[X]}{\varepsilon^2} \tag{2.83}$$

Equivalently, this can be stated as in the following theorem, which is called as Chebyshev's inequality.

Theorem 2.3 (Chebyshev's inequality) Let X be a nonnegative random variable with a finite expected value $E[X]$ and a finite variance $V[X]$. Then, for any positive ε

$$P\left(\left|X - E[X]\right| > \varepsilon\right) \leq \frac{V[X]}{\varepsilon^2} \tag{2.84}$$

□

As stated previously, Chebyshev's inequality can be used to present the first proof of the weak law of large numbers. Consider a sequence of independent and identically distributed (i.i.d.) random variables, X_i, $i = 1, 2, \ldots, n$. The arithmetic mean of the sequence can be viewed as a new random variable,

$$Y = \frac{1}{n} \sum_{i=1}^{n} X_i \tag{2.85}$$

From Theorem 2.1, it is seen that the expected value and variance of Y can be expressed as

$$E[Y] = E[X] \tag{2.86}$$

$$V[Y] = \frac{V[X]}{n} \tag{2.87}$$

Applying Chebyshev's inequality yields

$$P\left(\left|\frac{1}{n}\sum_{i=1}^{n} X_i - E[X]\right| > \varepsilon\right) \leq \frac{V[X]}{n\varepsilon^2} \tag{2.88}$$

As n increases, the right-hand side will tend to be zero, bounding the arithmetic mean near to the expected value. Stated differently,

$$\lim_{n\to\infty} P\left(\left|\frac{1}{n}\sum_{i=1}^{n} X_i - E[X]\right| < \varepsilon\right) = 1 \tag{2.89}$$

which gives the *weak law of large numbers*. This type of probabilistic convergence is often called *convergence in probability* and denoted by $\xrightarrow{p}$. Hence, the relation in (2.89) can be stated in the following theorem:

Theorem 2.4 (Weak law of large numbers) Let $X_1, X_2, \ldots, X_n$ be a set of i.i.d. random variables with finite expectation $E[X]$. Then the arithmetic mean $Y = \frac{1}{n}\sum_i X_i$ converges in probability to $E[X]$,

$$Y \xrightarrow{p} E[X], \quad n \to \infty. \tag{2.90}$$

□

It should be noted that the proof given above requires that the variance is finite, but the same result can be obtained without this restriction. This result is presented in the previous theorem.

To visualize the theorem, consider n consecutive rolls with a fair die, giving the results vector $\boldsymbol{x} = (x_1, \ldots, x_n)$. In Figure 2.4, the average value of such a test is shown where $1 \leq n \leq 500$. It is clearly seen that the average value $y_n = \frac{1}{n}\sum_i x_i$ of the results is approaching the expected value $E[X]$. It should be mentioned that the series of attempts are restarted for each value n.

Another way to view the results in Theorem 2.4 is to consider the number of outcomes of a certain value in a sequence. In the next example, the probability for the number of occurrences of ones and zeros is determined.

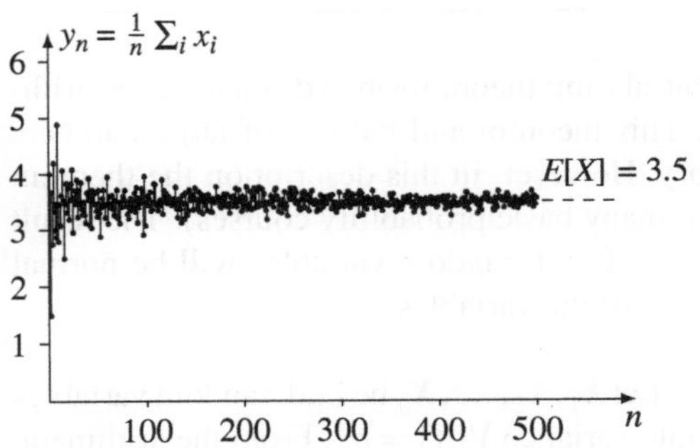

Figure 2.4 Average of n consecutive rolls with a fair die.

k	$P(k) = \binom{5}{k}\frac{2^k}{3^5}$
0	0.0041
1	0.0412
2	0.1646
3	0.3292
4	0.3292
5	0.1317

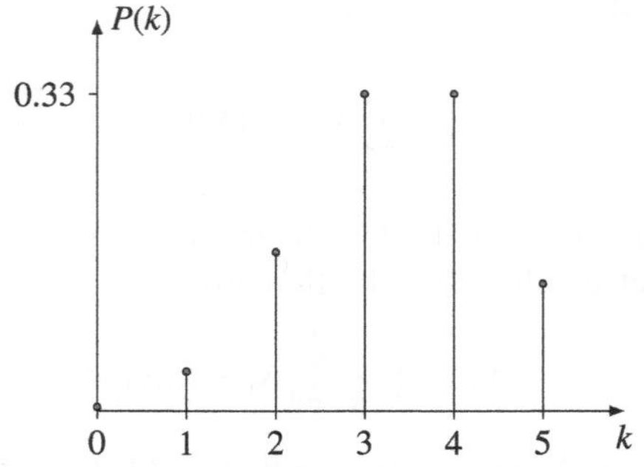

Figure 2.5 Probability distribution for k 1s in a vector of length 5, when $p(1) = 2/3$ and $p(0) = 1/3$.

Example 2.6 Let $\boldsymbol{X} = (X_1, X_2, \ldots, X_n)$ be a length n vector of i.i.d. binary random variables, where $p_X(0) = 1/3$ and $p_X(1) = 2/3$. According to the binomial distribution, the probability for a vector to have k ones is

$$P(k \text{ ones in } \boldsymbol{X}) = \binom{n}{k}\left(\frac{2}{3}\right)^k\left(\frac{1}{3}\right)^{n-k} = \binom{n}{k}\frac{2^k}{3^n}. \tag{2.91}$$

In Figure 2.5, the probability distribution of a number of ones in a vector of length $n = 5$ is shown both as a table and in a graphical version. It is most likely that a vector with three or four ones occur. Also, it is interesting to observe that it is less probable to get all five ones than three or four, even though the all one vector is the most probable vector.

In Figure 2.6, the distribution for the number of ones is shown when the length of the vector is increased to 10, 50, 100, and 500. With increasing length, it becomes more evident that the most likely outcome will be about $n \cdot E[X]$ ones. It also becomes more evident that although the all one vector is the most probable vector, the probability for the event of having all ones in a vector is much smaller than having about $\frac{2n}{3}$ ones. This is of course because there is only one vector with all ones and many more with $\frac{2n}{3}$ ones. For a large n, it becomes meaningless to consider specific vectors. Instead, vectors of a certain type, here the number of ones, should be considered.

It would be unfair not to mention the central limit theorem when discussing the arithmetic mean of i.i.d. random variables. This theorem and the law of large numbers are two main limits in probability theory. However, in this description the theorem is given without a proof (as also done in many basic probability courses). The result is that the arithmetic mean of a sequence of i.i.d. random variables will be normal distributed, independent of the distribution of the variables.

Theorem 2.5 (Central limit theorem) Let $X_1, X_2, \ldots, X_n$ be i.i.d. random variables with finite expectation $E[X] = m$ and finite variance $V[X] = \sigma^2$. From the arithmetic

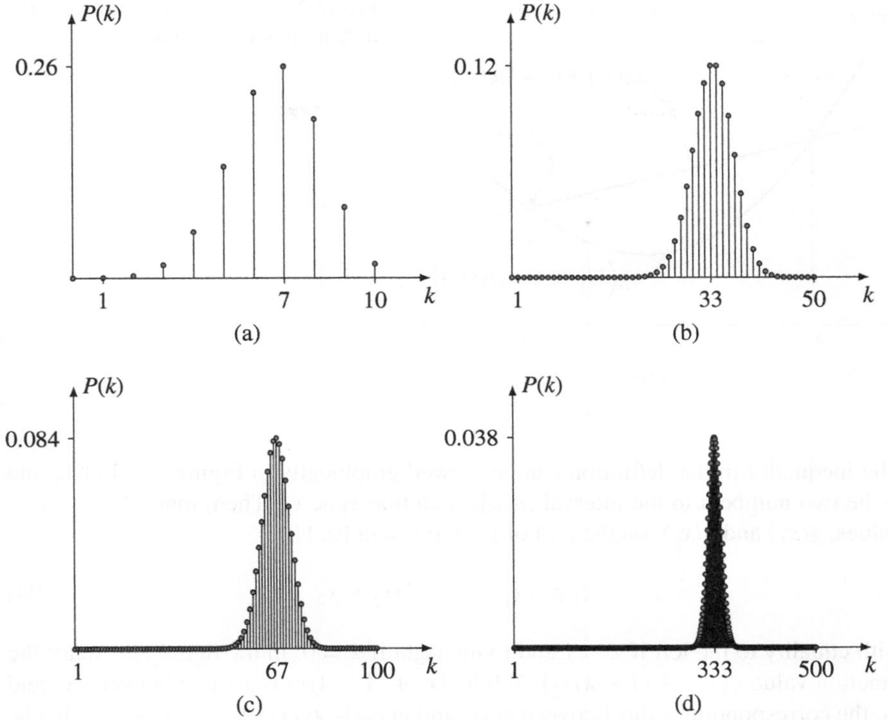

Figure 2.6 Probability distributions for k 1s in a vector of length 10 (a), 50 (b), 100 (c) and 500 (d), when $p(1) = 2/3$ and $p(0) = 1/3$.

mean as $Y = \frac{1}{n} \sum_i X_i$, then, as n increases to infinity, Y becomes distributed according to a normal distribution, i.e.,

$$\frac{Y - m}{\sigma/\sqrt{n}} \sim \mathrm{N}(0, 1), \quad n \to \infty. \tag{2.92}$$

□

2.5 JENSEN'S INEQUALITY

In many applications like, for example, optimization theory, convex functions are of special interest. A convex function is defined in the following definition:

Definition 2.4 (Convex function) A function $g(x)$ is *convex* in the interval $[a, b]$ if, for any x_1, x_2 such that $a \le x_1 \le x_2 \le b$, and any λ, $0 \le \lambda \le 1$,

$$g\big(\lambda x_1 + (1 - \lambda)x_2\big) \le \lambda g(x_1) + (1 - \lambda)g(x_2) \tag{2.93}$$

Similarly, a function $g(x)$ is *concave* in the interval $[a, b]$ if $-g(x)$ is convex in the same interval. □

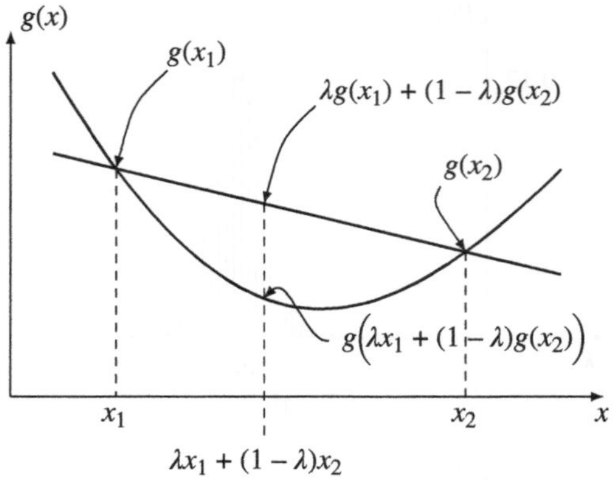

Figure 2.7 A graphical view of the definition of convex functions.

The inequality in the definition can be viewed graphically in Figure 2.7. Let x_1 and x_2 be two numbers in the interval $[a, b]$, such that $x_1 < x_2$. Then, mark the function values, $g(x_1)$ and $g(x_2)$, on the plot of $g(x)$. For λ in $[0, 1]$,

$$x_1 \leq \lambda x_1 + (1-\lambda)x_2 \leq x_2 \tag{2.94}$$

with equality to the left if $\lambda = 1$ and to the right if $\lambda = 0$. In the figure also mark the function value $g\big(\lambda x_1 + (1-\lambda)x_2\big)$. While $\lambda x_1 + (1-\lambda)x_2$ is a value between x_1 and x_2, the corresponding value between $g(x_1)$ and $g(x_2)$ is $\lambda g(x_1) + (1-\lambda)g(x_2)$. That is, the coordinates

$$\Big(\lambda x_1 + (1-\lambda)x_2, \lambda g(x_1) + (1-\lambda)g(x_2)\Big) \tag{2.95}$$

describe a straight line between $\big(x_1, g(x_1)\big)$ and $\big(x_2, g(x_2)\big)$ for $0 \leq \lambda \leq 1$. Since the inequality in (2.93) holds for all possible x_1 and x_2 in the interval, a convex function is typically shaped like a bowl. Similarly, a concave function is the opposite and looks like a hill.

Example 2.7 The functions x^2 and e^x are typical convex functions. On the other hand, the functions $-x^2$ and $\log x$ are concave functions. Similarly, $\sin(x)$ is concave at the interval $[0, \pi]$ and convex at the interval $[\pi, 2\pi]$.

In the literature, the names convex ∪ and convex ∩, as well as convex up and convex down, are also used. To determine whether a function is convex, or concave, the second derivative can be considered. Considering Figure 2.7, a convex function is bending upward at the whole interval. Then the derivative is constantly increasing, which means that the second derivative is positive, or at least, nonnegative. This can be stated as a theorem.

Theorem 2.6 A function $g(x)$ is convex in the interval $[a, b]$ if and only if its second derivative is nonnegative in the interval,

$$\frac{\partial^2}{\partial x^2} g(x) \geq 0, \quad a \leq x \leq b \tag{2.96}$$

Similarly, the function $g(x)$ is concave in the interval $[a, b]$ if and only if its second derivative is nonpositive in the interval,

$$\frac{\partial^2}{\partial x^2} g(x) \leq 0, \quad a \leq x \leq b \tag{2.97}$$

□

The convex functions in Example 2.7 have the nonnegative second derivatives:

$$\frac{\partial^2}{\partial x^2} x^2 = 2 > 0, \quad x \in \mathbb{R} \tag{2.98}$$

$$\frac{\partial^2}{\partial x^2} e^x = e^x \geq 0, \quad x \in \mathbb{R} \tag{2.99}$$

The second derivatives for the concave functions are nonpositive,

$$\frac{\partial^2}{\partial x^2} - x^2 = -2 < 0, \quad x \in \mathbb{R} \tag{2.100}$$

$$\frac{\partial^2}{\partial x^2} \log x = \frac{-1}{x^2 \ln 2} \leq 0, \quad x \in \mathbb{R}^+ \tag{2.101}$$

From the definition of the convex function follows *Jensen's inequality*.

Theorem 2.7 (Jensen's inequality) If $g(x)$ is a convex function and X a random variable,

$$E[g(X)] \geq g(E[X]) \tag{2.102}$$

If $g(x)$ is a concave function and X a random variable,

$$E[g(X)] \leq g(E[X]) \tag{2.103}$$

□

Jensen's inequality will be an important tool in the theory described later, and in the following a proof of the result is shown. Even though it will only be shown for the discrete case here, the result is also valid in the continuous case.

Proof: First, consider a binary random variable X with outcomes x_1 and x_2, and probabilities λ and $1 - \lambda$. Then $g\big(\lambda x_1 + (1 - \lambda)x_2\big)$ and $\lambda g(x_1) + (1 - \lambda)g(x_2)$ constitute the expected values $g(E[X])$ and $E[g(X)]$, respectively. That means, the result is true for the binary case.

To show that the theorem also holds for distributions with more than two outcomes, induction can be used. Assume that for a set of positive numbers $a_1, a_2, \ldots, a_n$, where $\sum_i a_i = 1$, it holds that

$$g\Big(\sum_i a_i x_i\Big) \leq \sum_i a_i g(x_i) \tag{2.104}$$

where $g(x)$ is a convex function.

Let $p_1, p_2, \ldots, p_{n+1}$ be a probability distribution for X with $n+1$ outcomes. Then

$$\begin{aligned}
g\big(E[X]\big) &= g\Big(\sum_{i=1}^{n+1} p_i x_i\Big) = g\Big(p_1 x_1 + \sum_{i=2}^{n+1} p_i x_i\Big) \\
&= g\Big(p_1 x_1 + (1-p_1)\sum_{i=2}^{n+1} \frac{p_i}{1-p_1} x_i\Big) \\
&\leq p_1 g(x_1) + (1-p_1) g\Big(\sum_{i=2}^{n+1} \frac{p_i}{1-p_1} x_i\Big) \\
&\leq p_1 g(x_1) + (1-p_1)\sum_{i=2}^{n+1} \frac{p_i}{1-p_1} g(x_i) \\
&= \sum_{i=1}^{n+1} p_i f(x_i) = E\big[f(X)\big]
\end{aligned} \tag{2.105}$$

where the first inequality follows from the convexity of $g(x)$, and the second from the induction assumption. What is left to show is that $a_i = \frac{p_i}{1-p_1}$, satisfying the requirements in the assumption. Clearly, since p_i is a probability, $a_i \geq 0$. The second requirement follows from

$$\sum_{i=2}^{n+1} \frac{p_i}{1-p_1} = \frac{1}{1-p_1}\sum_{i=2}^{n+1} p_i = \frac{1}{1-p_1}(1-p_1) = 1 \tag{2.106}$$

which completes the proof. ∎

Example 2.8 Since $g(x) = x^2$ is a convex function, it follows that $E[X^2] \geq E[X]^2$. Then $V[X] = E[X^2] - E[X]^2 \geq 0$ shows that the variance is a nonnegative function.

Clearly, the above example comes as no surprise since it is already evident from the definition of variance. A somewhat more interesting result from Jensen's inequality is the so-called *log-sum inequality*. The function $g(t) = t\log t$ is in fact a convex function, which is seen from the second derivative,

$$\frac{\partial^2}{\partial t^2} t\log t = \frac{1}{t\ln 2} \geq 0, \quad t \geq 0 \tag{2.107}$$

Then, if α_i forms a probability distribution, it follows from Jensen's inequality that

$$\sum_i \alpha_i t_i \log t_i \geq \Big(\sum_i \alpha_i t_i\Big) \log\Big(\sum_i \alpha_i t_i\Big) \tag{2.108}$$

This can be used to get

$$\begin{aligned} \sum_i a_i \log \frac{b_i}{a_i} &= \textstyle\sum_j b_j \displaystyle\sum_i \frac{b_i}{\sum_j b_j} \frac{a_i}{b_i} \log \frac{a_i}{b_i} \\ &\geq \textstyle\sum_j b_j \displaystyle\sum_i \frac{b_i}{\sum_j b_j} \frac{a_i}{b_i} \log \sum_i \frac{b_i}{\sum_j b_j} \frac{a_i}{b_i} \\ &= \sum_i a_i \log \frac{\sum_i a_i}{\sum_j b_j} \end{aligned} \tag{2.109}$$

where the identities $\alpha_i = \frac{a_i}{\sum_j b_j}$ and $t_i = \frac{a_i}{b_i}$ are used in (2.108). The above is summarised in the following theorem.

Theorem 2.8 (log-sum inequality) Let $a_1, \dots, a_n$ and $b_1, \dots, b_n$ be nonnegative numbers. Then

$$\sum_i a_i \log \frac{a_i}{b_i} \geq \Big(\sum_i a_i\Big) \log \frac{\sum_i a_i}{\sum_i b_i} \tag{2.110}$$

□

2.6 RANDOM PROCESSES

So far, it has been assumed that the generated symbols in a sequence are independent. It is often also useful to consider how symbols in a sequence depend on each other, i.e. a dynamic system. To show the essence, an example taken from Shannon's paper in 1948 [1] illustrates how the letters in a text depend on the surrounding letters. Shannon assumed an alphabet with 27 symbols, i.e., 26 letters and 1 space. To get the zero-order approximation, a sample text was generated with equal probability for the letters.

Example 2.9 [Zero-order approximation] Choose letters from the English alphabet with equal probability.

```
XFOML RXKHRJFFJUJ ZLPWCFWKCYJ FFJEYVKCQSGHYD
QPAAMKBZAACIBZLHJQD
```

Clearly, this text does not have much in common with normal written English. So, instead, count the number of occurrences per letter in normal English texts, and estimate the probabilities. The probabilities are given by Table 2.1.

TABLE 2.1 Probabilities in percent for the letters in English text.

X	P	X	P	X	P
A	8.167	J	0.153	S	6.327
B	1.492	K	0.772	T	9.056
C	2.782	L	4.025	U	2.758
D	4.253	M	2.406	V	0.978
E	12.702	N	6.749	W	2.360
F	2.228	O	7.507	X	0.150
G	2.015	P	1.929	Y	1.974
H	6.094	Q	0.095	Z	0.074
I	6.966	R	5.987		

Then, according to these probabilities, a sample text for the first-order approximation can be generated. Here, the text has a structure of more English text, but still far from readable.

Example 2.10 [First-order approximation] Choose the symbols according to their estimated probability (12% E, 2% W, etc.):

```
OCRO HLI RGWR NMIELWIS EU LL NBNESEBYA TH EEI
ALHENHTTPA OOBTTVA NAH BRL
```

The next step is to extend the distribution, so the probability depends on the previous letter, i.e. the probability for the letter at time t becomes $P(S_t|S_{t-1})$.

Example 2.11 [Second-order approximation] Choose the letters according to probabilities conditioned on the previous letter:

```
ON IE ANTSOUTINYS ARE T INCTORE ST BE S DEAMY ACHIN
D ILONASIVE TUCOOWE AT TEASONARE FUSO TIZIN ANDY
TOBE SEACE CTISBE
```

Similarly, the third-order approximation conditions on the two previous letters. Here the structure of such text becomes more like English.

Example 2.12 [Third-order approximation] Choose the symbols conditioned on the two previous symbols:

```
IN NO IST LAT WHEY CRATICT FROURE BIRS GROCID
PONDENOME OF DEMONSTRURES OF THE REPTAGIN IS
REGOACTIONA OF CRE
```

If instead of letters, the source of the text is generated from probabilities for words. The first-order approximation uses the unconditioned probabilities.

Example 2.13 [First-order word approximation] Choose words independently (but according to an estimated probability distribution):

```
REPRESENTING AND SPEEDILY IS AN GOOD APT OR COME
CAN DIFFERENT NATURAL HERE HE THE A IN CAME THE
TO OF TO EXPERT GRAY COME TO FURNISHES THE LINE
MESSAGE HAD BE THESE
```

If the probabilities for words are conditioned on the previous word, a much more readable text is obtained. Still, without any direct meaning, of course.

Example 2.14 [Second-order word approximation] Choose words conditioned on the two previous word:

```
THE HEAD AND IN FRONTAL ATTACK ON AN ENGLISH WRITER
THAT THE CHARACTER OF THIS POINT IS THEREFORE
ANOTHER METHOD FOR THE LETTERS THAT THE TIME OF
WHO EVER TOLD THE PROBLEM FOR AN UNEXPECTED
```

The above examples show that in many situations it is important to view sequences instead of individual symbols. In probability theory, this is called a *random process*, or a stochastic process. In a general form, a discrete time process can be defined as follows:

Definition 2.5 (Random process) A *discrete random process* is a sequence of random variables, $\{X_i\}_{i=1}^{n}$, defined as the same sample space. □

There can be an arbitrary dependency among the variables, and the process is characterized by the joint probability function

$$P\big(X_1, X_2, \dots, X_n = x_1, x_2, \dots, x_n\big) = p(x_1, x_2, \dots, x_n),\ n = 1, 2, \dots \quad (2.111)$$

As a consequence of introducing dependencies along the sequence, generalizations of the second-order moment and variance are needed. The *autocorrelation* function reflects the correlation in time and is defined as

$$r_{XX}(n, n+k) = E\big[X_n X_{n+k}\big] \quad (2.112)$$

If the mean and the autocorrelation function are time independent, i.e. for all n

$$E[X_n] = E[X] \quad (2.113)$$

$$r_{XX}(n, n+k) = r_{XX}(k) \quad (2.114)$$

the process is said to be *wide sense stationary* (WSS). The relation with the second-order moment function is that $r_{XX}(0) = E[X^2]$. The same relation for the variance comes with the *autocovariance function*, defined for a WSS process as

$$c_{XX}(k) = E\big[(X_n - E[X])(X_{n+k} - E[X])\big] \tag{2.115}$$

It is directly observed that $c_{XX}(k) = r_{XX}(k) - E[X]^2$ and that $c_{XX}(0) = V[X]$.

The class of WSS processes is a very powerful tool when modeling random behaviors. However, sometimes even stronger restriction on the time invariance is imposed. A process is *stationary* if the probability distribution does not depend on the time shift. That is, if

$$P\big(X_1, \dots, X_n = x_1, \dots, x_n\big) = P\big(X_{l+1}, \dots, X_{l+n} = x_1, \dots, x_n\big) \tag{2.116}$$

for all n and time shifts ℓ. Clearly, this is a subclass of WSS processes.

2.7 MARKOV PROCESS

A widely used class of the discrete stationary random processes is the class of Markov processes. The process has unit memory, i.e. the probability for a symbol depends only on the previous symbol. With this simplification, a system is achieved that is relatively easy to handle from a mathematical and computer implementation point of view, while still having time dependency in the sequence to be a powerful modeling tool.

Definition 2.6 (Markov chain) A *Markov chain*, or *Markov process*, is a stationary random process with a unit memory, i.e.

$$P\big(x_n | x_{n-1}, \dots, x_1\big) = P\big(x_n | x_{n-1}\big) \tag{2.117}$$

for all x_i. □

In this text, only time-invariant Markov chains will be considered. That is, the distribution for the conditional probabilities does not change over time,

$$P(X_n = x_a | X_{n-1} = x_b) = P(X_{n+\ell} = x_a | X_{n-1+\ell} = x_b) \tag{2.118}$$

for all relevant n, ℓ, x_a, and x_b.

Using the chain rule for probabilities (2.16), the joint probability function for a Markov chain can be written as

$$\begin{aligned} p(x_1, x_2, \dots, x_n) &= \prod_{i=1}^{n} p(x_i | x_{i-1}) \\ &= p(x_1)p(x_2|x_1)p(x_3|x_2)\cdots p(x_n|x_{n-1}) \end{aligned} \tag{2.119}$$

The unit memory property of a Markov chain results a process characterized by

- A finite set of *states*

$$X \in \{x_1, x_2, \dots, x_k\} \tag{2.120}$$

where the state determines everything about the past. This represents the unit memory of the chain.

- A *state transition matrix*

$$P = [p_{ij}]_{i,j\in\{1,\dots,k\}}, \text{ where } p_{ij} = p(x_j|x_i) \tag{2.121}$$

and $\sum_j p_{ij} = 1$.

In this description, the state is directly the output of the model. Sometimes, it is desirable to use a *hidden Markov chain*, where the output is a function of the state.

The behavior of a Markov chain can be visualized in a *state transition graph* consisting of states and edges, labeled with probabilities. The next example is intended to show the procedure.

Example 2.15 Consider a three-state Markov chain described by the three states

$$X \in \{x_1, x_2, x_3\} \tag{2.122}$$

and a state transition matrix

$$P = \begin{pmatrix} \frac{1}{3} & \frac{2}{3} & 0 \\ \frac{1}{4} & 0 & \frac{3}{4} \\ \frac{1}{2} & \frac{1}{2} & 0 \end{pmatrix} \tag{2.123}$$

From the matrix P, it is observed that, conditioned on the previous state x_1, the probability for x_1 is $1/3$, x_2 is $2/3$, and x_3 is 0. This can be viewed as transitions in a graph from state x_1 to the other states according to the said probabilities (see Figure 2.8).

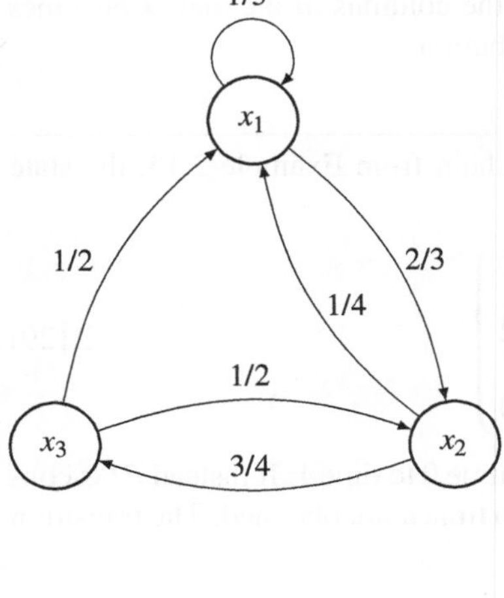

Figure 2.8 A state transition graph of a three-state Markov chain.

Similarly, the other rows in the state transition matrix describe the probabilities for transitions from the other states. The structure of the process is often more visible when viewed as a state transition graph.

For a Markov chain with k states, the transition probability matrix will be a $k \times k$ matrix where each row constitutes a probability distribution. Assume that the initial probabilities for the states at time 0 are

$$\boldsymbol{\pi}^{(0)} = \begin{pmatrix} \pi_1^{(0)} & \cdots & \pi_k^{(0)} \end{pmatrix} \tag{2.124}$$

where $\pi_i^{(0)} = P(X_0 = i)$. Then the probability for being in state j at time 1 becomes

$$\begin{aligned} \pi_j^{(1)} &= P(X_1 = j) = \sum_i P(X_1 = j | X_0 = i) P(X_0 = i) \\ &= \sum_i p_{ij} \pi_i^{(0)} = \begin{pmatrix} \pi_1^{(0)} \dots \pi_1^{(k)} \end{pmatrix} \begin{pmatrix} p_{1j} \\ \vdots \\ p_{kj} \end{pmatrix} \end{aligned} \tag{2.125}$$

That is, the vector describing the state probabilities at time 1 becomes

$$\boldsymbol{\pi}^{(1)} = \boldsymbol{\pi}^{(0)} P, \tag{2.126}$$

where P is the state transition matrix. Similarly, letting $\boldsymbol{\pi}^{(n)}$ be the state probabilities at time n,

$$\boldsymbol{\pi}^{(n)} = \boldsymbol{\pi}^{(n-1)} P = \boldsymbol{\pi}^{(n-2)} P^2 = \cdots = \boldsymbol{\pi}^{(0)} P^n \tag{2.127}$$

is needed. Hence, the matrix P^n describes the probabilities for the process to through state j at time n, conditioned on the starting state i at time 0,

$$P^n = \left[P(X_n = x_j | X_0 = x_i) \right]_{i,j \in \{1,2,\ldots,k\}} \tag{2.128}$$

In the next example, the state transition matrices from time 0 to time 1, 2, 4, 8, and 16, respectively, are derived. It is observed that the columns of the matrix becomes more and more independent of the starting distribution.

Example 2.16 Continuing with the Markov chain from Example 2.15, the state transition matrix is

$$P = \begin{pmatrix} \frac{1}{3} & \frac{2}{3} & 0 \\ \frac{1}{4} & 0 & \frac{3}{4} \\ \frac{1}{2} & \frac{1}{2} & 0 \end{pmatrix} \tag{2.129}$$

It shows the probabilities for the transitions from time 0 to time 1. If instead P^n is considered, the transition probabilities from time 0 to time n are obtained. The transition probabilities to time 2, 4, 8, and 16 are as follows:

$$P^2 = PP = \begin{pmatrix} \frac{20}{72} & \frac{16}{72} & \frac{36}{72} \\ \frac{33}{72} & \frac{39}{72} & 0 \\ \frac{21}{72} & \frac{24}{72} & \frac{27}{72} \end{pmatrix} \approx \begin{pmatrix} 0.2778 & 0.2222 & 0.5000 \\ 0.4583 & 0.5417 & 0 \\ 0.2917 & 0.3333 & 0.3750 \end{pmatrix} \tag{2.130}$$

$$P^4 = P^2P^2 = \begin{pmatrix} \frac{1684}{5184} & \frac{1808}{5184} & \frac{1692}{5184} \\ \frac{1947}{5184} & \frac{2049}{5184} & \frac{1188}{5184} \\ \frac{1779}{5184} & \frac{1920}{5184} & \frac{1485}{5184} \end{pmatrix} \approx \begin{pmatrix} 0.3248 & 0.3488 & 0.3264 \\ 0.3756 & 0.3953 & 0.2292 \\ 0.3432 & 0.3704 & 0.2865 \end{pmatrix} \tag{2.131}$$

$$P^8 = P^4P^4 \approx \begin{pmatrix} 0.3485 & 0.3720 & 0.2794 \\ 0.3491 & 0.3721 & 0.2788 \\ 0.3489 & 0.3722 & 0.2789 \end{pmatrix} \tag{2.132}$$

$$P^{16} = P^8P^8 \approx \begin{pmatrix} 0.3488 & 0.3721 & 0.2791 \\ 0.3488 & 0.3721 & 0.2791 \\ 0.3488 & 0.3721 & 0.2791 \end{pmatrix}. \tag{2.133}$$

Already for 16 steps in the process, numbers in each column are equal up to four decimals (actually, this is true already for P^{12}). This means that for 16 steps or more in the graph the probability for a state is independent of the starting state. For any starting distribution $\boldsymbol{\pi}^{(0)}$, the probabilities for the states after 16 steps are

$$\boldsymbol{\pi}^{(16)} = \boldsymbol{\pi}^{(0)}P^{16} = \begin{pmatrix} 0.3488 & 0.3721 & 0.2791 \end{pmatrix} \tag{2.134}$$

To get higher accuracy in the derivations, a higher order of the exponent is required, but eventually it will stabilize.

As presented in Example 2.16, the *asymptotic distribution* $\boldsymbol{\pi} = (\pi_1, \dots, \pi_k)$ is reached by

$$\lim_{n\to\infty} P^n = \begin{pmatrix} \boldsymbol{\pi} \\ \boldsymbol{\pi} \\ \vdots \\ \boldsymbol{\pi} \end{pmatrix} = \begin{pmatrix} \pi_1 & \cdots & \pi_k \\ \pi_1 & \cdots & \pi_k \\ \vdots & \ddots & \vdots \\ \pi_1 & \cdots & \pi_k \end{pmatrix} \tag{2.135}$$

It can be shown [8] that if there exists an n_0 for which P^{n_0} has only strictly positive entrences, the limit in (2.135) exists. The requirement of P^{n_0} being strictly positive means that, with a positive probability, there is a path from every state to every other state in n_0 steps. If this is true for n_0 steps, it will also be true for n steps, where $n \geq n_0$. Eventually, the asymptotic distribution will be reached.

Assuming the asymptotic distribution exists, then as $n \to \infty$,

$$\begin{pmatrix} \boldsymbol{\pi} \\ \vdots \\ \boldsymbol{\pi} \end{pmatrix} = P^n = P^{n+1} = P^n P = \begin{pmatrix} \boldsymbol{\pi} \\ \vdots \\ \boldsymbol{\pi} \end{pmatrix} P \tag{2.136}$$

Consider one row in the left matrix to conclude

$$\boldsymbol{\pi} = \boldsymbol{\pi} P \tag{2.137}$$

which is the *stationary distribution* of the system. The next theorem establishes the relationship between the stationary and the asymptotic distributions.

Theorem 2.9 Let $\boldsymbol{\pi} = (\pi_1 \ \pi_2 \ \dots \ \pi_r)$ be an asymptotic distribution of the state probabilities. Then

- $\boldsymbol{\pi}$ is a *stationary distribution*, i.e., $\boldsymbol{\pi} P = \boldsymbol{\pi}$.
- $\boldsymbol{\pi}$ is a unique stationary distribution for the source.

□

The first property is already discussed above, but the second, on uniqueness, still needs some clarification.

Assume that $\boldsymbol{\nu} = (\nu_1 \cdots \nu_k)$ is a stationary distribution, i.e. it fulfills $\sum_i \nu_i = 1$ and $\boldsymbol{\nu} P = \boldsymbol{\nu}$. Then, as $n \to \infty$, the equation $\boldsymbol{\nu} = \boldsymbol{\nu} P^n$ can be written as

$$\begin{pmatrix} \nu_1 & \dots & \nu_k \end{pmatrix} = \begin{pmatrix} \nu_1 & \dots & \nu_k \end{pmatrix} \begin{pmatrix} \pi_1 & \cdots & \pi_j & \cdots & \pi_k \\ \vdots & & \vdots & & \vdots \\ \pi_1 & \cdots & \pi_j & \cdots & \pi_k \end{pmatrix} \tag{2.138}$$

This implies that

$$\begin{aligned} \nu_j &= \begin{pmatrix} \nu_1 & \dots & \nu_k \end{pmatrix} \begin{pmatrix} \pi_j \\ \vdots \\ \pi_j \end{pmatrix} \\ &= \nu_1 \pi_j + \cdots + \nu_k \pi_j = \pi_j \underbrace{(\nu_1 + \cdots + \nu_k)}_{=1} = \pi_j \end{aligned} \tag{2.139}$$

That is, $\boldsymbol{\nu} = \boldsymbol{\pi}$, which proves uniqueness.

To derive the stationary distribution, start with the equation $\boldsymbol{\pi} P = \boldsymbol{\pi}$. Equivalently, it can be written as

$$\boldsymbol{\pi}(P - I) = \mathbf{0} \tag{2.140}$$

However, since $\boldsymbol{\pi} \neq \mathbf{0}$ it is seen that the matrix $P - I$ cannot have full rank and at least one more equation is needed to solve the equation system. It is natural to use $\sum_j \pi_j = 1$, which gives

$$\begin{cases} \boldsymbol{\pi}(P - I) = \mathbf{0} \\ \sum_j \pi_j = 1. \end{cases} \tag{2.141}$$

Example 2.17 Again use the state transition matrix from Example 2.15,

$$P = \begin{pmatrix} \frac{1}{3} & \frac{2}{3} & 0 \\ \frac{1}{4} & 0 & \frac{3}{4} \\ \frac{1}{2} & \frac{1}{2} & 0 \end{pmatrix} \tag{2.142}$$

Starting with $\boldsymbol{\pi}(P - I) = \mathbf{0}$

$$\boldsymbol{\pi}\left(\begin{pmatrix} \frac{1}{3} & \frac{2}{3} & 0 \\ \frac{1}{4} & 0 & \frac{3}{4} \\ \frac{1}{2} & \frac{1}{2} & 0 \end{pmatrix} - I\right) = \boldsymbol{\pi}\begin{pmatrix} -\frac{2}{3} & \frac{2}{3} & 0 \\ \frac{1}{4} & -1 & \frac{3}{4} \\ \frac{1}{2} & \frac{1}{2} & -1 \end{pmatrix} = \mathbf{0} \tag{2.143}$$

In $P - I$ column 2 plus column 3 equals column 1. Therefore, exchange column 1 with the equation $\sum_j w_j = 1$,

$$\boldsymbol{\pi}\begin{pmatrix} 1 & \frac{2}{3} & 0 \\ 1 & -1 & \frac{3}{4} \\ 1 & \frac{1}{2} & -1 \end{pmatrix} = \begin{pmatrix} 1 & 0 & 0 \end{pmatrix} \tag{2.144}$$

This is solved by

$$\begin{aligned} \boldsymbol{\pi} &= \begin{pmatrix} 1 & 0 & 0 \end{pmatrix}\begin{pmatrix} 1 & \frac{2}{3} & 0 \\ 1 & -1 & \frac{3}{4} \\ 1 & \frac{1}{2} & -1 \end{pmatrix}^{-1} \\ &= \begin{pmatrix} 1 & 0 & 0 \end{pmatrix}\begin{pmatrix} \frac{15}{43} & \frac{16}{43} & \frac{12}{43} \\ \frac{42}{43} & -\frac{24}{43} & -\frac{18}{43} \\ \frac{36}{43} & \frac{4}{43} & -\frac{40}{43} \end{pmatrix} = \begin{pmatrix} \frac{15}{43} & \frac{16}{43} & \frac{12}{43} \end{pmatrix} \end{aligned} \tag{2.145}$$

Derived with four decimals, the same result is obtained as earlier when the asymptotic distribution was discussed,

$$\boldsymbol{\pi} \approx \begin{pmatrix} 0.3488 & 0.372 & 0.2791 \end{pmatrix} \tag{2.146}$$

PROBLEMS

2.1 **(a)** Let X be a binary stochastic variable with $P(X = 0) = P(X = 1) = \frac{1}{2}$, and let Y be another independent binary stochastic variable with $P(Y = 0) = p$ and $P(Y = 1) = 1 - p$. Consider the modulo two sum $Z = X + Y \bmod 2$. Show that Z is independent of Y for all values of p.

(b) Let X be a stochastic variable uniformly distributed over $\{1, 2, \dots, M\}$. Let Y be independent of X, with an arbitrary probability function over $\{1, 2, \dots, M\}$. Consider the sum $Z = X + Y$, mod M. Show that Z is independent of Y.

2.2 Two cards are drawn from an ordinary deck of cards. What is the probability that neither of them is a heart?

2.3 Two persons flip a fair coin n times each. What is the probability that they have the same number of Heads?

2.4 The random variable X denotes the outcome of a roll with a five-sided fair die and Y the outcome from a roll with an eight-sided fair die.

(a) What is the distribution of $Z_a = X + Y$?

(b) What is the distribution of $Z_b = X - Y$?

(c) What is the distribution of $Z_c = |X - Y|$?

2.5 Flip a fair coin until Heads comes up and denote the number of flips by X.

(a) What is the probability distribution of the number of coin flips, X?

(b) What is the expected value of the number of coin flips, $E[X]$?

(c) Repeat (a) and (b) for an unfair coin with $P(\text{head}) = p$ and $P(\text{tail}) = q = 1 - p$.

2.6 Let X be Poisson-distributed, $X \sim \text{Po}(\lambda)$ (see Appendix A). Show that the expectation and variance are $E[X] = V[X] = \lambda$.

2.7 Let X be exponentially distributed, $X \sim \text{Exp}(\lambda)$ (see Appendix A). Show that the expectation and variance are $E[X] = \frac{1}{\lambda}$ and $V[X] = \frac{1}{\lambda^2}$.

2.8 Show that the second-order moment around a point c is minimized by the variance, i.e.,

$$E\big[(X-c)^2\big] \geq E\big[(X-m)^2\big]$$

with equality if and only if $c = m$, where $m = E[X]$.

2.9 Consider a binary vector of length $N = 10$ where the bits are i.i.d. with $P(X = 0) = p = 0.2$. Construct a table where you list, for each possible number of zeros in the vector, the number of vectors with that number of zeros, the probability for each vector and the probability for the number of zeros.

2.10 An urn has 10 balls, seven white and three black. Six times after each other a ball is drawn from the urn. What is the probability of the number of black balls drawn in the series, if

(a) the ball is replaced in the urn after each draw?

(b) drawn balls are not replaced?

2.11 Use Jensen's inequality to show

$$(x_1 x_2)^{\frac{1}{2}} \leq \frac{x_1 + x_2}{2}, \qquad x_1 x_2 \in \mathbb{Z}^+.$$

Hint: The logarithm is a concave function.

2.12 At some places in the textbook, Stirling's approximation is used to relate the binomial function with the binary entropy, defined in Chapter 3. There are different versions of this approximation in the literature, with different accuracy (and difficulty). Here, one of the basic versions is derived.

(a) Consider the logarithm of the faculty function

$$y(n) = \ln n!$$

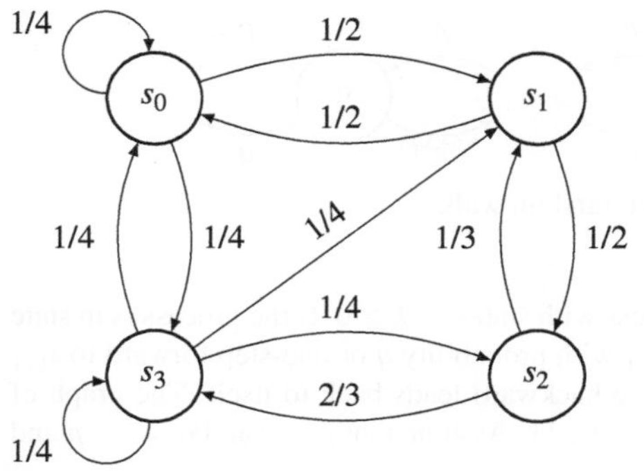

Figure 2.9 A state transition graph for a Markov process.

View $y(n)$ as a sum and interprete it as a trapezoid approximation of an integral. Use this to show that

$$\ln n! \approx n \ln n - n + 1 + \frac{1}{2} \ln n$$

or, equivalently,

$$n! \approx e\sqrt{n}\left(\frac{n}{e}\right)^n$$

(b) To improve the approximation for large n, use the limit value

$$\lim_{n\to\infty} n!\left(\frac{e}{n}\right)^n \frac{1}{\sqrt{n}} = \sqrt{2\pi}$$

Show how this gives the more common version of Stirling's approximation

$$n! \approx \sqrt{2\pi n}\left(\frac{n}{e}\right)^n$$

(c) Use the result in (b) to estimate the approximation error in (a) for large n.

2.13 A Markov process is defined by the state transition graph in Figure 2.9.

(a) Give the state transition matrix, P.

(b) Derive the steady state distribution, $\boldsymbol{\pi} = (\pi_0\ \pi_1\ \pi_2\ \pi_3)$.

2.14 Often in communication systems, the transmission is distorted by bursty noise. One way to model the noise bursts is through the so called Gilbert–Elliott channel model. It consists of a time discrete Markov model with two states, *Good* and *Bad*. In the *Good* state, the transmission is essentially error free, whereas in the *Bad* state the error probability is high, e.g., 0.5. The probability for transition from *Good* to *Bad* is denoted by P_{gb}, and from *Bad* to *Good* is P_{gb} (see Figure 2.10).

(a) Derive the steady-state distribution for the Markov model.

(b) What is the expected time duration for a burst?

(c) What is the expected time between two consecutive bursts?

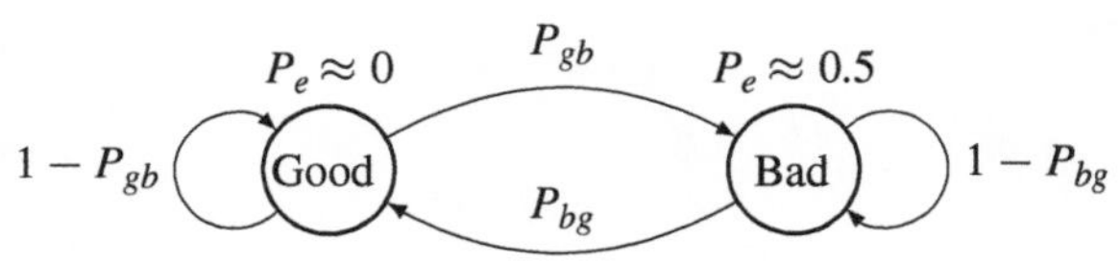

Figure 2.10 Gilbert–Elliott channel model for bursty noise.

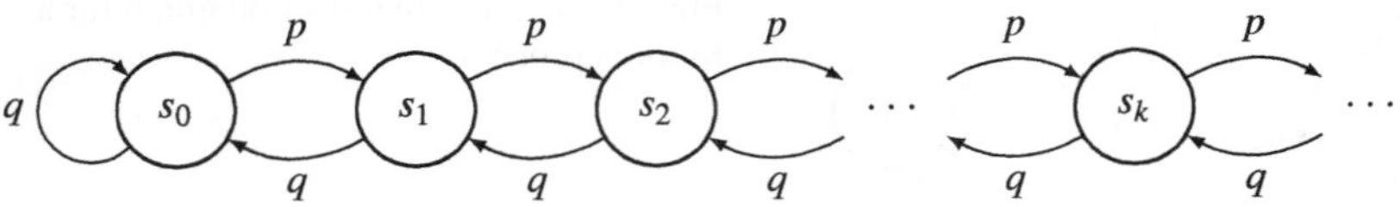

Figure 2.11 Markov chain for an infinite state random walk.

2.15 Consider an infinite random walk process with states s_k, $k \geq 0$. If the process is in state s_k it will take one-step backward to s_{k-1} with probability q or one-step forward to s_{k+1} with probability p. For state s_0, the step backward leads back to itself. The graph of the Markov process is presented in Figure 2.11. Assume that $p < q$ and $q = 1 - p$ and derive the steady-state distribution.

CHAPTER 3

INFORMATION MEASURES

INFORMATION THEORY is a mathematical theory of communication, based on probability theory. It was introduced by Claude Shannon in his landmark paper *A Mathematical Theory of Communications* in 1948 [1]. The theory is centered around two fundamentally important questions:

- What is information?
- What is communication?

Even if common knowledge gives an interpretation and understanding of information and communication, it is not the same as defining it in a mathematical context. As an example, one can consider an electronic copy of Shannon's paper in pdf format. The one considered here has the file size of 2.2 MB. It contains a lot of information about the subject, but the aim here is to measure the information contents of the file.

One way to look at the information content is to compress the file as much as possible. This size can serve as a measure of the amount of information required to reproduce the file. In an experimental test using the zip format, the same file can be stored with 713 kB. To quantify the amount of information in the paper, the pdf version contains at least 1.5 MB data that is not necessary to describe the text. Is then the number 713 kB a measure of the contained information? From a mathematical point of view, it will get close to the truth. However, the information measure here is related to describing the file and not the actual content in it. In comparison, a text of the same size containing only randomly chosen letters, uniformly chosen, will have more or less the same size of the compressed files. The question is then, do they contain the same amount of information? These semantic doubts are not considered in the mathematical model. Instead, the question answered is the amount of data needed to describe the text.

3.1 INFORMATION

In his paper, Shannon sets up a mathematical theory for information and communication, based on probability theory. He gave a quantitative measure of the amount of information stored in a variable and gave limits of how much information can be transmitted from one place to another over a given communication channel.

Information and Communication Theory, First Edition. Stefan Höst.

Already 20 years earlier, in 1928, Hartley stated that a symbol can contain information only if it has multiple choices [4]. That is, he realized the important fact that the symbol must be a random variable. If a random symbol, X, has k alternatives, a vector of n independent such symbols $(X_1, \dots, X_n)$ has k^n alternatives. To form a measure of information, Hartley noticed that if the symbol X has the information I, then the vector should have the information nI, since there are n variables. The conclusion of this observation was that an appropriate information measure should be based on the logarithm of the number of alternatives,

$$I_H(X) = \log k \tag{3.1}$$

In that way,

$$I_H(X_1, \dots, X_n) = \log k^n = n \log k = nI_H(X) \tag{3.2}$$

Example 3.1 Consider an outcome of a throw with a fair die. It has six alternatives, and hence, the information according to Hartley is[1]

$$I_H(\text{Die}) = \log_2 6 \approx 2.585 \text{ bit} \tag{3.3}$$

In this example, Hartley's information measure makes sense, since it is the number of bits needed to point out one of the six alternatives. But there can be other situations where it is less intuitive, like in the next example.

Example 3.2 Let the variable X be the outcome of a counterfeit coin, with the probabilities $P(X = \text{Head}) = p$ and $P(X = \text{Tail}) = 1 - p$. According to Hartley, the information is

$$I_H(X) = \log_2(2) = 1 \text{ bit} \tag{3.4}$$

In the case when the outcomes from the coin flip are equally likely, i.e. $p = \frac{1}{2}$, the measure is intuitive, since 1 bit is the amount of data needed to describe the outcome.

If the probability for Head is small, say $p = \frac{1}{4}$, it is expected that the result is Tail. That means, when the result is Tail, the result is as expected, and there is not much information in it. On the other hand, when the result is Head, the moment of surprise is higher and there should be more information in the event. That is, the uncertainty is connected to our expectations, or the distribution, the outcomes and should also be reflected by the information measure. So, even if Hartley's information measure was groundbreaking at that time, it did not considered the probability distribution of the experiment.

---

[1] Hartley did not specify the basis of the logarithm. Using the binary base, the information measure has the unit bits. In this way, it specifies the number of bits required to distinguish the alternatives.

When Shannon 20 years later introduced his information measure, it is based on a probabilistic view about how two random variables are related. Consider two outcomes, A and B, the probability $P(A)$ describes the probability that outcome A occurs and $P(A|B)$ the probability that outcome A occurs if it isknown that outcome B has occurred. The difference between these two probabilities describes how the knowledge about that B has occurred, affecting the probability of outcome A. This probability change for A then reflects an information propagation from event B to event A.

Definition 3.1 The mutual information between event A and event B, denoted as $I(A; B)$, is

$$I(A; B) = \log_2 \frac{P(A|B)}{P(A)} \tag{3.5}$$

where it is assumed $P(A) \neq 0$ and $P(B) \neq 0$ □

If nothing is stated, the logarithmic base 2 will be utilized to achieve the unit *bit* (binary digit). In this text, the binary logarithm will be denoted as $\log x$. The binary unit bit was first used in Shannon's paper, but it is also stated that it was John W. Tukey who coined the expression. The base in the logarithm can be changed through

$$x = a^{\log_a x} = b^{\log_b x} = a^{\log_a b \log_b x} \tag{3.6}$$

where the last equality follows from $b = a^{\log_a b}$. This leads to

$$\log_a x = \log_a b \log_b x \quad \Rightarrow \quad \log_b x = \frac{\log_a x}{\log_a b} \tag{3.7}$$

Especially, it is convenient to use

$$\log_2 x = \frac{\ln x}{\ln 2} = \frac{\log_{10} x}{\log_{10} 2} \tag{3.8}$$

It is also worth noting that $\log_2 e = \frac{1}{\ln 2}$. In, e.g., Matlab, the binary logarithm can be derived by the command log2(n).

Example 3.3 The outcome of a die is reflected by the two random variables:

$X =$ Number and

$Y =$ Odd or even number

The information achieved about the event $X = 3$ from the event $Y = \text{Odd}$ is

$$I(X = 3; Y = \text{Odd}) = \log \frac{P(X = 3|Y = \text{Odd})}{P(X = 3)}$$

$$= \log \frac{1/3}{1/6} = \log 2 = 1 \text{ bit} \tag{3.9}$$

In other words, by knowing that the number is odd, the set of outcomes is split into two halves, which means that 1 bit of information is given about the event that the number is 3.

The symmetry of the information measure follows from

$$I(A;B) = \log\frac{P(A|B)}{P(A)} = \log\frac{P(A,B)}{P(A)P(B)} = \log\frac{P(B|A)}{P(B)} = I(B;A) \tag{3.10}$$

That is, the information gained about event A by observing event B is the same as the information about event B by observing event A. This is the reason why it is called *mutual information.*

Example 3.4 The information from the event $X = 3$ about the event $Y = \text{Odd}$ is

$$\begin{aligned} I(Y = \text{Odd}; X = 3) &= \log\frac{P(Y = \text{Odd}|X = 3)}{P(\text{Odd})} \\ &= \log\frac{1}{1/2} = \log 2 = 1 \text{ bit} \end{aligned} \tag{3.11}$$

The knowledge about $X = 3$ gives full knowledge about the outcome of Y, which is a binary choice with two equally sized parts. To specify one of the two outcomes of Y, it is required 1 bit.

The mutual information between the events A and B can be bounded by

$$-\infty \le I(A;B) \le \min\{-\log P(A), -\log P(B)\} \tag{3.12}$$

The bound follows from the varying $P(A|B)$ between 0 and 1. Since the logarithm is a strictly increasing function, the two end cases give the following result:

$$P(A|B) = 0 \Rightarrow I(A;B) = \log\frac{0}{P(A)} = \log 0 = -\infty \tag{3.13}$$

$$P(A|B) = 1 \Rightarrow I(A;B) = \log\frac{1}{P(A)} = -\log P(A) \tag{3.14}$$

Similarly, by letting $P(B|A) = 1$, the information is $I(A;B) = -\log P(B)$. If $P(A)$ and $P(B)$ are not equal, there are two bounds, $-\log P(A)$ and $-\log P(B)$, where the minimum should be used. Notice that since $0 \le P(A) \le 1$ the value $-\log P(A)$ is positive. If $I(A;B) = 0$, the events A and B are statistically independent since it implies

$$\frac{P(A|B)}{P(A)} = 1 \quad \Rightarrow \quad P(A|B) = P(A). \tag{3.15}$$

To get a measure for the information related to event A, consider the mutual information between A and A. That is, the amount of information achieved about the event by observing the same event. This quantity is called the *self-information.*

Definition 3.2 The *self-information* in the event A is defined as

$$I(A) = I(A;A) = \log \frac{P(A|A)}{P(A)} = -\log P(A). \tag{3.16}$$

□

That is, $-\log P(A)$ is the amount of information needed to determine that the event A has occurred. The self-information is always a positive quantity, and as long as the outcome is not deterministic, i.e., $P(A) = 1$ for some event A, it is strictly positive.

3.2 ENTROPY

The above quantities deal with information related to specific events. An interesting measure then is the average required information to determine the outcome of a random variable. This is directly achieved from the expected value of the self-information, as stated in the following important definition:

Definition 3.3 The *entropy* of a random variable X is

$$H(X) = E_X[-\log p(X)] = -\sum_x p(x)\log p(x). \tag{3.17}$$

□

In the derivations, the convention that $0 \log 0 = 0$ is used, which follows from the corresponding limit value. It is sometimes convenient to use the notation

$$H(p_1, p_2, \dots, p_k) = -\sum_{i=1}^{k} p_i \log p_i, \tag{3.18}$$

when considering the entropy function for a probability distribution given as a vector $\boldsymbol{p} = (p_1, p_2, \dots, p_k)$.

The entropy is the average amount of information needed to determine the outcome of a random variable. As such, it can also be interpreted as the *uncertainty* of the outcome. Since the self-information is nonnegative so is its average,

$$H(X) \geq 0 \tag{3.19}$$

In other words, the uncertainty cannot be negative.

In many cases, a random variable describes a binary choice, e.g., a flip of a coin. The entropy function for this case is so widely used that it often gets a definition of its own.

Definition 3.4 The *binary entropy function* for the probability p is defined as

$$h(p) = -p\log p - (1-p)\log(1-p) \tag{3.20}$$

□

The binary entropy function has its maximum for $h(\frac{1}{2}) = 1$. In Figure 3.1, a plot of the function is shown, where the maximum value is depicted. It can be seen from the

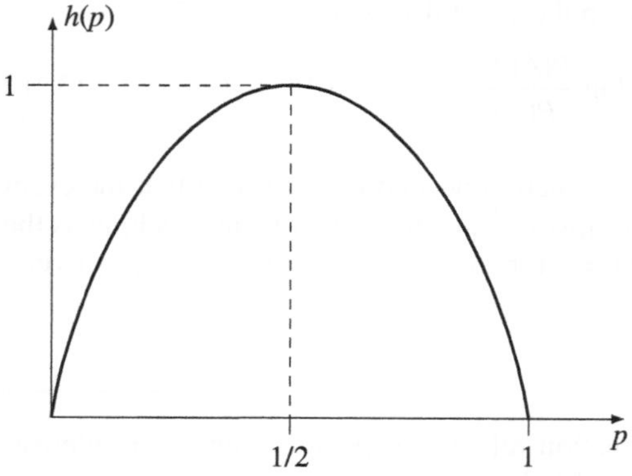

Figure 3.1 The binary entropy function.

figure that the function is symmetric in p, i.e. $h(p) = h(1-p)$, which is also evident directly from the definition. In the case of a coin flip, the uncertainty is maximized for a fair coin, i.e., $P(\text{Head}) = p = \frac{1}{2}$. If p increases, a natural initial guess is that the outcome would be Head, and the uncertainty decreases. Similarly, if p decreases the uncertainty should also decrease. At the end points, where $p = 0$ or $p = 1$, the outcome is known and the uncertainty is zero, corresponding to $h(0) = h(1) = 0$.

Example 3.5 Let X be the outcome of a fair die. Then $P(X = x) = 1/6$, $x = 1, 2, \dots, 6$. The entropy is

$$H(X) = -\sum_x \frac{1}{6} \log \frac{1}{6} = \log 6 = 1 + \log 3 \approx 2.5850 \text{ bit} \tag{3.21}$$

Let Y be the outcome of a die with a small weight inside, such that the probabilities are $P(Y = 1) = 0$, $P(Y = 6) = 1/3$, and $P(Y = i) = 1/6$, $i = 2, 3, 4, 5$. Then, the corresponding uncertainty of the outcome is

$$H(Y) = -\frac{1}{3} \log \frac{1}{3} - \frac{4}{6} \log \frac{1}{6} = \frac{4}{6} + \log 3 \approx 2.2516 \text{ bit} \tag{3.22}$$

Again, since there is a high probability for the outcome to be 6, there is less information is this outcome. Furthermore, since $X = 1$ will not happen, there is only five possible outcomes compared to six for the fair die. So, in total the uncertainty of the outcome has decreased compared to the fair die.

The definition of the entropy is also valid for vectorized random variables, such as (X, Y) with the joint probability function $p(x, y)$.

Definition 3.5 The *joint entropy* for a pair of random variables with the joint distribution $p(x, y)$ is

$$H(X, Y) = E_{XY}[-\log p(X, Y)] = -\sum_{x,y} p(x, y) \log p(x, y) \tag{3.23}$$

□

Similarly, in the general case with an n-dimensional vector $\boldsymbol{X} = (X_1, \dots, X_n)$, the joint entropy function is

$$H(X_1, \dots, X_n) = E_{\boldsymbol{X}}[-\log p(\boldsymbol{X})] = -\sum_{\boldsymbol{x}} p(\boldsymbol{x}) \log p(\boldsymbol{x}) \tag{3.24}$$

Example 3.6 Let X and Y be the outcomes from two independent fair dice. Then the joint probability is $P(X, Y = x, y) = 1/36$ and the joint entropy

$$H(X, Y) = -\sum_{x,y} \frac{1}{36} \log \frac{1}{36} = \log 36 = 2 \log 6 \approx 5.1699 \tag{3.25}$$

Clearly, the uncertainty of the outcome of two dice is twice the uncertainty of one die.

Let Z be the sum of the dice, $Z = X + Y$. The probability distribution can be derived as the convolution of the distributions for the two dice, as shown in the following table:

Z:	2	3	4	5	6	7	8	9	10	11	12
$P(Z)$:	$\frac{1}{36}$	$\frac{2}{36}$	$\frac{3}{36}$	$\frac{4}{36}$	$\frac{5}{36}$	$\frac{6}{36}$	$\frac{5}{36}$	$\frac{4}{36}$	$\frac{3}{36}$	$\frac{2}{36}$	$\frac{1}{36}$

The entropy of Z is

$$\begin{aligned} H(Z) &= H\left(\tfrac{1}{36}, \tfrac{2}{36}, \tfrac{3}{36}, \tfrac{4}{36}, \tfrac{5}{36}, \tfrac{6}{36}, \tfrac{5}{36}, \tfrac{4}{36}, \tfrac{3}{36}, \tfrac{2}{36}, \tfrac{1}{36}\right) \\ &= -2\frac{1}{36} \log \frac{1}{36} - 2\frac{2}{36} \log \frac{2}{36} - 2\frac{3}{36} \log \frac{3}{36} \\ &\quad - 2\frac{4}{36} \log \frac{4}{36} - 2\frac{5}{36} \log \frac{5}{36} - \frac{6}{36} \log \frac{6}{36} \\ &= \frac{23}{18} + \frac{5}{3} \log 3 - \frac{5}{18} \log 5 \approx 3.2744 \end{aligned} \tag{3.26}$$

The uncertainty of the sum of the dice is less than the outcomes of the pair of dice. This is natural, since several outcomes of the pair X, Y give the same sum Z.

In (3.19), it was seen that the entropy function is a nonnegative function. To achieve an upper bound, the following inequality will help.

Lemma 3.1 (IT-inequality) For every positive real number r

$$\log(r) \le (r - 1) \log(e) \tag{3.27}$$

with equality if and only if $r = 1$. □

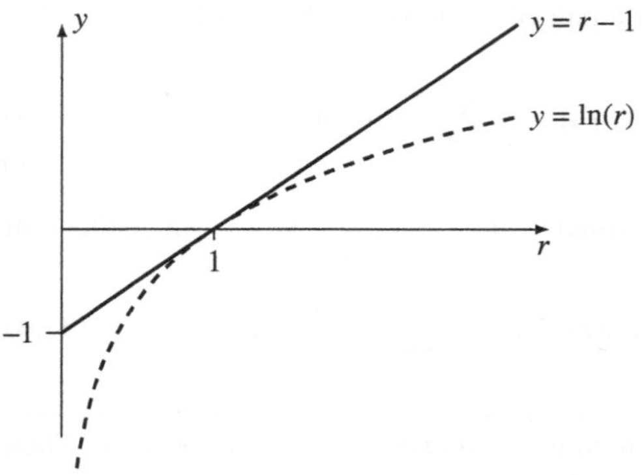

Figure 3.2 Graphical interpretation of the IT-inequality.

Proof: Consider the two functions $y_1 = r - 1$ and $y_2 = \ln r$ as shown in Figure 3.2. It is easy to verify that at $r = 1$ the two curves coincide, $\ln r = r - 1$. To show that in all other cases $\ln r < r - 1$, notice that the derivative of $r - 1$ is always 1 and the derivative of $\ln r$ is

$$\frac{d}{dr}\ln r = \frac{1}{r} = \begin{cases} >1, & r < 1 \Rightarrow \ln r < r - 1 \\ < 1, & r > 1 \Rightarrow \ln r < r - 1 \end{cases} \tag{3.28}$$

Hence, for $0 < r < 1$ the curve for $\ln r$ is steeper than $r - 1$ and for $r > 1$ it is flatter. So in all cases but $r = 1$, the curve for $\ln r$ must be strictly lower than $r - 1$. Hence, it has been shown that $\ln r \leq r - 1$ with equality if and only if $r = 1$. Rewriting into the binary logarithm completes the proof. ■

From the previous examples, a natural guess is that the maximum value of the entropy occurs when the outcomes have equal probabilities. In that case, a random variable X with k outcomes, $\{x_1, \dots, x_k\}$, has the probabilities $P(X = x_i) = \frac{1}{k}$. The entropy is

$$H(X) = -\sum_i \frac{1}{k}\log\frac{1}{k} = \log k.$$

To show that this value is a maximum over all distributions, consider

$$\begin{aligned} H(X) - \log k &= -\sum_x p(x)\log p(x) - \sum_x p(x)\log k \\ &= \sum_x p(x)\log\frac{1}{p(x)k} \\ &\leq \sum_x p(x)\Big(\frac{1}{p(x)k} - 1\Big)\log e \\ &= \Big(\sum_x \frac{1}{k} - \sum_x p(x)\Big)\log e \\ &= (1-1)\log e = 0 \end{aligned} \tag{3.29}$$

where the inequality follows from the IT-inequality with $r = \frac{1}{p(x)k}$, implying equality if and only if $\frac{1}{p(x)k} = 1$. In other words, it is shown that

$$H(X) \leq \log k \tag{3.30}$$

with equality if and only if $p(x) = \frac{1}{k}$. Combining (3.19) and (3.30), the following theorem can be stated.

Theorem 3.1 If X is a random variable with k outcomes, then

$$0 \leq H(X) \leq \log k \tag{3.31}$$

with equality to the left if and only if there exists some i where $p(x_i) = 1$, and with equality to the right if and only if $p(x_i) = 1/k$ for all $i = 1, 2, \dots, k$. □

To be able to consider also dependencies between random variables, the conditional entropy is introduced. This is defined as the average self-information for the event A conditioned on the event B, that is $E[-\log p(X|Y)]$. Since there are two random variables, the average needs to be taken over the joint probability distribution.

Definition 3.6 The *conditional entropy* for X conditioned on Y is

$$H(X|Y) = E_{XY}[-\log p(X|Y)] = -\sum_{x,y} p(x, y) \log p(x|y) \tag{3.32}$$

where $p(x, y)$ and $p(x|y)$ are the joint and conditional probability functions, respectively. □

Using the chain rule for probabilities, $p(x, y) = p(x|y)p(y)$, the formula can be rewritten as

$$\begin{aligned} H(X|Y) &= -\sum_x \sum_y p(x, y) \log p(x|y) = -\sum_x \sum_y p(x|y)p(y) \log p(x|y) \\ &= \sum_y p(y)\Big(-\sum_x p(x|y) \log p(x|y)\Big) \end{aligned} \tag{3.33}$$

By introducing the entropy of X conditioned on the event $Y = y$,

$$H(X|Y = y) = -\sum_x p(x|y) \log p(x|y) \tag{3.34}$$

the conditional entropy can be written as

$$H(X|Y) = \sum_y H(X|Y = y)p(y) \tag{3.35}$$

Example 3.7 The joint distribution of the random variables X and Y is given by

$P(X, Y)$

X	$Y=0$	$Y=1$
0	0	$\frac{3}{4}$
1	$\frac{1}{8}$	$\frac{1}{8}$

The marginal distributions of X and Y can be derived as $p(x) = \sum_y p(x, y)$ and $p(y) = \sum_x p(x, y)$ giving

X	$P(X)$
0	$\frac{3}{4}$
1	$\frac{1}{4}$

Y	$P(Y)$
0	$\frac{1}{8}$
1	$\frac{7}{8}$

The individual entropies are

$$H(X) = h\left(\tfrac{1}{4}\right) \approx 0.8113 \tag{3.36}$$

$$H(Y) = h\left(\tfrac{1}{8}\right) \approx 0.5436 \tag{3.37}$$

and the joint entropy is

$$H(X, Y) = H\left(0, \tfrac{3}{4}, \tfrac{1}{8}, \tfrac{1}{8}\right) = \frac{9}{4} - \frac{3}{4}\log 3 \approx 1.0613 \tag{3.38}$$

To calculate the conditional entropy $H(X|Y)$, first consider the conditional probabilities, derived by $p(x|y) = \frac{p(x,y)}{p(y)}$.

$P(X|Y)$

X	$Y=0$	$Y=1$
0	0	$\frac{6}{7}$
1	1	$\frac{1}{7}$

Then

$$H(X|Y=0) = h(0) = 0 \tag{3.39}$$

$$H(X|Y=1) = h\left(\tfrac{1}{7}\right) \approx 0.5917 \tag{3.40}$$

Putting things together, the conditional entropy becomes

$$H(X|Y) = H(X|Y=0)P(Y=0) + H(X|Y=1)P(Y=1)$$
$$= h(0)\frac{1}{8} + h\left(\tfrac{1}{7}\right)\frac{7}{8} = 0.5177 \quad (3.41)$$

Similarly, the probability for Y conditioned on X is given by

$P(Y|X)$

X	$Y=0$	$Y=1$
0	0	1
1	$\frac{1}{2}$	$\frac{1}{2}$

Then

$$H(Y|X=0) = h(0) = 0 \quad (3.42)$$
$$H(Y|X=1) = h\left(\tfrac{1}{2}\right) = 1 \quad (3.43)$$

and the conditional entropy becomes

$$H(Y|X) = H(Y|X=0)P(X=0) + H(Y|X=1)P(X=1)$$
$$= h(0)\frac{1}{2} + h\left(\tfrac{1}{2}\right)\frac{1}{4} = \frac{1}{4} \quad (3.44)$$

The chain rule for probabilities can also be used to achieve a corresponding chain rule for entropies,

$$H(X,Y) = -\sum_{x,y} p(x,y)\log p(x,y) = -\sum_{x,y} p(x,y)\log p(x|y)p(y)$$
$$= -\sum_{x,y} p(x,y)\log p(x|y) - \sum_{y} p(y)\log p(y)$$
$$= H(X|Y) + H(Y) \quad (3.45)$$

Rewriting the result yields $H(X|Y) = H(X,Y) - H(Y)$. That is, the conditional entropy is the difference between the uncertainty of the pair (X,Y) and the information gained by observing Y. A more general version of (3.45) can be stated as the chain rule for entropies in the following theorem. It follows directly from the general chain rule for probabilities (see (2.16)).

Theorem 3.2 Let $X_1, \dots, X_n$ be an n-dimensional random variable drawn according to $p(x_1, \dots, x_n)$. Then the chain rule for entropies states that

$$H(X_1, \dots, X_n) = \sum_{i=1}^{n} H(X_i|X_1, \dots, X_{i-1}) \quad (3.46)$$

□

Example 3.8 [Continued from Example 3.7] The joint entropy can alternatively be derived as

$$H(X, Y) = H(X|Y) + H(Y)$$
$$= \frac{7}{8}h\left(\tfrac{1}{7}\right) + h\left(\tfrac{1}{8}\right) = \frac{9}{4} - \frac{3}{4}\log 3 \approx 1.0613 \tag{3.47}$$

or

$$H(X, Y) = H(Y|X) + H(X)$$
$$= \frac{1}{4} + h\left(\tfrac{1}{4}\right) = \frac{9}{4} - \frac{3}{4}\log 3 \approx 1.0613 \tag{3.48}$$

3.3 MUTUAL INFORMATION

The entropy was obtained by averaging the self-information for a random variable. Similarly, the average mutual information between the random variables X and Y can be defined as follows:

Definition 3.7 The *mutual information* between the random variables X and Y is defined as

$$I(X; Y) = E_{X,Y}\left[\log \frac{p(X, Y)}{p(X)p(Y)}\right] = \sum_{x,y} p(x, y) \log \frac{p(x, y)}{p(x)p(y)} \tag{3.49}$$

□

Utilizing that $p(x, y) = p(x)p(y|x) = p(y)p(x|y)$, the function can also be written as the fraction between the conditional and the nonconditional probabilities,

$$I(X; Y) = E_{X,Y}\left[\log \frac{p(X|Y)}{p(X)}\right] = E_{X,Y}\left[\log \frac{p(Y|X)}{p(Y)}\right] \tag{3.50}$$

The mutual information describes the strength in the relation between two variables. From the definition, it is clear that it is a symmetric measure,

$$I(X; Y) = I(Y; X). \tag{3.51}$$

Splitting the logarithm argument in the definition, it is possible to derive the mutual information from the entropies as

$$\begin{aligned}
I(X;Y) &= E_{X,Y}\left[\log \frac{p(X,Y)}{p(X)p(Y)}\right] \\
&= E_{X,Y}[\log p(X,Y) - \log p(X) - \log p(Y)] \\
&= E_{X,Y}[\log p(X,Y)] - E_X[\log p(X)] - E_Y[\log p(Y)] \\
&= H(X) + H(Y) - H(X,Y) \\
&= H(X) - H(X|Y) \\
&= H(Y) - H(Y|X) \qquad (3.52)
\end{aligned}$$

where the last two equalities follows from the chain rule of entropies.

Example 3.9 [Continued from Example 3.7] The mutual information can be derived as

$$\begin{aligned}
I(X;Y) &= H(X) + H(Y) - H(X,Y) \\
&\approx 0.5436 + 0.8113 - 1.0613 = 0.2936 \qquad (3.53)
\end{aligned}$$

Alternatively,

$$I(X;Y) = H(X) - H(X|Y) \approx 0.5436 - \frac{1}{4} = 0.2936 \qquad (3.54)$$

The mutual information between two random variables X and Y can be affected by observing a third variable Z. This is reflected in the conditional mutual information.

Definition 3.8 The *conditional mutual information* between the random variables X and Y, when Z is observed, is defined as

$$\begin{aligned}
I(X;Y|Z) &= E_{X,Y,Z}\left[\log \frac{p(X,Y|Z)}{p(X|Z)p(Y|Z)}\right] \\
&= \sum_{x,y,z} p(x,y,z) \log \frac{p(x,y|z)}{p(x|z)p(y|z)} \qquad (3.55)
\end{aligned}$$

□

Similar to the unconditional case, the conditional mutual information can be derived from the entropies as

$$\begin{aligned}
I(X;Y|Z) &= H(X|Z) + H(Y|Z) - H(X,Y|Z) \\
&= H(X|Z) - H(X|YZ) \\
&= H(Y|Z) - H(Y|XZ). \qquad (3.56)
\end{aligned}$$

Both the entropy and the mutual information are important measures of information. The entropy states how much information is needed to determine the outcome of a random variable. It will be shown later that this is equivalent to the number of bits needed to describe the variable on average. In other words, this is a limit of how much a symbol can be compressed without any information being lost.

The mutual information, on the other hand, describes the amount of information achieved about the variable X by observing the variable Y. In a communication system, a symbol, X, is transmitted. The received symbol Y is then a distorted version of X and used by a receiver to estimate X. The mutual information is a measure of how much information about X can be obtained from Y, i.e. how much information that can be transmitted to the receiver. It will lead to the concept of the channel capacity.

To get more knowledge about these quantities, introduce the relative entropy. It was first considered by Kullback and Leibler in 1951 [9].

Definition 3.9 Given two probability distributions $p(x)$ and $q(x)$ for the same sample set $\mathcal{X}$. The *relative entropy*, or the *Kullback–Leibler divergence*, is defined as

$$D(p||q) = E_p\left[\log \frac{p(X)}{q(X)}\right] = \sum_x p(x) \log \frac{p(x)}{q(x)} \tag{3.57}$$

□

In the derivations, it is used that $0 \log 0 = 0$ and $0 \log \frac{0}{0} = 0$.

In the next example, the Poisson distribution is considered. Assume that the result from a random experiment is Poisson distributed with intensity λ. Then if the experimentally estimated intensity is λ_0, the example shows the relative entropy from the true distribution to the estimated. It will later be shown that this value reflects the penalty at a compression rate due to the estimation mismatch.

Example 3.10 Consider a random variable that is Poisson distributed, i.e. the probability function is

$$p(k) = \frac{\lambda^k e^{-\lambda}}{k!}, \qquad k = 0, 1, 2, \ldots \tag{3.58}$$

Then compare this distribution with another Poisson distribution with the parameter λ_0, i.e.

$$p_0(k) = \frac{\lambda_0^k e^{-\lambda_0}}{k!}, \qquad k = 0, 1, 2, \ldots \tag{3.59}$$

The relative entropy from $p(k)$ to $p_0(k)$ is then

$$D(p||p_0) = \sum_k \frac{\lambda^k e^{-\lambda}}{k!} \log \frac{\frac{\lambda^k e^{-\lambda}}{k!}}{\frac{\lambda_0^k e^{-\lambda_0}}{k!}}$$

$$= \sum_k \frac{\lambda^k e^{-\lambda}}{k!} \Big(k \log \frac{\lambda}{\lambda_0} + \lambda_0 \log e - \lambda \log e \Big)$$

$$= \log \frac{\lambda}{\lambda_0} \sum_k k \frac{\lambda^k e^{-\lambda}}{k!} + (\lambda_0 - \lambda) \log e \sum_k \frac{\lambda^k e^{-\lambda}}{k!}$$

$$= \lambda \log \frac{\lambda}{\lambda_0} + \frac{\lambda_0 - \lambda}{\ln 2} \tag{3.60}$$

where in the last equality it is used that $E[k] = \lambda$, $\sum_k p(k) = 1$ and $\log e = \frac{1}{\ln 2}$.

The relative entropy can be regarded as a divergence from one distribution to another. However, one should be careful here since it is not a distance in a natural meaning. From a mathematical point of view, a distance should be seen as a metric. That is, if a function $g(x, y)$ is a metric between x and y, it should hold that

- $g(x, y) \geq 0$ with equality if and only if $x = y$ (i.e., nonnegative),
- $g(x, y) = g(y, x)$ (i.e., symmetry), and
- $g(x, y) + g(y, z) \geq g(x, z)$ (i.e., triangular inequality).

A divergence does not have to fulfill the symmetry and triangular inequality criteria, which is the case for the relative entropy. In Theorem 3.4, it will be stated that the relative entropy is nonnegative. The other two criteria are easily shown not to hold. In the next example, it is shown that the relative entropy is not symmetric.

Example 3.11 Consider a binary random variable, $X \in \{0, 1\}$, and compare two distributions. First assume that the values are equally probable,

$$p(0) = p(1) = 1/2$$

and, second, assume a skew distribution,

$$q(0) = 1/4 \text{ and } q(1) = 3/4.$$

The relative entropy from p to q is then

$$D(p||q) = \frac{1}{2} \log \frac{1/2}{1/4} + \frac{1}{2} \log \frac{1/2}{3/4}$$

$$= 1 - \frac{1}{2} \log 3 \approx 0.2075 \tag{3.61}$$

On the other hand, the relative entropy from q to p is

$$D(q||p) = \frac{1}{4}\log\frac{1/4}{1/2} + \frac{3}{4}\log\frac{3/4}{1/2}$$
$$= \frac{3}{4}\log 3 - 1 \approx 0.1887 \tag{3.62}$$

That is, the relative entropy is not a symmetric measure.

While the previous example shows that the relative entropy is, in general, nonsymmetric, the next example shows that the triangular inequality does not hold.

Example 3.12 Consider three binary distributions p, q, and r defined by

$$p(0) = 1/2 \qquad p(1) = 1/2$$
$$q(0) = 1/3 \qquad q(1) = 2/3$$
$$r(0) = 1/6 \qquad r(1) = 5/6$$

Then $D(p||q) = 0.085$, $D(q||r) = 0.119$, and $D(p||r) = 0.424$. Hence,

$$D(p||q) + D(q||r) \not\geq D(p||r) \tag{3.63}$$

which contradicts the triangle inequality. Hence, in general, the relative entropy does not satisfy the triangle inequality.

In statistics, it is important to find a measure describing the symmetric relation between two distributions. In Problem 3.16, it is shown that modified versions of the relative entropy can be used for this purpose.

Even though the relative entropy cannot be viewed as a measure of difference between distributions, it is often very helpful in deriving properties for the other information measures. The mutual information can be expressed as a special case of the relative entropy as

$$I(X;Y) = E_{XY}\left[\frac{p(X,Y)}{p(X)p(Y)}\right] = D(p(X,Y)||p(X)p(Y)) \tag{3.64}$$

The mutual information is the information divergence from the joint distribution to the independent case, i.e. the information divergence describes the relation between X and Y.

Another aspect of the relative entropy to consider is the relationship with the entropy function. Consider a random variable with k possible outcomes and probability distribution $p(x)$. Let $u(x) = 1/k$ be the uniform distribution for the same set of outcomes. Then,

$$H(X) = -\sum_x p(x)\log p(x) = \log k - \sum_x p(x)\log p(x)k$$

$$= \log k - \sum_x p(x)\log\frac{p(x)}{u(x)} = \log k - D(p||u) \tag{3.65}$$

That is, the relative entropy from $p(x)$ to $u(x)$ is the difference between the entropy based on the true distribution and the maximum value of the entropy. Since the maximum value is achieved by the uniform distribution, the relative entropy is a measure of how much $p(x)$ diverges from the uniform distribution.

By using the IT-inequality in Lemma 3.1, it can be shown that the relative entropy is a nonnegative function. The outcomes with $p(x) = 0$ will not give any contribution to the sum, but to avoid any problems in the derivations let S_p be the set of x such that $p(x) > 0$, i.e., $S_p = \{x|p(x) > 0\}$. Then

$$D(p||q) = \sum_{x\in S_p} p(x)\log\frac{p(x)}{q(x)} \tag{3.66}$$

$$= -\sum_{x\in S_p} p(x)\log\frac{q(x)}{p(x)}$$

$$\geq -\sum_{x\in S_p} p(x)\Big(\frac{q(x)}{p(x)} - 1\Big)\log e$$

$$= \Big(\sum_{x\in S_p} p(x) - \sum_{x\in S_p} q(x)\Big)\log e \geq (1-1)\log e = 0 \tag{3.67}$$

with equality if and only if $\frac{q(x)}{p(x)} = 1$, i.e. when $p(x) = q(x)$ for all x. The result is expressed in the next theorem.

Theorem 3.3 Given two probability distributions $p(x)$ and $q(x)$ for the same sample set, the relative entropy is nonnegative

$$D(p||q) \geq 0 \tag{3.68}$$

with equality if and only if $p(x) = q(x)$ for all x. □

An alternative proof of the above theorem can be derived from the log-sum inequality, Theorem 2.8, as

$$D(p||q) = \sum_x p(x)\log\frac{p(x)}{q(x)} \geq \sum_x p(x)\log\frac{\sum_x p(x)}{\sum_x q(x)} = 1\log\frac{1}{1} = 0 \tag{3.69}$$

Since the mutual information can be expressed as the relative entropy, the following corollary follows immediately.

Corollary 3.1 The mutual information for two random variables, X and Y, is a nonnegative function,

$$I(X; Y) \geq 0 \tag{3.70}$$

with equality if and only if X and Y are independent. □

By knowing the mutual information is nonnegative, the relation between the conditional and unconditioned entropy can be further examined. Since

$$I(X; Y) = H(X) - H(X|Y) \geq 0 \tag{3.71}$$

the conditioned entropy cannot exceed the unconditioned. The result can be stated as a corollary.

Corollary 3.2 The conditional entropy is upper bounded by the unconditional entropy, i.e.

$$H(X|Y) \leq H(X) \tag{3.72}$$

with equality if and only if X and Y are independent. □

Interpreted as uncertainty, it means the uncertainty about a random variable X will not increase by viewing the side information Y. If this side information does not have anything to do with the considered variable, meaning that they are independent, the uncertainty will not change. From (3.72), together with (3.45), the joint entropy does not exceed the sum of the individual entropies,

$$H(X, Y) = H(X|Y) + H(Y) \leq H(X) + H(Y) \tag{3.73}$$

Generalization of n-dimensional vectors gives the following theorem.

Theorem 3.4 Let $\boldsymbol{X} = (X_1, \dots, X_n)$ be an n-dimensional random vector drawn according to $p(x_1, \dots, x_n)$. Then

$$H(X_1, \dots, X_n) \leq \sum_{i=1}^{n} H(X_i) \tag{3.74}$$

with equality if and only if all X_i are independent. □

That is, the uncertainty is minimized when considering a random vector as a whole, instead of individual variables. In other words, the relationship between the variables should be taken into account when minimizing the uncertainty.

3.3.1 Convexity of Information Measures

In Definition 2.4, the terminology of convex functions was introduced. It is a class of function with special interest in, for example, optimization since there are no local optima in the interval. In this section, the convexity of the information measures will be investigated. First, the relative entropy will be shown to be convex. With this as a tool, the entropy can be shown to be concave and then the convexity of the mutual information is investigated.

Our previous definition of a convex function is stated for one-dimensional functions. Therefore, to start with a generalization of the definition is given. A straightforward way is to say that a multidimensional function is convex if it is convex in all dimensions. For the two-dimensional case, the function surface resembles a bowl. Comparing with Figure 2.7, the two-dimensional argument $\lambda(x_1, y_1) + (1 - \lambda)(x_2, y_2)$, for $0 \leq \lambda \leq 1$, describes a straight line between the points (x_1, y_1) and (x_2, y_2) in the argument plane. The coordinates for this line can be rewritten as $(\lambda x_1 + (1 - \lambda)x_2, \lambda y_1 + (1 - \lambda)y_2)$ for λ between 0 and 1. Considering the two-dimensional function $g(x, y)$, the values corresponding to the endpoints are $z_1 = g(x_1, y_1)$ and $z_2 = g(x_2, y_2)$. If the function value along the argument line, $g(\lambda(x_1, y_1) + (1 - \lambda)(x_2, y_2))$ never exceeds the corresponding value at the line, $\lambda g(x_1, y_1) + (1 - \lambda)g(x_2, y_2)$, the function $g(x, y)$ is a convex function. That is, $g(x, y)$ is convex over the region $\mathcal{A}$ if

$$g(\lambda(x_1, y_1) + (1 - \lambda)(x_2, y_2)) \leq \lambda g(x_1, y_1) + (1 - \lambda)g(x_2, y_2) \tag{3.75}$$

for all λ such that $0 \leq \lambda \leq 1$ and all $(x_1, y_1), (x_2, y_2) \in \mathcal{A}$. Here $\mathcal{A}$ denotes a two-dimensional convex region, i.e. a straight line between two points in the region should never be outside the region. The regions considered in this text are easily verified to satisfy this criterion. The above reasoning for convexity of functions can easily be generalized for n-dimensional functions.

Definition 3.10 Let $\boldsymbol{x}^{(1)} = (x_1^{(1)}, \dots, x_n^{(1)})$ and $\boldsymbol{x}^{(2)} = (x_1^{(2)}, \dots, x_n^{(2)})$ be two n-dimensional vectors in the region $\mathcal{A}$ and $g(\boldsymbol{x})$ an n-dimensional function. Then, $g(\boldsymbol{x})$ is a convex function in $\mathcal{A}$ if

$$\lambda g\big(\boldsymbol{x}^{(1)} + (1 - \lambda)\boldsymbol{x}^{(2)}\big) \leq g\big(\lambda \boldsymbol{x}^{(1)}\big) + (1 - \lambda)g\big(\boldsymbol{x}^{(2)}\big) \tag{3.76}$$

for all λ such that $0 \leq \lambda \leq 1$ and all $\boldsymbol{x}^{(1)}, \boldsymbol{x}^{(2)} \in \mathcal{A}$. □

The relative entropy is a two-dimensional function in the probability pair (p, q) and can thus be checked for convexity. Then, consider the four probability distributions $p_1(x)$, $p_2(x)$, $q_1(x)$, and $q_2(x)$ over the same sample space $\mathcal{X}$. For λ between 0 and 1, two new distributions can be formed as

$$p_\lambda(x) = \lambda p_1(x) + (1 - \lambda)p_1(x) \tag{3.77}$$

$$q_\lambda(x) = \lambda q_1(x) + (1 - \lambda)q_1(x) \tag{3.78}$$

Considering the relative entropy from p_λ to q_λ, it can be seen that

$$\begin{aligned}
D\big(p_\lambda||q_\lambda\big) &= D\big(\lambda p_1 + (1-\lambda)p_2||\lambda q_1 + (1-\lambda)q_2\big) \\
&= \sum_x (\lambda p_1(x) + (1-\lambda)p_2(x)) \log \frac{\lambda p_1(x) + (1-\lambda)p_2(x)}{\lambda q_1(x) + (1-\lambda)q_2(x)} \\
&\le \sum_x \lambda p_1(x) \log \frac{\lambda p_1(x)}{\lambda q_1(x)} + \sum_x (1-\lambda)p_1(x) \log \frac{(1-\lambda)p_1(x)}{(1-\lambda)q_1(x)} \\
&= \lambda \sum_x p_1(x) \log \frac{p_1(x)}{q_1(x)} + (1-\lambda) \sum_x p_1(x) \log \frac{p_1(x)}{q_1(x)} \\
&= \lambda D\big(p_1||q_1\big) + (1-\lambda)D\big(p_2||q_2\big)
\end{aligned} \tag{3.79}$$

where the inequality is a direct application of the log-sum inequality in Theorem 2.8. Hence, as stated in the next theorem, the relative entropy is a convex function.

Theorem 3.5 The relative entropy is convex in (p, q). □

From (3.65), the entropy can be expressed as $H_p(X) = \log k - D(p||u)$, where u is uniformly distributed. Again using $p_\lambda(x) = \lambda p_1(x) + (1-\lambda)p_1(x)$ to get

$$\begin{aligned}
H_{p_\lambda}(X) &= \log k - D\big(\lambda p_1 + (1-\lambda)p_2||u\big) \\
&\ge \log k - \lambda D\big(p_1||u\big) - (1-\lambda)D\big(p_2||u\big) \\
&= \lambda(\log k - D(p_1||u)) + (1-\lambda)(\log k - D(p_2||u)) \\
&= \lambda H_{p_1}(X) + (1-\lambda)H_{p_2}(X)
\end{aligned} \tag{3.80}$$

where the inequality follows from the convexity of the relative entropy. The above result is stated in the following theorem.

Theorem 3.6 The entropy is concave in p. □

The mutual information can be written as $I(X;Y) = H(Y) - H(Y|X)$. Hence, it consists of two parts that needs to be treated separately. The first case to consider is two distributions on X, $p_1(x)$ and $p_2(x)$, while the conditional probability on Y, $p(y|x)$, is fixed. Then, again form $p_\lambda(x) = \lambda p_1(x) + (1-\lambda)p_2(x)$. The unconditional probability on Y then becomes

$$\begin{aligned}
p_\lambda(y) &= \sum_x p_\lambda(x)p(y|x) \\
&= \lambda \sum_x p_1(x)p(y|x) + (1-\lambda) \sum_x p_2(x)p(y|x) \\
&= \lambda p_1(y) + (1-\lambda)p_2(y)
\end{aligned} \tag{3.81}$$

Meaning that introducing $p_\lambda(x)$ gives

$$H_{p_\lambda(y)}(Y) \geq \lambda H_{p_1(y)}(Y) + (1-\lambda)H_{p_2(y)}(Y) \tag{3.82}$$

since the entropy is concave. On the other hand, the conditional entropy is

$$\begin{aligned} H_{p_\lambda(x)}(Y|X) &= -\sum_{x,y} p_\lambda(x)p(y|x)\log p(y|x) \\ &= -\lambda \sum_{x,y} p_1(x)p(y|x)\log p(y|x) \\ &\quad - (1-\lambda)\sum_{x,y} p_2(x)p(y|x)\log p(y|x) \\ &= \lambda H_{p_1(x)}(Y|X) + (1-\lambda)H_{p_2(x)}(Y|X) \end{aligned} \tag{3.83}$$

Putting things together concludes

$$\begin{aligned} I_{p_\lambda(x)}(X;Y) &= H_{p_\lambda(y)}(Y) - H_{p_\lambda(x)}(Y|X) \\ &\geq \lambda H_{p_1(y)}(Y) + (1-\lambda)H_{p_2(y)}(Y) \\ &\quad - \lambda H_{p_1(x)}(Y|X) - (1-\lambda)H_{p_2(x)}(Y|X) \\ &= \lambda I_{p_1(x)}(X;Y) + (1-\lambda)I_{p_2(x)}(X;Y) \end{aligned} \tag{3.84}$$

That is, for fixed $p(y|x)$ the mutual information $I(X;Y)$ is concave in $p(x)$.

Similarly, if $p(x)$ is fixed and considering two distributions on the conditional probability, $p_1(y|x)$ and $p_2(y|x)$, introduce

$$p_\lambda(y|x) = \lambda p_1(y|x) + (1-\lambda)p_2(y|x) \tag{3.85}$$

The corresponding joint and marginal probabilities are

$$p_\lambda(x,y) = p(x)p_\lambda(y|x) = \lambda p_1(x,y) + (1-\lambda)p_2(x,y) \tag{3.86}$$

and

$$p_\lambda(y) = \sum_x p(x)p_\lambda(y|x) = \lambda p_1(y) + (1-\lambda)p_2(y) \tag{3.87}$$

where $p_i(x,y) = p(x)p_i(y|x)$ and $p_i(y) = \sum_x p(x)p_i(y|x)$. Then by writing the mutual information as the relative entropy and using its convexity gives

$$\begin{aligned} I_{p_\lambda(y|x)}(X;Y) &= D(p_\lambda(X,Y)||p(X)p_\lambda(Y)) \\ &\leq \lambda D(p_1(X,Y)||p(X)p_1(Y)) \\ &\quad + (1-\lambda)D(p_2(X,Y)||p(X)p_2(Y)) \\ &= \lambda I_{p_1(y|x)}(X;Y) + (1-\lambda)I_{p_2(y|x)}(X;Y) \end{aligned} \tag{3.88}$$

That is, for fixed $p(x)$ the mutual information is convex in $p(y|x)$. The convexity of the mutual information can be summarized in the following theorem.

Theorem 3.7 The mutual information $I(X;Y)$ is

- concave in $p(x)$ if $p(y|x)$ is fixed.
- convex in $p(y|x)$ if $p(x)$ is fixed. □

3.4 ENTROPY OF SEQUENCES

In the previous section, the information measures are defined for random variables. Often it is desirable to use as well for random processes where the variables in a sequence are statistically dependent. Then, to generalize the entropy measure complete sequences must be considered. In this section, first a famous result on data processing will be derived, called the data-processing lemma. After this, the entropy rate will be defined, which is the corresponding entropy measure for random processes.

For the first part, consider a Markov chain with three variables X, Y, and Z. Their dependencies are described in Figure 3.3. The process A transforms X into Y, and process B transforms Y into Z. These processes are very general and can, for example, represent preprocessing, postprocessing, or transmission of data.

The assumed Markov property gives that X and Z are independent when conditioned on Y, i.e.

$$P(XZ|Y) = P(X|Y)P(Z|XY) = P(X|Y)P(Z|Y) \tag{3.89}$$

where the second equality follows from the Markov condition. Then the mutual information between the end points can be derived and bounded in two ways,

$$\begin{aligned} I(X;Z) &= H(X) - H(X|Z) \\ &\leq H(X) - H(X|YZ) = H(X) - H(X|Y) = I(X;Y) \end{aligned} \tag{3.90}$$

and

$$\begin{aligned} I(X;Z) &= H(Z) - H(Z|X) \\ &\leq H(Z) - H(Z|XY) = H(Z) - H(Z|Y) = I(Z;Y) \end{aligned} \tag{3.91}$$

This result is stated as the data-processing lemma.

Lemma 3.2 (Data-Processing Lemma) Let the random variables X, Y, and Z form a Markov chain, $X \rightarrow Y \rightarrow Z$. Then

$$I(X;Z) \leq I(X;Y) \tag{3.92}$$

$$I(X;Z) \leq I(Y;Z) \tag{3.93}$$

□

An interpretation of the lemma can be viewed in the following way. Assume first that X is transformed into Y by process A. This, for example, can be a transmission of data over a channel distorting the signals (e.g., wired or wireless

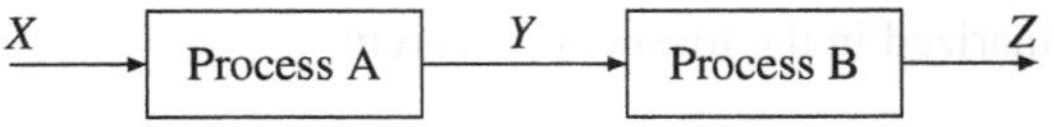

Figure 3.3 The dependencies used in the data-processing lemma.

communication or writing and reading of a CD, DVD, or flash memory). The aim of the receiver is then to get as much information about X by observing Y. It is common to perform postprocessing, which in this model is represented by process B. The data-processing lemma states that the information about X by viewing Z cannot exceed the information about X by viewing Y. In other words, the information about X will not increase by postprocessing, it can only decrease. In practice, however, postprocessing is often used to transform the information into another representation where the information is easier accessible for interpretation. For example, it is easier to understand an image when viewed on a screen than it is from the data received.

Similarly, process A can represent preprocessing and process B the transmission. Then, the data-processing lemma states that the information cannot increase by the preprocessing. Still, in practice it is common to use preprocessing in communication systems to transform data into appropriate representations. Summarizing, the lemma states that the information cannot increase by neither pre- nor postprocessing. The information can only decrease in the processing.

3.4.1 Entropy Rate

Next, the description will go to a more general description of information measure for sequences. In many cases, there is a dependency between symbols in a sequence, which can be modeled by a random process. In this section, two natural generalizations of the entropy function will be introduced. It turns out that these two definitions are in fact equivalent. The measure can in many cases be used and interpreted in the same way for a random process as the entropy for random variables.

A natural way to define the entropy per symbol for a sequence is by treating the sequence as a multidimensional random variable and averaging over the number of symbols. As the length of the sequence tends to be infinity, the following definition is obtained.

Definition 3.11 The *entropy rate* of a random process is

$$H_\infty(X) = \lim_{n\to\infty} \frac{1}{n} H(X_1X_2 \dots X_n) \tag{3.94}$$

□

To see that this is a natural generalization of the entropy function where the variables in a sequence are considered independent, consider a sequence of i.i.d. variables as in the next example.

Example 3.13 Consider a sequence of i.i.d. random variables with entropy $H(X)$. Then the entropy rate equals the entropy function since

$$\begin{aligned} H_\infty(X) &= \lim_{n\to\infty} \frac{1}{n} H(X_1 \dots X_n) = \lim_{n\to\infty} \frac{1}{n} \sum_i H(X_i|X_1 \dots X_{i-1}) \\ &= \lim_{n\to\infty} \frac{1}{n} \sum_i H(X_i) = \lim_{n\to\infty} H(X)\frac{1}{n} \sum_i 1 = H(X) \end{aligned} \tag{3.95}$$

An alternative definition for the entropy of one symbol in a random process is to consider the entropy of the nth variable in the sequence, conditioned on all the previous case. By letting $n \to \infty$, it is the entropy of one symbol conditioned on an infinite sequence.

Definition 3.12 The *alternative entropy rate* of a random process is

$$H(X|X^{\infty}) = \lim_{n\to\infty} H(X_n|X_1X_2\dots X_{n-1}) \tag{3.96}$$

□

Clearly, for the case of i.i.d. symbols in the sequence this alternative definition also gives the entropy $H(X)$. To see how the two definitions relates, rewrite the entropy with the chain rule,

$$\frac{1}{n}H(X_1\dots X_n) = \frac{1}{n}\sum_i H(X_i|X_1\dots X_{i-1}) \tag{3.97}$$

The right-hand side is the arithmetic mean of $H(X_i|X_1\dots X_{i-1})$. By the law of large numbers, as $n \to \infty$ this will approach $H(X|X^{\infty})$. Hence, asymptotically as the length of the sequence grows to infinity, the two definitions for the entropy rate are equal. This important result is stated in the next theorem. In the continuation of the text, the notation from the first definition will be adopted.

Theorem 3.8 The entropy rate and the alternative entropy rate are equivalent, i.e.

$$H_{\infty}(X) = H(X|X^{\infty}) \tag{3.98}$$

□

Consider a stationary random process, then

$$H(X_n|X_1\dots X_{n-1}) \le H(X_n|X_2\dots X_{n-1}) = H(X_{n-1}|X_1\dots X_{n-2}) \tag{3.99}$$

where the last equality follows since from the stationarity of the process. Hence, $H(X_n|X_1\dots X_{n-1})$ is a decreasing function in n and a lower bound for the entropy function is obtained from

$$H(X_n|X_1\dots X_{n-1}) \le \cdots \le H(X_2|X_1) \le H(X_1) = H(X) \le \log k \tag{3.100}$$

Finally, since the entropy is a nonnegative function, the following relation between the entropy rate and the entropy can be concluded.

Theorem 3.9 For a stationary random process, the entropy rate is bounded by

$$0 \le H_{\infty}(X) \le H(X) \le \log k \tag{3.101}$$

□

In Figure 3.4, the relation between $\log k$, $H(X)$, $H(H_n|X_1\dots X_{n-1})$ and $H_{\infty}(X)$ is shown as a function of n. One natural conclusion is that the uncertainty of the sequence is less, if the dependency between symbols is taken into consideration.

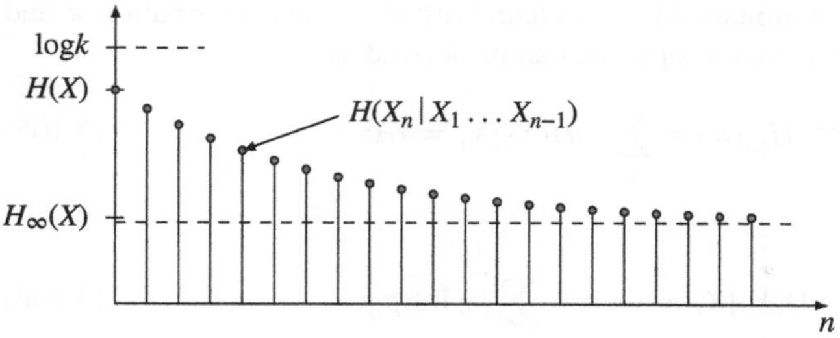

Figure 3.4 The relation between $H_\infty(X)$ and $H(X)$.

3.4.2 Entropy Rate of Markov Models

So far the entropy rate has been treated for the class of stationary random processes. If the theory is limited to the often used Markov chains, it is possible to be more specific on derivations of the entropy rate. From the unit memory property and stationarity of a Markov process, the conditional entropy can be written as $H(X_n|X_1 \dots X_{n-1}) = H(X_n|X_{n-1})$. Then, the entropy rate is

$$\begin{aligned} H_\infty(X) &= \lim_{n\to\infty} H(X_n|X_1 \dots X_{n-1}) \\ &= \lim_{n\to\infty} H(X_n|X_{n-1}) = H(X_2|X_1) \\ &= \sum_{i,j} P(X_1 = x_i, X_2 = x_j) \log P(X_2 = x_j|X_1 = x_i) \\ &= \sum_i P(X_1 = x_i) \sum_j P(X_2 = x_j|X_1 = x_i) \log P(X_2 = x_j|X_1 = x_i) \\ &= \sum_i H(X_2|X_1 = x_i) P(X_1 = x_i) \end{aligned} \tag{3.102}$$

where

$$H(X_2|X_1 = x_i) = \sum_j P(X_2 = x_j|X_1 = x_i) \log P(X_2 = x_j|X_1 = x_i) \tag{3.103}$$

In (3.102), the transition probability is given by the state transition matrix for the Markov chain

$$P = [p_{ij}]_{i,j\in\{1,2,\dots,k\}}, \tag{3.104}$$

where $p_{ij} = P(X_2 = x_j|X_1 = x_i)$ and k is the number of states. With the stationary distribution given by $\pi_i = P(X_1 = x_i)$, the entropy rate for a Markov chain can be derived as stated in the next theorem.

Theorem 3.10 For a stationary Markov chain with stationary distribution $\boldsymbol{\pi}$ and transition matrix $P = [p_{ij}]$, the entropy rate can be derived as

$$H_\infty(X) = \sum_i \pi_i H(X_2|X_1 = x_i) \tag{3.105}$$

where

$$H(X_2|X_1 = x_i) = -\sum_j p_{ij} \log p_{ij}. \tag{3.106}$$

□

In the next example, the Markov chain used in Chapter 2 is reused.

Example 3.14 The Markov chain shown in Figure 2.8 has the state transition matrix

$$P = \begin{pmatrix} \frac{1}{3} & \frac{2}{3} & 0 \\ \frac{1}{4} & 0 & \frac{3}{4} \\ \frac{1}{2} & \frac{1}{2} & 0 \end{pmatrix} \tag{3.107}$$

In Example 2.17, the steady-state distribution was derived as

$$\boldsymbol{\pi} = \begin{pmatrix} \frac{15}{43} & \frac{16}{43} & \frac{12}{43} \end{pmatrix} \tag{3.108}$$

The conditional entropies are derived row-wise in P,

$$H(X_2|X_1 = s_1) = h\left(\tfrac{1}{3}\right) = \log 3 - \tfrac{2}{3} \tag{3.109}$$

$$H(X_2|X_1 = s_2) = h\left(\tfrac{1}{4}\right) = 2 - \tfrac{3}{4}\log 3 \tag{3.110}$$

$$H(X_2|X_1 = s_3) = h\left(\tfrac{1}{2}\right) = 1 \tag{3.111}$$

and the entropy rate becomes

$$\begin{aligned} H_\infty(X) &= \pi_1 H(X_2|X_1 = s_1) + \pi_2 H(X_2|X_1 = s_2) + \pi_3 H(X_2|X_1 = s_3) \\ &= \tfrac{15}{43} h\left(\tfrac{1}{3}\right) + \tfrac{16}{43} h\left(\tfrac{1}{4}\right) + \tfrac{12}{43} h\left(\tfrac{1}{2}\right) \\ &= \tfrac{3}{43}\log 3 + \tfrac{34}{43} \approx 0.9013 \text{ bit/symbol} \end{aligned} \tag{3.112}$$

In this example, the entropy of the stationary distribution is

$$H\left(\frac{15}{43}, \frac{16}{43}, \frac{12}{43}\right) \approx 1.58 \tag{3.113}$$

which can be seen as a measure of the entropy when the relations between the symbols are neglected. The uncertainty per symbol is lower when the relations in the sequence are considered.

PROBLEMS

3.1 The so-called IT-inequality is in the text described as a consequence of the fact that the functions $\ln x$ lower than $x - 1$, with equality if and only if $x = 1$. Show that this relation

$$x - 1 \geq \log_b x, \quad b > 1$$

only holds for the natural base, i.e., when $b = e$.

3.2 Use the IT-inequality to show that, for all positive x, $\ln x \geq 1 - \frac{1}{x}$ with equality if and only if $x = 1$.

3.3 The outcome of a throw with a fair die is denoted by X. Then, let Y be Even if X is even and Odd otherwise. Determine

(a) $I(X = 2; Y = \text{Even})$, $I(X = 3; Y = \text{Even})$, $I(X = 2 \text{ or } X = 3; Y = \text{Even})$.

(b) $I(X = 4)$, $I(Y = \text{Odd})$.

(c) $H(X)$, $H(Y)$.

(d) $H(X, Y)$, $H(X|Y)$, $H(Y|X)$.

(e) $I(X; Y)$.

3.4 Let X_1 and X_2 be two variables describing the outcome of a throw with two dice and let $Y = X_1 + X_2$ be the total number.

(a) What is the probability function for the stochastic variable Y?

(b) Determine $H(X_1)$ and $H(Y)$.

(c) Determine $I(Y; X_1)$.

3.5 The joint probability of X and Y is given by

$P(X, Y)$

X	$Y = a$	$Y = b$	$Y = c$
0	$\frac{1}{12}$	$\frac{1}{6}$	$\frac{1}{3}$
1	$\frac{1}{4}$	0	$\frac{1}{6}$

Calculate

(a) $P(X)$, $P(Y)$, $P(X|Y)$, and $P(Y|X)$

(b) $H(X)$ and $H(Y)$

(c) $H(X|Y)$ and $H(Y|X)$

(d) $H(X, Y)$

(e) $I(X, Y)$.

3.6 The joint probability of X and Y is given by

	$P(X,Y)$		
X	$Y=a$	$Y=b$	$Y=c$
A	$\frac{1}{12}$	$\frac{1}{6}$	0
B	0	$\frac{1}{9}$	$\frac{1}{5}$
C	$\frac{1}{18}$	$\frac{1}{4}$	$\frac{2}{15}$

Calculate

(a) $P(X)$, $P(Y)$, $P(X|Y)$, and $P(Y|X)$

(b) $H(X)$ and $H(Y)$

(c) $H(X|Y)$ and $H(Y|X)$

(d) $H(X,Y)$

(e) $I(X,Y)$.

3.7 In an experiment, there are two coins. The first is a fair coin, while the second has Heads on both sides. Choose with equal probability one of the coins, and flip it twice. How much information do you get about the identity of the coin by studying the number of Heads from the flips?

3.8 An urn has 18 balls; ten blue, five red, and three green. Someone draws one ball from the urn and puts it in a box without looking. Let the random variable X denote the color of this first ball. Next, you draw a ball from the urn and let Y denote the color of this second ball.

(a) What is the uncertainty of X?

(b) What is the uncertainty of Y if you first open the box to get the color of the first ball?

(c) What is the uncertainty of Y if you do not open the box?

(d) Assume that you do not open the box. How much information about X do you get from Y?

3.9 Consider two dice where the first has equal probability for all six numbers and the second has a small weight close to the surface of number 1. Let X be the outcome of a roll with one of the dice, then the corresponding probability distributions for the dice are given below.

x :	1	2	3	4	5	6
$p(x)$:	$\frac{1}{6}$	$\frac{1}{6}$	$\frac{1}{6}$	$\frac{1}{6}$	$\frac{1}{6}$	$\frac{1}{6}$
$q(x)$:	$\frac{1}{14}$	$\frac{1}{7}$	$\frac{1}{7}$	$\frac{1}{7}$	$\frac{1}{7}$	$\frac{5}{14}$

(a) What is the entropy of a throw with the fair die and the manipulated die, respectively?

(b) What is $D(p||q)$?

(c) What is $D(q||p)$?

3.10 The joint distribution of X and Y is given by

$$p(x,y) = k^2 2^{-(x+y)}, \qquad x,y = 0,1,2,\dots$$

(a) Determine k.

(b) Derive $P(X < 4, Y < 4)$.

(c) Derive the joint entropy.

(d) Derive the conditional probability $H(X|Y)$.

3.11 The two distributions $p(x,y)$ and $q(x,y)$ are defined over the same set of outcomes. Verify that

$$\begin{aligned} D\big(p(x,y)||q(x,y)\big) &= D\big(p(x)||q(x)\big) + \sum_x D\big(p(y|x)||q(y|x)\big)p(x) \\ &= D\big(p(y)||q(y)\big) + \sum_y D\big(p(x|y)||q(x|y)\big)p(y) \end{aligned}$$

and that, if X and Y are independent,

$$D\big(p(x,y)||q(x,y)\big) = D\big(p(x)||q(x)\big) + D\big(p(y)||q(y)\big).$$

3.12 Sometimes a function called *Cross Entropy*, closely related to the relative entropy, is used. It is defined as

$$H(p,q) = -\sum_x p(x)\log q(x).$$

Show that

$$H(p,q) = D\big(p||q\big) - H_p(X).$$

3.13 **(a)** Show that if α, β, and γ form a probability distribution, then

$$H(\alpha,\beta,\gamma) = h(\alpha) + (1-\alpha)h\left(\tfrac{\beta}{1-\alpha}\right).$$

(b) Show that if $p_1, p_2, p_3, \dots, p_n$ form a probability distribution, then

$$H(p_1,p_2,\dots,p_n) = h(p_1) + (1-p_1)H\left(\tfrac{p_2}{1-p_1}, \tfrac{p_3}{1-p_1}, \dots, \tfrac{p_n}{1-p_1}\right).$$

3.14 Consider two urns, numbered 1 and 2. Urn 1 has four white balls and three black balls, while Urn 2 has three white balls and seven black. Choose one of the urns with equal probability, and draw one ball from it. Let X be the color of that ball and Y the number of the chosen urn.

(a) Derive the uncertainty of X.

(b) How much information is obtained about Y when observing X?

(c) Introduce a third urn, Urn 3, with only one white ball (and no black). Redo problems (a) and (b) for this case.

3.15 Show that

$$I(X;Y,Z) = I(X;Y) + I(X;Z|Y)$$

3.16 In statistics, sometimes it is desirable to compare distributions and have a measure of how different they are. One way is, of course, to use the relative entropy $D(p||q)$ as a measure. However, the relative entropy is not a symmetric measure. Since, symmetry

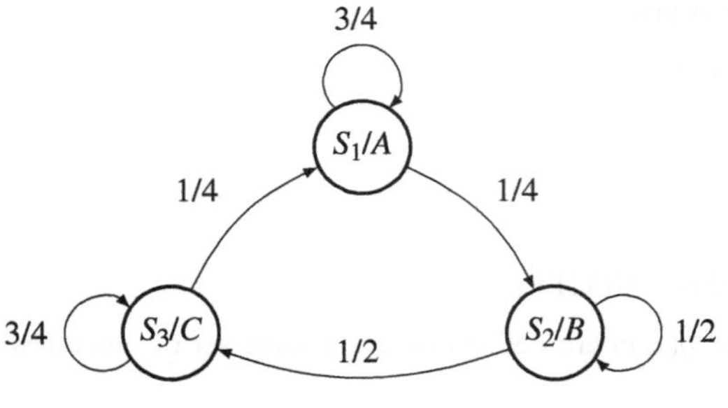

Figure 3.5 A Markov graph for the source in Problem 3.18.

is one of the basic criterion for a metric this property is desirable. Below are given two symmetric measures based on the relative entropy.

(a) One direct way to get a symmetric measurement of the difference between two distributions is the Jeffrey's divergence [9]

$$D_J(p||q) = D(p||q) + D(q||p)$$

named after the statistician Harold Jeffreys. Show that it can be written as (for discrete distributions)

$$D_J(p||q) = \sum_x (p(x) - q(x)) \log \frac{p(x)}{q(x)}.$$

(b) To get around the problem that there can occur infinite values in the Jeffrey's divergence, Lin introduced in 1991 the so called Jensen–Shannon [10] divergence,

$$D_{JS}(p||q) = \frac{1}{2}D\left(p||\frac{p+q}{2}\right) + \frac{1}{2}D\left(q||\frac{p+q}{2}\right).$$

Show that an alternative way to write this is

$$D_{JS}(p||q) = H\left(\frac{p+q}{2}\right) - \frac{H(p) + H(q)}{2}. \tag{3.114}$$

3.17 Let $p(x)$ and $q(x)$ be two probability functions for the random variable X. Use the relative entropy to show that

$$\sum_x \frac{p^2(x)}{q(x)} \geq 1$$

with equality if and only if $p(x) = q(x)$ for all x.

3.18 A Markov source with output symbols $\{A, B, C\}$, is characterized by the graph in Figure 3.5.

(a) What is the stationary distribution for the source?

(b) Determine the entropy of the source, H_∞.

(c) Consider a memory-less source with the same probability distribution as the stationary distribution calculated in (a). What is the entropy for the memory-less source?

3.19 The engineer Inga is going to spend her vacation in an archipelago with four main islands. The islands are connected with four different boat lines, and one sightseeing tour around the largest island (see the map in Figure 3.6).

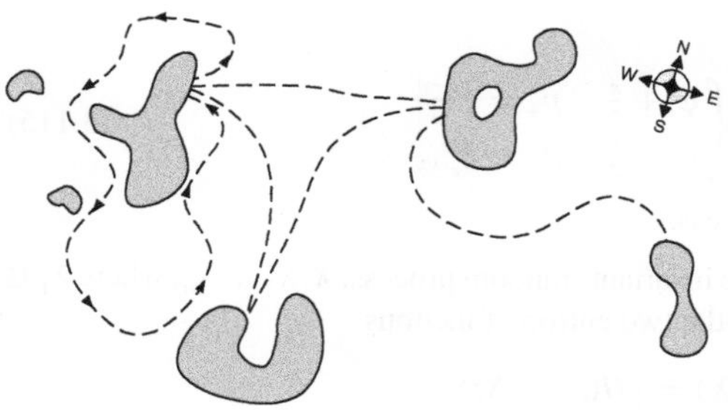

Figure 3.6 A map over the islands and the boat connections.

To avoid planning the vacation route too much, she decides to take a boat every day. She will choose one of the boat lines going out from the island with equal probability. All the boat lines are routed both ways every day, except the sightseeing tour that is only one way.

(a) When Inga has traveled around in the archipelago for a long time, what is the probabilities for being on each of the islands?

(b) Inga has promised to write home and tell her friends about her travel. How many bits, on average, does she need to write per day to describe her route? Assume that she will choose a starting island for her vacation according to the distribution in (a).

3.20 A man climbs an infinitely long ladder. At each time instant, he tosses a coin. If he gets Head he takes a step up on the ladder but if he gets Tail he drops down to the ground (step 0). The coin is counterfeit with $P(\text{Head}) = p$ and $P(\text{Tail}) = 1 - p$. The sequence of where on the ladder the man stands forms a Markov chain.

(a) Construct the state transition matrix for the process and draw the state transition graph.

(b) What is the entropy rate of the process?

(c) After the man has taken many steps according to the process you call him and ask if he is on the ground. What is the uncertainty about his answer? If he answers that he is not on the ground, what is the uncertainty of which step he is on? (You can trust that he is telling the truth.)

3.21 Four points are written on the unit circle (see Figure 3.7). A process moves from the current point to one of its neighbors. If the current point is $\Phi = \varphi_i$, the next point is

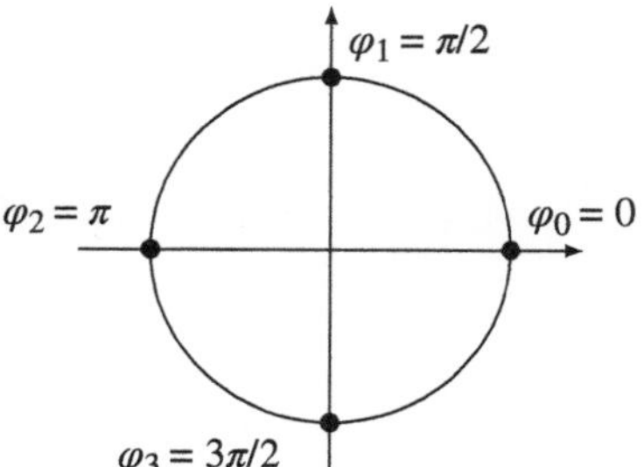

Figure 3.7 Four points on the unit circle.

chosen with probabilities

$$\Phi^+ = \begin{cases} \varphi_i + \frac{\pi}{2}, & p_+ = \frac{2\pi - \varphi_i}{2\pi} \\ \varphi_i - \frac{\pi}{2}, & p_- = \frac{\varphi_i}{2\pi} \end{cases} \tag{3.115}$$

Derive the entropy rate for the process.

3.22 Consider a discrete stationary (time invariant) random process, $X_1X_2 \dots X_n$, where $X_i \in \{x_1, x_2, \dots, x_k\}$ and k finite. Define the two entropy functions

$$H_n(X) = \frac{1}{n}H(X_1 \dots X_n)$$
$$H(X|X^n) = H(X_n|X_1 \dots X_{n-1})$$

As $n \to \infty$, these functions approach the entropy rate functions $H_\infty(X)$ and $H(X|X^\infty)$. In the course, it has been shown that these limits exist and are equal, $H_\infty(X) = H(X|X^\infty)$. In this problem, the same result is derived in an alternative way.

(a) Show that

$$H_n(X) \geq H(X|X^n).$$

(b) Show that $H_n(X)$ is a decreasing function, i.e. that

$$H_n(X) \leq H_{n-1}(X).$$

Remark: This shows $0 \leq H_\infty(X) \leq H_n(X) \leq H_{n-1}(X) \leq H(X) \leq \log k$, and that the limit exists.

(c) Show that for any fixed integer μ, in $1 \leq \mu < n$,

$$H_n(X) \leq \frac{\mu}{n}H_\mu(X) + \frac{n-\mu}{n}H(X|X^\mu).$$

(d) Use the results above to show that for all $\mu \ll n$

$$H(X|X^\infty) \leq H_\infty(X) \leq H(X|X^\mu).$$

Remark: By letting $\mu \to \infty$ this gives $H_\infty(X) = H(X|X^\infty)$.

CHAPTER 4

OPTIMAL SOURCE CODING

IN PRACTICE, IMAGES or texts contain redundant data. A text, for example, is most often still readable if every fourth letter is replaced with an erasure symbol, e.g., ★. That means these letters can be removed, and the text only takes about three fourth of the space in the memory. The same reasoning holds, e.g., for images or video where there is a lot of dependencies between the pixels. The aim of source coding is to remove as much redundancy as possible to achieve only the information in a file.

In the previous chapter, the entropy was interpreted as the average amount of information needed to describe the random variable. This should then mean that the entropy is a lower bound for how much the data, or the outcome from a random variable, can be compressed. One of the main results in Shannon's paper is the source coding theorem, where it is shown that for a vector of length n the compressed codeword length per symbol can approach the entropy as n grows to infinity. In this chapter, it will be shown that with the requirement of unique decompression, it is possible to achieve the average codeword length arbitrarily close to the entropy, but not less. A simple algorithm will also be given to construct a code, the Huffman code, that is optimal in terms of codeword length for a source with independent and identically distributed (i.i.d.) symbols. The chapter is concluded with a description of arithmetic coding.

4.1 SOURCE CODING

In Figure 4.1, a block model for a source coding system is shown. The symbols from the source is fed to the source encoder where the redundancy is removed, i.e., the sequence is compressed. The source decoder is the inverse function, and the source data are reconstructed. In the figure, X is a random variable with k outcomes. In general, X can be a vector, where the length n is also a random variable. The corresponding codeword, $\boldsymbol{Y}$, is an ℓ-dimensional vector of random variables Y_j with D outcomes. The codeword length ℓ is viewed as a random variable, and the average codeword length is $L = E[\ell]$. Normally, the code symbols will be considered to be drawn from the alphabet $\mathbb{Z}_D = \{0, 1, \dots, D-1\}$. Since the algorithms often are implemented in computers, it is common to use binary vectors, i.e., $D = 2$.

Information and Communication Theory, First Edition. Stefan Höst.

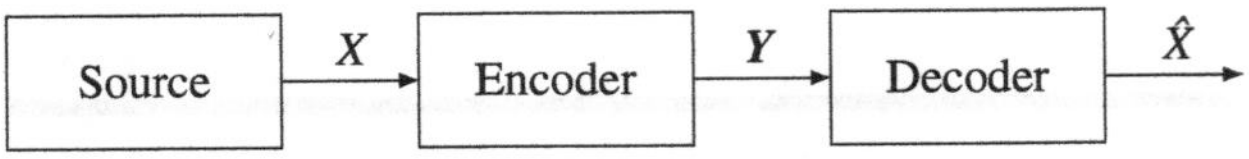

Figure 4.1 Block model for source coding system.

The decoder estimates the source vector as $\hat{X}$. In case of *lossless* source coding, the aim is to have $\hat{X} = X$. This is typically the case when the source is a text or a program code that needs to be flawlessly reconstructed at the receiver side. A typical example of this is zip compression and variants thereof. If it is acceptable with a certain limit of distortion *lossy* source coding can be used, which is the case in image, video, or audio coding, e.g., jpeg, mpeg, and mp3. Then the reconstructed message should be similar enough to the original, $\hat{\boldsymbol{X}} \approx \boldsymbol{X}$. What "similar enough" means varies from case to case. For example, the same sound quality is not expected in a phone as for a stereo equipment. Then the phone audio can be much compressed than the sound on a music file. In this section, only lossless source coding will be considered. Lossy compression will be revisited in Chapter 11 where the concept of distortion is introduced.

In the previous chapter, the entropy was considered to be a measure of the uncertainty of the outcome of a random variable. This can be interpreted as the amount of information needed to determine the outcome. It is then natural to view the entropy as a limit for how much it is possible to compress the data, in view of this statistics. In the following, it will not only be shown that indeed the entropy is a lower bound on the codeword length but also that it is a reachable bound. The latter is shown by Shannon's source coding theorem, stated in Shannon's paper from 1948 [1]. However, there it was shown using the law of large numbers and asymptotic equipartition property (AEP) whereas in this chapter a more natural construction will be used.

In Chapter 6, AEPs are introduced and both the source coding theorem and the channel coding theorem derived from it. In this chapter, the source symbols will be fixed in size, which most often is one letter from a fixed alphabet, and the codewords varying length vectors of D-ary symbols. This means the source symbols can be considered as a random variable, and source coding can be defined as follows.

Definition 4.1 (Source coding) A source code is a mapping from the outcomes of a random variable $X \in \{x_1, \dots, x_k\}$ to a vector of variable length, $\boldsymbol{y} = (y_1, \dots, y_\ell)$, where $y_i \in \{0, 1, \dots, D-1\}$. The length of the codeword $\boldsymbol{y}$ corresponding to the source symbol x is denoted as ℓ_x. □

Since the source symbols are assumed to have fixed length, the efficiency of the code can be measured by the average codeword length,

$$L = E\big[\ell_x\big] = \sum_x p(x)\ell_x \tag{4.1}$$

In a more general description, the input $\boldsymbol{X}$ can be viewed as a random vector with a random length n. In that case, both the input vector and the codeword vector are allowed to vary in length. In that case, a more suitable measure of the code efficiency can be expressed as

$$R = \frac{E[n]}{E[\ell]} \tag{4.2}$$

This will be often used in the next chapter while considering Lempel–Ziv codes.

4.2 KRAFT INEQUALITY

In Table 4.1, the probability function for a random variable X is given. There are also five different examples of codes for this variable. The first code, C_1, is an example of a direct binary mapping with equal lengths, not taking the distribution into account. With four source symbols, it is needed at least $\log 4 = 2$ bits in a vector to have unique codewords. This can be used as a reference code, representing the uncoded case. Since all codewords are equal in length, the average codeword length will also be $L(C_1) = 2$. In the other codes, symbols with high probability are mapped to short codewords, since this will give a low average codeword length.

The second code, C_2, has two symbols, x_1 and x_2, that are mapped to the same codeword. This means that it is impossible to find a unique decoding, and even though it has a short average length such code should be avoided.

The third code in Table 4.1, C_3, has an average codeword length

$$L(C_2) = \sum_x p(x)\ell_x = 1.25 \text{ bit} \tag{4.3}$$

which is considerably lower than for the reference code C_1. All the code sequences are distinct, which is classified as a *nonsingular code*. The generated codeword sequence is a direct concatenation of the codewords, without any spaces or other separating symbols. Consider then the code sequence

$$y = 00110\ldots \tag{4.4}$$

Even though all codewords are distinct, there might be a problem with decoding here as well. Since no codeword contain any double zeros, it is easy to see that the first zero corresponds to the symbol x_1. But then the next pair of bits, 01, can either mean the combination x_1, x_2, or the single symbol x_3. It is not possible to make a clear decision between the alternatives, which means the code is not uniquely decodable.

TABLE 4.1 Some different codes.

x	$p(x)$	C_1	C_2	C_3	C_4	C_5
x_1	1/2	00	0	0	0	0
x_2	1/4	01	0	1	01	10
x_3	1/8	10	1	01	011	110
x_4	1/8	11	11	10	0111	111

The problem has occurred because there are unequal lengths in the codewords, and no separator between the codewords.

The fourth code, C_4, has unique codewords for all symbols, hence it is nonsingular. Since all codewords starts with a zero, and this is the only occurrence of zeros, any code sequence can be uniquely decoded, which is normally classified as a *uniquely decodable code*. The average codeword length is $L(C_4) = 1.875$. The only flaw with the code is that it is not possible see the end of a codeword until the start of the next codeword. Hence, to decode one codeword one must look ahead to see the start of the next codeword. In general, the requirement for a uniquely decodable code is that any codeword sequence should have a unique mapping to a sequence of source symbols. That means all codeword sequences can be uniquely decoded. A drawback is that it might imply that the complete codeword sequence is read before the decoding starts.

Finally, the fifth code, C_5, is both nonsingular and uniquely decodable. Apart from this, it has the property that no codeword is a prefix to any other codeword, which denotes a *prefix code*. For decoding, as soon as a complete codeword is found in the code sequence, it can be mapped to the corresponding source symbol. For example, if the code sequence is

$$y = 01011010111\dots \tag{4.5}$$

the first 0 corresponds to x_1. Then 10 gives x_2, and after that 110 gives x_3. Continuing, the sequence can be decoded as follows:

$$y = \underbrace{0}_{x_1}\underbrace{10}_{x_2}\underbrace{110}_{x_3}\underbrace{10}_{x_2}\underbrace{111}_{x_4}\dots \tag{4.6}$$

To summarize, the following classes of codes can be distinguished:

- *Nonsingular codes.* Each source symbol is mapped to a distinct code vector.
- *Uniquely decodable codes.* Each sequence of source symbols is mapped to a sequence of code symbols, different from any other valid code sequence that might appear.
- *Prefix codes.*[1] No codeword is a prefix to any other codeword.

From the above reasoning, it can be concluded that prefix codes are desirable since they are easily decoded. Clearly, the class of prefix codes is a subclass of the uniquely decodable codes. One basic criterion for a code to be uniquely decodable is that the set of codewords is nonoverlapping. That is, the class of uniquely decodable codes is a subclass of the nonsingular code. Finally, the class of nonsingular codes is a subclass of all codes. In Figure 4.2, a graphical representation of the relation between the classes is shown.

[1] The notation of prefix code is a bit misleading since the code should *not* contain prefixes. However, it is today the most common notation of this class of codes (see, e.g., [11, 12]) and will therefore be adopted in this text. Sometimes in the literature, it is mentioned as a prefix condition code (e.g., [13]), a prefix-free code (e.g., [14]), or an instantaneous code (e.g., [15]).

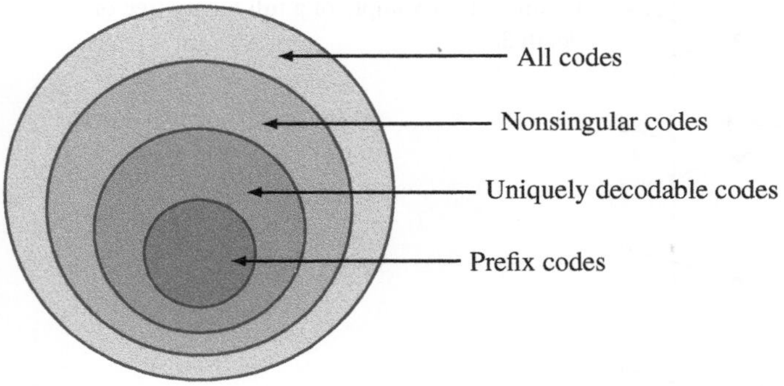

Figure 4.2 All prefix codes are uniquely decodable, and all uniquely decodable codes are nonsingular.

In the continuation of this section, mainly prefix codes will be considered. In the coming pages, it will be shown that to optimize the average codeword length, there is no advantage with uniquely decodable codes compared with prefix codes. That is, given a uniquely decodable code, the same set of codeword lengths can be used to build a prefix code. For this analysis, a tree structure is needed to represent the codewords. A general tree has a root node, which may have one or more child nodes. Each child node may also have one or more child nodes and so on. A node that does not have any child nodes is called a leaf. In a D-ary tree, each node has either zero or D child nodes. In Figure 4.3, two examples of D-ary trees are shown, the left with $D = 2$ and the right with $D = 3$. Often a 2-ary tree is mentioned as a binary tree. Notice that the trees grow to the right from the root. Normally, in computer science trees grow downwards, but in many topics related to information theory and communication theory they are drawn from left to right. The branch labels are often read as codewords, and then it is natural to read from left to right.

The depth of a node is the number of branches back to the root node in the tree. Starting with the root node, it has depth 0. In the left tree of Figure 4.3, the node labeled A has depth 2 and the node labeled B has depth 4. A tree is said to be *full* if all leaves are located at the same depth. In Figure 4.4, a full binary tree of depth 3 is shown. In a full D-ary tree of depth d, there are D^d leaves.

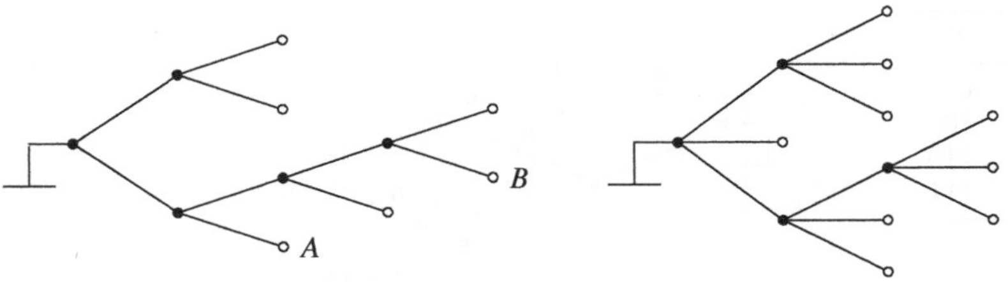

(a) Binary ($D = 2$) tree (b) 3-ary tree

Figure 4.3 Examples of binary and 3-ary tree.

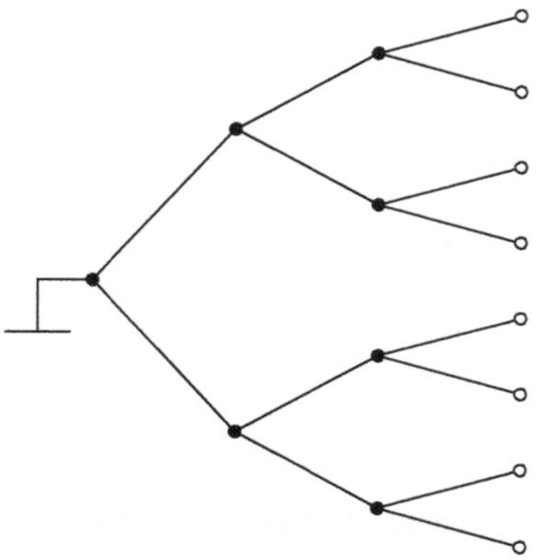

Figure 4.4 Example of a full binary tree of depth 3.

A prefix codes of alphabet size D can be represented in a D-ary tree. The first letter of the codeword is represented by a branch stemming from the root. The second letter is represented by a branch stemming from a node at depth 1 and so on. That is, the labels of the branches represent the symbols of the codeword, and the codeword is read from the root to the leaf. In this way, a structure is built where each sequence from root to leaf in the tree gives a codeword. In that way, a codeword cannot be a prefix of another codeword.

In Figure 4.5, the prefix code C_5 from Table 4.1 is shown in a binary tree representation. In this representation, the probabilities for each source symbol is also added. The labeling of the tree nodes is the sum of the probabilities for the source symbols stemming from that node, i.e. the probability that a codeword passess through that node. Among the codes in Table 4.1, the reference code C_1 is also a prefix code. In Figure 4.6, a representation of this code is shown in a tree. Since all codewords are of equal length, it is a full binary tree.

There are many advantages with the tree representation. One is that it gives a graphical interpretation of the code, which in many occasions is a great help for the intuitive understanding. It is also a helpful tool in the derivations of the code properties. For example, according to the next lemma, the average codeword length can be derived from the tree representation.

x	$p(x)$	C_5
x_1	1/2	0
x_2	1/4	10
x_3	1/8	110
x_4	1/8	111

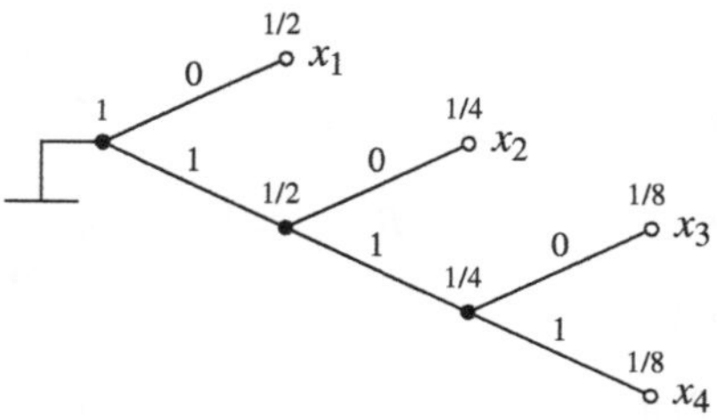

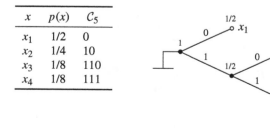

Figure 4.5 Representation of the prefix code C_5 in a binary tree.

x	$p(x)$	$\mathcal{C}_1$
x_1	1/2	00
x_2	1/4	01
x_3	1/8	10
x_4	1/8	11

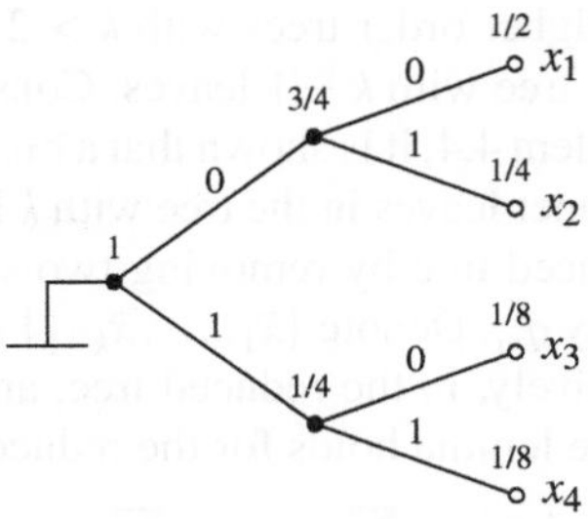

Figure 4.6 Representation of the code $\mathcal{C}_1$ in a (full) binary tree.

Lemma 4.1 **(Path length lemma)** In a tree representation of a prefix code, the average codeword length $L = E[\ell_x]$ equals the sum of probabilities for the inner nodes, including the root. □

Before stating a proof of the lemma, the next example describes the general idea of why the lemma works.

Example 4.1 Consider again the prefix code $\mathcal{C}_5$. The average codeword length can be derived as

$$L = \sum_x p(x)\ell_x = \tfrac{1}{2}1 + \tfrac{1}{4}2 + \tfrac{1}{8}3 + \tfrac{1}{8}3 = 1.75 \tag{4.7}$$

The derivations can be rewritten as

$$\begin{aligned} L &= \underbrace{\tfrac{1}{2}}_{x_1} + \underbrace{\tfrac{1}{4} + \tfrac{1}{4}}_{x_2} + \underbrace{\tfrac{1}{8} + \tfrac{1}{8} + \tfrac{1}{8}}_{x_3} + \underbrace{\tfrac{1}{8} + \tfrac{1}{8} + \tfrac{1}{8}}_{x_4} \\ &= \underbrace{\tfrac{1}{2} + \tfrac{1}{4} + \tfrac{1}{8} + \tfrac{1}{8}}_{1} + \underbrace{\tfrac{1}{4} + \tfrac{1}{8} + \tfrac{1}{8}}_{1/2} + \underbrace{\tfrac{1}{8} + \tfrac{1}{8}}_{1/4} = 1.75 \end{aligned} \tag{4.8}$$

That is, by rearranging the terms of the sum in equality two, it can be seen that each leaf probability is present in each of the nodes on the path to the leaf. Hence, by summing up all inner node probabilities the contribution from a leaf probability is the probability times its depth.

A more formal proof for Lemma 4.1 is as follows, based on induction. The proof only deals with binary trees, but the lemma is also true for arbitrary D-ary trees.

Proof: First assume a code tree with two leaves at depth 1 and one inner node, i.e., the root, with probability 1. Clearly, the lemma is true for this case since the root probability is one and the depth of the leaves is one. This is the only tree with two leaves that is considered. All other trees with two leaves can be viewed as higher order trees with all but two leaf probabilities set to zero.

To show the lemma for higher order trees with $k > 2$ leaves, first assume that the lemma holds for any binary tree with $k-1$ leaves. Consider a tree with k leaves denoted by $\{x_1, \dots, x_k\}$. In Problem 4.4, it is shown that a binary tree with k leaves has $k-1$ inner nodes. Denote the inner leaves in the tree with k leaves by $\{q_1, \dots, q_{k-1}\}$. From this tree, construct a reduced tree by removing two sibling leaves, say x_i and x_j, and denote the parent node by q_m. Denote $\{\tilde{x}_1, \dots, \tilde{x}_{k-1}\}$ and $\{\tilde{q}_1, \dots, \tilde{q}_{k-2}\}$ as the leaves and inner nodes, respectively, in the reduced tree, and let $\tilde{x}_n = q_m$. From the assumption presented above, the lemma holds for the reduced tree, and

$$\widetilde{L} = E\big[\ell_{\tilde{x}}\big] = \sum_{\tilde{x}} p_{\tilde{x}} \ell_{\tilde{x}} = \sum_{\tilde{q}} p_{\tilde{q}} \tag{4.9}$$

Since the parent node probability equals the sum of the sibling node probabilities, $p_{\tilde{x}_n} = p_{q_m} = p_{x_i} + p_{x_j}$ and $\ell_{\tilde{x}_n} = \ell_{q_m} = \ell_{x_i} - 1 = \ell_{x_j} - 1$, the average codeword length for the tree with k leaves can be written as

$$\begin{aligned}
L = E\big[\ell_x\big] &= \sum_x p_x \ell_x \\
&= \sum_{x \neq x_i, x_j} p_x \ell_x + p_{x_i} \ell_{x_i} + p_{x_j} \ell_{x_j} \\
&= \sum_{x \neq x_i, x_j} p_x \ell_x + \underbrace{(p_{x_i} + p_{x_j})}_{p_{\tilde{x}_n}} \underbrace{(\ell_{x_i} - 1)}_{\ell_{\tilde{x}_n}} + \underbrace{p_{x_i} + p_{x_j}}_{p_{q_m}} \\
&= \sum_{\tilde{x} \neq \tilde{x}_n} p_{\tilde{x}} \ell_{\tilde{x}} + p_{\tilde{x}_n} \ell_{\tilde{x}_n} + p_{q_m} \\
&= \sum_{\tilde{x}} p_{\tilde{x}} \ell_{\tilde{x}} + p_{q_m} = \sum_{\tilde{q}} p_{\tilde{q}} + p_{q_m} = \sum_{q \neq q_m} p_q + p_{q_m} = \sum_q p_q
\end{aligned} \tag{4.10}$$

Hence, the lemma is true for the tree with k leaves, which completes the proof. ■

Since the sum of all leaf probabilities equals the root probability, an alternative formulation can be given where the average length is the sum of all node probabilities except the root. This is used later when considering adaptive compression methods.

It is now time to state and show a famous result, first derived in a master thesis's project by Leon Kraft in 1949 [16]. It meets the requirement of the codeword lengths used to form a prefix code.

Theorem 4.1 (Kraft inequality) There exists a *prefix D-ary code* with codeword lengths $\ell_1, \ell_2, \dots, \ell_k$ if and only if

$$\sum_{i=1}^{k} D^{-\ell_i} \leq 1 \tag{4.11}$$

□

To show this, consider a D-ary prefix code where the longest codeword length is $\ell_{\max} = \max_x \ell_x$. This code can be represented in a D-ary tree. A full D-ary tree of

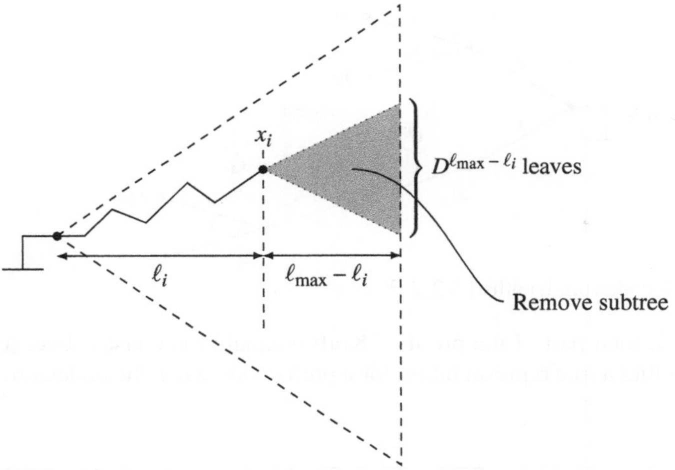

Figure 4.7 Tree construction for a general prefix code.

depth ℓ_{max} has $D^{\ell_{max}}$ leaves. In the tree representation, the codeword corresponding to symbol x_i is a length ℓ_i path from the root. Since it should be a prefix code, the subtree spanned with x_i as a root is not allowed to be used by any other codeword, hence it should be removed from the tree (see Figure 4.7). So, when including the codeword for symbol x_i, $D^{\ell_{max}-\ell_i}$ leaves are removed from the tree. The number of removed leaves cannot exceed the total number of leaves in the full D-ary tree of depth ℓ_{max},

$$\sum_i D^{\ell_{max}-\ell_i} \leq D^{\ell_{max}} \tag{4.12}$$

By cancelling $D^{\ell_{max}}$ on both sides, this proves that for a prefix code

$$\sum_i D^{-\ell_i} \leq 1 \tag{4.13}$$

To show that if the inequality is fulfilled, it is possible to construct a prefix code, start by assuming $\sum_i D^{\ell_i} \leq 1$ and order the codeword lengths as $\ell_1 \leq \ell_2 \leq \cdots \leq \ell_k$, where $\ell_k = \ell_{max}$. Then use the same construction as above; start with the shortest codeword and remove the subtree. After $m < k$ steps, the number of nonused leaves left at depth ℓ_{max} in the tree is

$$D^{\ell_{max}} - \sum_{i=1}^{m} D^{\ell_{max}-\ell_i} = D^{\ell_{max}}\left(1 - \sum_{i=1}^{m} D^{-\ell_i}\right) > 0 \tag{4.14}$$

where the last inequality comes from the assumption, i.e., $\sum_{i=1}^{m} D^{-\ell_i} < 1$. In other words, as long as $i < k$, there is at least one leaf left at depth ℓ_{max}. The last codeword only needs one leaf since it is of maximum length, which shows that it is always possible to construct a prefix code if the inequality is fulfilled.

x	C_1	ℓ
x_1	00	2
x_2	01	2
x_3	10	2
x_4	110	3
x_5	1110	4

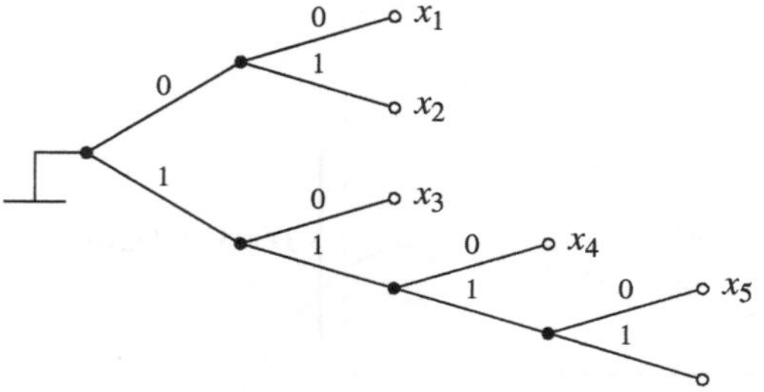

Figure 4.8 A code with codeword lengths $\{2, 2, 2, 3, 4\}$.

Notice that the second part of the proof of Kraft inequality is constructive. It describes a method to find a tree representation for a prefix code when the codeword lengths are given.

Example 4.2 To construct a binary prefix code with codeword lengths $\ell \in \{2, 2, 2, 3, 4\}$, first check that it is possible according to Kraft inequality. The derivation

$$\sum_i 2^{-\ell_i} = 2^{-2} + 2^{-2} + 2^{-2} + 2^{-3} + 2^{-4} = 3\tfrac{1}{4} + \tfrac{1}{8} + \tfrac{1}{16} = \tfrac{15}{16} < 1 \tag{4.15}$$

shows that there exists such code. For example, the code and code tree in Figure 4.8 can be used.

Notice that the tree in Figure 4.8 has one unused leaf. That means the codeword for x_5 is unnecessarily long and can be shortened to 111. With this improvement, the set of codeword lengths becomes $\ell^{(2)} \in \{2, 2, 2, 3, 3\}$, which also satisfies Kraft inequality,

$$\sum_i 2^{-\ell_i^{(2)}} = 2^{-2} + 2^{-2} + 2^{-2} + 2^{-3} + 2^{-3} = 3\tfrac{1}{4} + 2\tfrac{1}{8} = 1 \tag{4.16}$$

An attempt to shorten one of the shorter codewords, say x_1, instead of x_5, results in the set of codeword lengths $\ell^{(3)} \in \{1, 2, 2, 3, 4\}$. Checking for Kraft inequality gives

$$\sum_i 2^{-\ell_i^{(3)}} = 2^{-1} + 2^{-2} + 2^{-2} + 2^{-3} + 2^{-4} = \tfrac{1}{2} + 2\tfrac{1}{4} + \tfrac{1}{8} + \tfrac{1}{16} = \tfrac{19}{16} > 1 \tag{4.17}$$

Since Kraft inequality is not fulfilled, it is not possible to construct a prefix code with these codeword lengths.

In the last part of the previous example, the set of codeword lengths does not fulfill Kraft inequality, and it is not possible to construct a prefix code. It should then be noted that Kraft inequality can be extended to hold for uniquely decodable codes, a result due to McMillan [17]. That means, any set of codeword lengths that can be implemented with a uniquely decodable code can also be implemented with a prefix code.

Theorem 4.2 (McMillan inequality) There exists a *uniquely decodable D-ary code* with codeword lengths $\ell_1, \ell_2, \dots, \ell_k$ if and only if

$$\sum_{i=1}^{k} D^{-\ell_i} \leq 1 \tag{4.18}$$

□

Since a prefix code is also uniquely decodable, the existence of such a code follows directly from Theorem 4.1. To show that a uniquely decodable code satisfies the inequality, assume that the codeword lengths in the code are $\ell_1, \dots \ell_k$ and that the maximum length is $\ell_{\max} = \max_i \ell_i$. The sum can be rewritten as

$$\sum_{i=1}^{k} D^{-\ell_i} = \sum_{j=1}^{\ell_{\max}} w_j D^{-j} \tag{4.19}$$

where w_j denotes the total number of codewords of length j. The nth power of this sum equals

$$\begin{aligned}
\Big(\sum_{i=1}^{k} D^{-\ell_i}\Big)^n &= \Big(\sum_{j=1}^{\ell_{\max}} w_j D^{-j}\Big)^n \\
&= \sum_{j_1=1}^{\ell_{\max}} \cdots \sum_{j_n=1}^{\ell_{\max}} (w_{j_1} \cdots w_{j_n}) D^{-(j_1+\cdots+j_n)} \\
&= \sum_{\tau=n}^{n\ell_{\max}} W_\tau D^{-\tau}
\end{aligned} \tag{4.20}$$

where

$$W_\tau = \sum_{j_1+\cdots+j_n=\tau} (w_{j_1} \cdots w_{j_n}) \tag{4.21}$$

is the total number of code sequences of length τ obtained by concatenating n codewords. Since the code is uniquely decodable, all vectors must be distinct, and the number of length τ code vectors cannot exceed the total number length τ vectors, $W_\tau \leq D^\tau$. In other words, $W_\tau D^{-\tau} \leq 1$ and

$$\Big(\sum_{i=1}^{k} D^{-\ell_i}\Big)^n = \sum_{\tau=n}^{n\ell_{\max}} W_\tau D^{-\tau} \leq \sum_{\tau=n}^{n\ell_{\max}} 1 \leq n\ell_{\max} \tag{4.22}$$

Taking the nth root of this expression gives $\sum_{i=1}^{k} D^{-\ell_i} \leq \sqrt[n]{n\ell_{\max}}$, which should hold for any number of concatenated codewords n. Considering infinitely long code sequences completes the proof of McMillans inequality,

$$\sum_{i=1}^{k} D^{-\ell_i} \leq \lim_{n\to\infty} \sqrt[n]{n\ell_{\max}} = 1 \tag{4.23}$$

In the continuation of this chapter, results based on Kraft inequality are often stated for prefix codes, although McMillan inequality shows that they also hold for the larger class of codes, the uniquely decodable codes. The reason is that it is often much easier to construct and make use of prefix codes, and in light of McMillan's inequality there is no gain in the average codeword length to consider uniquely decodable codes. That is, given a uniquely decodable code, it is always possible to construct a prefix code with the same set of codeword lengths.

4.3 OPTIMAL CODEWORD LENGTH

With Kraft inequality, a mathematical foundation needed to consider an optimization function for the codeword lengths is obtained. One standard method for minimization of a function with some side criterion is the Lagrange multiplication method (see, e.g., [18]). Assign an optimization function to optimize the average codeword length with the side condition that Kraft inequality applies,

$$J = \sum_i p(x_i)\ell_i + \lambda\Big(\sum_i D^{-\ell_i} - 1\Big) \tag{4.24}$$

Setting the derivative of J equal zero for each ℓ_i gives an equation system,

$$\frac{\partial}{\partial \ell_i} J = p(x_i) - \lambda D^{-\ell_i} \ln D = 0 \tag{4.25}$$

or, equivalently,

$$D^{-\ell_i} = \frac{p(x_i)}{\lambda \ln D} \tag{4.26}$$

The condition from Kraft inequality gives

$$\sum_i D^{-\ell_i} = \sum_i \frac{p(x_i)}{\lambda \ln D} = \frac{1}{\lambda \ln D} = 1 \tag{4.27}$$

Combine (4.26) and (4.27) to get

$$D^{-\ell_i} = p(x_i)\frac{1}{\lambda \ln D} = p(x_i) \tag{4.28}$$

Thus, the optimal codeword length for codeword i is obtained as

$$\ell_i^{(\text{opt})} = -\log_D p(x_i) \tag{4.29}$$

The optimal average codeword length for a prefix code then becomes[2]

$$L_{\text{opt}} = \sum_i p(x_i)\ell_i^{(\text{opt})} = -\sum_i p(x_i)\log_D p(x_i) = H_D(X) = \frac{H(X)}{\log D} \tag{4.30}$$

[2] The notation

$$H_D(X) = \sum_x p(x)\log_D p(x) = \sum_x p(x)\frac{\log p(x)}{\log D} = \frac{H(X)}{\log D}$$

is used for the entropy when derived over the base D instead of base 2.

To verify that the entropy $H_D(X)$ is a lower bound, i.e. that it is a global minimum, consider an arbitrary set of codeword lengths $\ell_1, \dots, \ell_k$ such that Kraft inequality is satisfied. Then,

$$\begin{aligned}
H_D(X) - L &= -\sum_i p(x_i)\log_D p(x_i) - \sum_i p(x_i)\ell_i \\
&= \sum_i p(x_i)\log_D \frac{1}{p(x_i)} - \sum_i p(x_i)\log_D D^{\ell_i} \\
&= \sum_i p(x_i)\log_D \frac{D^{-\ell_i}}{p(x_i)} \\
&\leq \sum_i p(x_i)\Big(\frac{D^{-\ell_i}}{p(x_i)} - 1\Big)\log_D e \\
&= \sum_i \big(D^{-\ell_i} - p(x_i)\big)\log_D e \\
&= \Big(\underbrace{\sum_i D^{-\ell_i}}_{\leq 1} - \underbrace{\sum_i p(x_i)}_{=1}\Big)\log_D e \leq 0
\end{aligned} \tag{4.31}$$

which implies that $L \geq H_D(X)$ for all prefix codes. In the first inequality the IT-inequality was used and in the second Kraft inequality. Both inequalities are satisfied with equality if and only if $\ell_i = -\log_D p(x_i)$. When deriving the optimal codeword lengths, the logarithmic function in general will not give integer values, so in practice it might not be possible to reach this lower limit. The above result is summarized in the following theorem.

Theorem 4.3 The average codeword length $L = E[\ell_x]$ for a prefix code is lower bounded by the entropy of the source,

$$L \geq H_D(X) = \frac{H(X)}{\log D} \tag{4.32}$$

with equality if and only if $\ell_x = -\log_D p(x)$. □

From the previous theorem on the optimal codeword length and the construction method in the proof of Kraft inequality, it is possible to find an algorithm for a code design. The codeword length for source symbol x_i is $\ell_i^{(\text{opt})} = -\log_D p(x_i)$. To assure that the codeword length is an integer use instead

$$\ell_i = \lceil -\log_D p(x_i) \rceil \tag{4.33}$$

From the following derivation it can be seen that Kraft inequality is still satisfied, which shows that it is possible to construct a prefix code,

$$\sum_i D^{-\ell_i} = \sum_i D^{-\lceil -\log_D p(x_i)\rceil} \leq \sum_i D^{-(-\log_D p(x_i))} = \sum_i p(x_i) = 1 \tag{4.34}$$

This code construction is often named the *Shannon–Fano code*. Since the codeword length is the upper integer part of the optimal length, it can be bounded as

$$-\log_D p(x_i) \leq \ell_i < -\log_D p(x_i) + 1 \tag{4.35}$$

Taking the expectation of the above gives

$$E\big[-\log_D p(X)\big] \leq E\big[\ell\big] \leq E\big[-\log_D p(X) + 1\big] \tag{4.36}$$

which can also be expressed as

$$H_D(X) \leq L \leq H_D(X) + 1 \tag{4.37}$$

That is, the above code construction gives a code with average codeword length $L < H_D(X) + 1$. It is reasonable to define an optimum prefix code as a prefix code with the minimum average codeword length over all prefix codes for that random variable. Such optimum code can clearly not exceed the Shannon–Fano average codeword length. Together with the result in Theorem 4.3, the following theorem is obtained.

Theorem 4.4 The average codeword length, $L = E[\ell_x]$, for an optimal D-ary prefix code satisfies

$$H_D(X) \leq L < H_D(X) + 1 \tag{4.38}$$

□

Even though the Shannon–Fano code construction is used to bound an optimal code, it is important to note that in general this code construction is not optimal. But since an optimal code cannot have a longer average codeword length, it can be used as an upper bound. In the next example, it is seen that the Shannon–Fano code construction can give a suboptimal code.

Example 4.3 Consider a random variable with four outcomes according to the table below. In the table, it is also listed the optimal codeword lengths and the lengths for the codewords in a Shannon–Fano code.

x	$p(x)$	$-\log p(x)$	$\ell = \lceil -\log p(x) \rceil$
x_1	0.45	1.152	2
x_2	0.25	2	2
x_3	0.20	2.32	3
x_4	0.10	3.32	4

Since Kraft inequality is fulfilled,

$$\sum_{\ell} 2^{-\ell} = \frac{1}{4} + \frac{1}{4} + \frac{1}{8} + \frac{1}{16} = \frac{11}{16} < 1 \tag{4.39}$$

it is possible to construct a prefix code with the listed codeword lengths by following the procedure described earlier. The binary tree representation and the codeword list are given in Figure 4.9.

x	$p(x)$	$\mathcal{C}$
x_1	0.45	00
x_2	0.25	01
x_3	0.20	100
x_4	0.10	1010

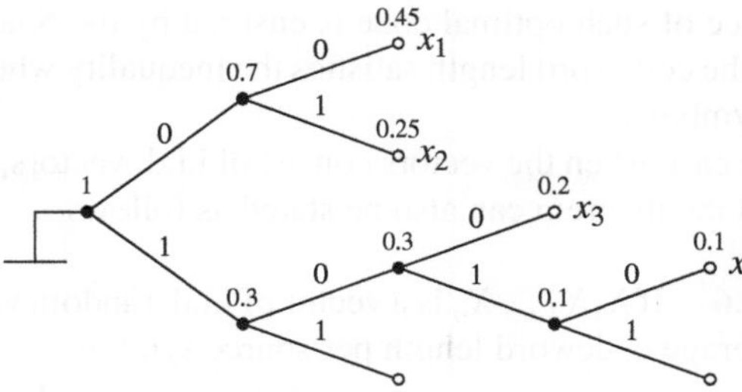

Figure 4.9 Code tree and list for a Shannon–Fano code.

In the tree following the paths for 11 and 1011, there are unused leaves. By moving the label for x_4 to the leaf at 11, the length for this codeword will decrease from 4 to 2, which will result in a code with lower average codeword length. Hence, the obtained Shannon–Fano code does not have optimal codeword length. As a comparison, the entropy for the random variable is

$$H(X) = H(0.45, 0.25, 0.2, 0.1) = 1.815 \tag{4.40}$$

With use of the path length lemma, the average codeword length for the code is

$$L = 1 + 0.7 + 0.3 + 0.3 + 0.1 = 2.4 \tag{4.41}$$

which lies in between $H(X) = 1.815$ and $H(X) + 1 = 2.815$.

To extend the results to a random process, a sequence of length n source symbols $\boldsymbol{X} = X_1X_2 \dots X_n$ is considered. By constructing an optimal code for this vector, the optimal codeword length will satisfy

$$H(X_1X_2 \dots X_n) \leq E[\ell_{\boldsymbol{x}}] \leq H(X_1X_2 \dots X_n) + 1 \tag{4.42}$$

Equivalently, the average codeword length per source symbol is

$$\frac{1}{n}H(X_1X_2 \dots X_n) \leq \frac{1}{n}E[\ell_{\boldsymbol{x}}] \leq \frac{1}{n}H(X_1X_2 \dots X_n) + \frac{1}{n} \tag{4.43}$$

As the length of the vector, n, grows the term $\frac{1}{n}$ will tend to zero and both sides of the bound will approach the entropy rate $\frac{1}{n}H(\boldsymbol{X}) \to H_\infty(X)$. More formally, the following theorem can be formulated.

Theorem 4.5 If $X_1X_2 \dots X_n$ forms an stationary ergodic random process, there exists a code with the average codeword length per source symbol

$$H_\infty(X) \leq E\big[\ell_{\boldsymbol{x}}\big] \leq H_\infty(X) + \frac{1}{n} \tag{4.44}$$

where $H_\infty(X)$ is the entropy rate for the process and the gap $\frac{1}{n}$ can be made arbitrarily small for large n. □

The existence of such optimal code is ensured by the Shannon–Fano code construction where the codeword length satisfies the inequality when length n vectors are used as source symbols.

In the case when the vectors consist of i.i.d. vectors, the entropy rate equals the entropy and the theorem can also be stated as follows

Theorem 4.6 If $X_1X_2 \dots X_n$ is a vector of i.i.d. random variables, there exists a code with the average codeword length per source symbol

$$H(X) \leq E[\ell_x] \leq H(X) + \frac{1}{n} \tag{4.45}$$

where $H(X)$ is the entropy of the variables and the gap $\frac{1}{n}$ can be made arbitrarily small for large n. □

Theorem 4.6 is closely related to Shannon's source coding theorem. In his original paper from 1948, it was shown using the law of large numbers and AEP. In Chapter 6, AEP will be introduced and used to show both the source coding theorem and the channel coding theorem. The generalization to ergodic processes was first published in [19].

4.4 HUFFMAN CODING

In 1951, Robert Fano at MIT lectured the first ever course on information theory. One of the students, David Huffman, developed as part of a class assignment, an algorithm for constructing a source code. He also showed that the code was in fact optimal. The year after, in 1952 the construction was published [20] and the produced code is normally mentioned as a *Huffman code*. In the following, the algorithm is described for the binary case and then shown that the procedure generates an optimal code. In this description, the algorithm will be described for binary codes, with an extension at the end of the chapter to D-ary codes. The algorithm will also be generalized to construct D-ary codes.

The idea of the Huffman code construction is to list all possible source symbols as nodes. Then, iteratively find the two least probable nodes and merge in a binary tree, letting the root represent a new node replacing the two merged nodes. Written as an algorithm the following is obtained.

Algorithm 4.1 (Binary Huffman code)

To construct the binary code tree for a random variable with k outcomes:

1. Sort the symbols according to their probabilities.
2. Let x_i and x_j, with probabilities p_i and p_j, respectively, be the two least probable symbols in the list
   - Remove x_i and x_j from the list and connect them in a binary tree.
   - Add the root node $\{x_i, x_j\}$ as one symbol with probability $p_{ij} = p_i + p_j$ to the list.

3. If only one symbol left in the list
   STOP
   Else
   GOTO 2

Before showing the optimality of the constructed code, two examples are studied. The first is a small example to go through the details, and then follows a slightly more complex example.

Example 4.4 Consider a random variable X with four outcomes, x_1, x_2, x_3, and x_4, with the probabilities

x :	x_1	x_2	x_3	x_4
$p(x)$:	1/2	1/4	1/8	1/8

In Figure 4.10a, each of the symbols are represented with one node. To start the algorithm, find the two nodes with least probabilities. In this case, it is x_3 and x_4, each with probability $1/8$. Merge these nodes in a binary tree and add the root as a node to the list. Now, the list contains three nodes, x_1, x_2, and x_3x_4, where the last one represents the newly added tree (see Figure 4.10b). Since there are more than one node left, continue from the beginning and find the two least probable nodes. It is the nodes x_2 and x_3x_4 that should be merged (see Figure 4.10c). Then there are two nodes left, x_1 and $x_2x_3x_4$, and they represent the two least probable nodes. Merge and let the root represent a new node to get Figure 4.10d. After this, there is only one node left and the algorithm stops.

The average codeword length in the constructed code tree is

$$L = 1 + \frac{1}{2} + \frac{1}{4} = 1.75 \text{ bit} \tag{4.46}$$

In Figure 4.11, a redrawn tree is shown together with the code table.

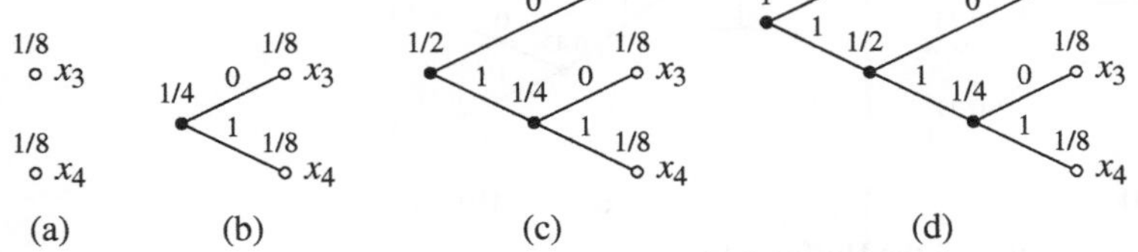

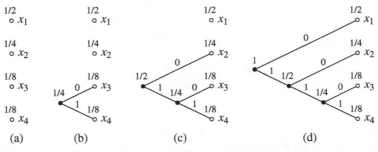

Figure 4.10 Construction of the Huffman tree.

x	$p(x)$	y
x_1	1/2	0
x_2	1/4	10
x_3	1/8	110
x_4	1/8	111

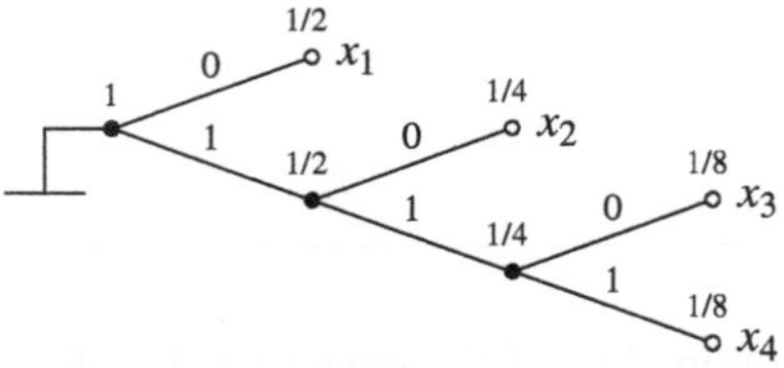

Figure 4.11 The tree and the code table for the constructed code.

From Theorem 4.3, the codeword length is lower bounded by the entropy of the source. So, to compare derive the entropy as

$$H\left(\tfrac{1}{2}, \tfrac{1}{4}, \tfrac{1}{8}, \tfrac{1}{8}\right) = 1.75 \text{ bit} \tag{4.47}$$

Since there is equality between the codeword length and the entropy, it can directly be seen that the constructed code is optimal.

The next example is slightly more complex. Later it will be shown that Huffman codes are indeed optimal. In that respect, the example also shows that the optimal codeword length is not necessarily equal to entropy.

Example 4.5 The random variable X have six outcomes with probabilities according to

$x:$	x_1	x_2	x_3	x_4	x_5	x_5
$p(x):$	0.05	0.10	0.15	0.20	0.23	0.27

In Figure 4.12a, the Huffman tree is constructed. The labeling of the inner nodes represents the order in which they are constructed in the algorithm. In Figure 4.12b, the same tree redrawn to better see the structure.

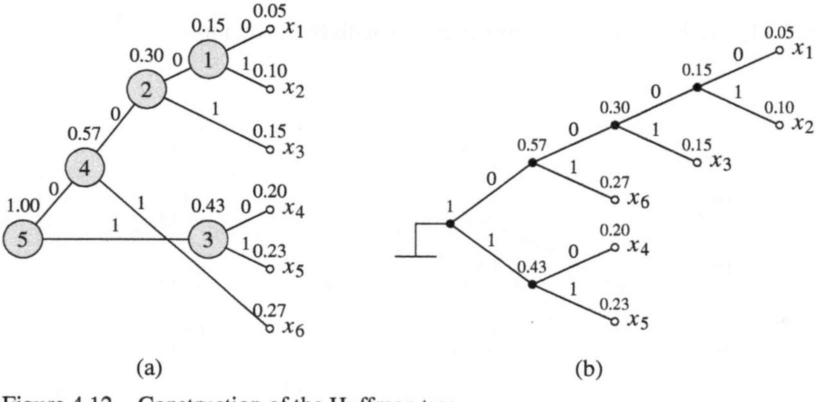

Figure 4.12 Construction of the Huffman tree.

The code can be written in a table as

$x:$	x_1	x_2	x_3	x_4	x_5	x_5
$p(x):$	0000	0001	001	10	11	01

and the average codeword length calculated as

$$L = 1 + 0.57 + 0.3 + 0.15 + 0.43 = 2.45 \text{ bit} \tag{4.48}$$

The entropy of the source is

$$H(X) = H(0.05, 0.10, 0.15, 0.20, 0.23, 0.27) = 2.42 \text{ bit} \tag{4.49}$$

In the previous example, the Huffman code does not give equality in the lower bound for the average codeword length. It is shown below that in fact the Huffman procedure gives an optimal code. In the previous example, that means there are no other prefix code that gives equality in the bound either. Thus, the optimal codeword lengths from Theorem 4.3 are not integers and it is not possible to meet the bound.

To show that Huffman's algorithm will give an optimal code, a couple of observations that apply to an optimal code are needed. Then it can be shown that by applying these observations the constructed code has to be optimal and that it all boils down to the construction given by Huffman.

The first observation is that in an optimal code the lengths for the codewords must follow the probabilities for the source symbols. That is, codewords corresponding to source symbols with low probability should not be shorter than codewords corresponding to more probable source symbols, $p_i < p_j \Rightarrow \ell_i \geq \ell_j$. Intuitively, it is clear that if a codeword with a length shorter than a codeword of higher probability, the average length will decrease if the codewords are swapped. To show it mathematically, assume that the source symbols x_i and x_j have probabilities p_i and p_j, where $p_i < p_j$ and the corresponding codeword lengths ℓ_i and ℓ_j satisfy $\ell_i \leq \ell_j$. Notice that the codeword length is chosen such that the lowest probability is mapped to the shortest length codeword. Then by swapping the codewords the average codeword length will decrease,

$$\begin{aligned}
L &= \sum_{\tau} p_\tau \ell_\tau = p_i\ell_i + p_j\ell_j + \sum_{\tau \neq i,j} p_\tau \ell_\tau \\
&= p_i\ell_i + p_j\ell_j + p_i\ell_j + p_j\ell_i - p_i\ell_j - p_j\ell_i + \sum_{\tau \neq i,j} p_\tau \ell_\tau \\
&= \underbrace{(p_j - p_i)}_{>0}\underbrace{(\ell_j - \ell_i)}_{\geq 0} + \underbrace{p_i\ell_j + p_j\ell_i + \sum_{\tau \neq i,j} p_\tau \ell_\tau}_{\widetilde{L}} + \geq \widetilde{L}
\end{aligned} \tag{4.50}$$

where $\widetilde{L}$ is the length when the codewords are swapped. This shows that by swapping the codewords $L \geq \widetilde{L}$, with equality when $\ell_i = \ell_j$. Hence, in an optimal code, codewords corresponding to less probable symbols are not shorter than codewords corresponding to more probable symbols.

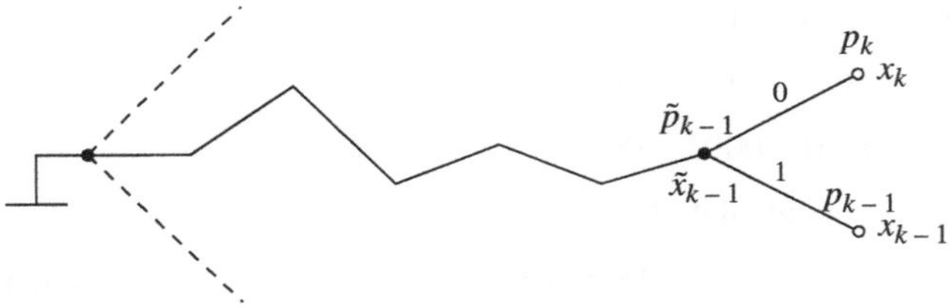

Figure 4.13 A code tree for an optimal code.

The second observation is that in a tree corresponding to an optimal code, there are no unused leaves. This can be seen by assuming a code with an unused leaf in the tree. Then the parent node of this leaf has only one branch. By removing this last branch and placing the leaf at the parent node, a shorter codeword is obtained, and therefore the average length decreases. Hence, the assumed code cannot be optimal.

The third observation is that the two least probable codewords are of the same length, and that a code tree can be constructed such that they differ only in the last bit. This can be seen by assuming that the least probable symbol has a codeword according to $y_1 \dots y_{\ell-1}0$. Then since there are no unused leaves, there exists a codeword $y_1 \cdots y_{\ell-1}1$. If this is not the second least probable codeword, it follows from the first observation that the second least probable codeword has the same length. So, this codeword can be swapped with $y_1 \dots y_{\ell-1}1$ and to get the desired result.

Then a binary code for a random variable X with k outcomes corresponds to a binary tree with k leaves, since there are no unused leaves. The two least probable symbols, x_k and x_{k-1}, corresponding to probabilities p_k and p_{k-1}, can be assumed to be located as siblings in the tree (see Figure 4.13). The parent node for these leaves has the probability $\tilde{p}_{k-1} = p_k + p_{k-1}$.

Then, a new code tree with $k-1$ leaves can be constructed by replacing the codewords for x_k and x_{k-1} by its parent node, called $\tilde{x}_{k-1}$ in the figure. Denote by L the average length in the original code with k codewords and $\tilde{L}$ the average length in the new code. From the path length lemma

$$L = \tilde{L} + \tilde{p}_{k-1} = \tilde{L} + (p_k + p_{k-1}) \tag{4.51}$$

Since p_k and p_{k-1} are the two least probabilities for the random variable, the code with k codewords can only be optimal if the code with $k-1$ elements is optimal. Continuing this reasoning until there are only two codewords left, the optimal code has the codewords 0 and 1. The steps taken here to construct an optimal code are exactly the same steps used in the Huffman algorithm. Hence, it is concluded that a Huffman code is an optimal code.

Theorem 4.7 A binary Huffman code is an optimal prefix code. □

On many occasions, there can be more than one way to merge the nodes in the algorithm. For exampl,e if there are more than two nodes with the same least probability. That means the algorithm can produce different codes depending on which merges

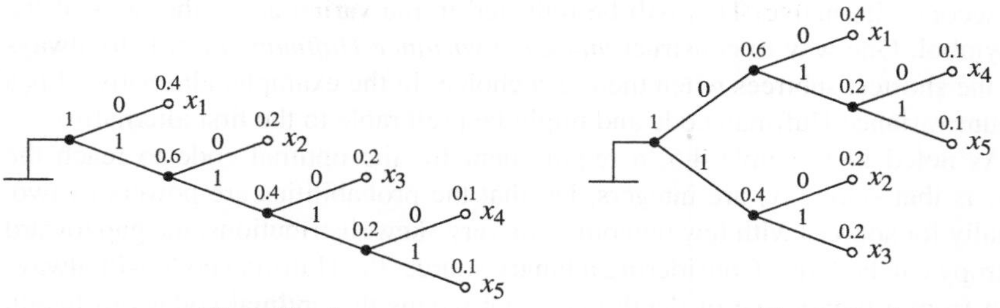

Figure 4.14 Two alternative Huffman trees for one source.

are chosen. Independent of which code is considered, the codeword length remains minimal, as mentioned in the following example.

Example 4.6 The random variable X with five outcomes has the probabilities

$x:$	x_1	x_2	x_3	x_4	x_5
$p(x):$	0.4	0.2	0.2	0.1	0.1

The Huffman algorithm can produce two different trees, and thus two different codes for this statistics. In Figure 4.14, the two trees are shown. The difference in the construction becomes evident after the first step of the algorithm. Then there are three nodes with least probability, x_2, x_3, and x_1x_2. In the first alternative, the nodes x_3 and x_1x_2 are merged into $x_3x_4x_5$ with probability 0.4, and in the second alternative the two nodes x_2 and x_3 are merged to x_2x_3.

The average codeword length for the both alternatives will give the same calculation

$$L_1 = L_2 = 1 + 0.6 + 0.4 + 0.2 = 2.2 \text{ bit} \tag{4.52}$$

In Example 4.6, both codes give the same codeword lengths and both codes are optimal. However, the difference can be of importance from another perspective. Source coding gives variations in the length of the coded symbols, i.e. the rate of the symbol varies from the encoder. In, for example, video coding, this might be an important design factor when choosing codes. For standard definition (SD), the rate is approximately 2–3 Mb/s on average but the peak levels can go up to as high as 6–8 Mb/s. For high definition (HD), the problem is even more pronounced.

In most communication schemes, the transmission is done with a fixed maximum rate. To handle this mismatch, the transmitter and receiver is often equipped with buffers. At the same time, the delays in the system should be kept as small as possible, and therefore the buffer sizes should also be small. This implies that the variations in the rates from the source encoder should be as small as possible. In the example, for the first alternative of the code tree, the variation in length is larger than in the second alternative. This will be reflected in the variations in the rates of the code symbol. One way to construct *minimum variance Huffman codes* is to always merge the shortest subtrees when there is a choice. In the example, alternative 2 is a minimum variance Huffman code and might be preferable to the first alternative.

As noted in Example 4.5, a requirement for the optimal code to reach the entropy is that $-\log p(x)$ are integers, i.e. that the probabilities are powers of two. Especially for sources with few outcomes, or very skew distributions, the gap toward the entropy can be large. Considering a binary source, the Huffman code will always be built from a binary tree of depth one. That means the optimal codeword length is still one, and the gap toward the entropy becomes $1 - h(p)$, where p is the probability for one of the outcomes. To circumvent this obstacle and force the codeword length toward the entropy, the source sequence can be viewed as a sequence of vectors of length n instead of symbols. Then there is a distribution for the vectors instead, on which a Huffman code can be built. This is the idea of Theorem 4.6, where the upper bound of the average codeword length is $H(X) + \frac{1}{n}$. In the next example, such construction will be shown.

Example 4.7 Consider a source with three outcomes distributed according to $P \in \{0.6, 0.25, 0.15\}$. Clearly, a Huffman code can be constructed as

x	$p(x)$	$y(x)$
1	0.6	1
2	0.25	01
3	0.15	00

The average codeword length for this code is $L = 1.4$ bit, whereas the entropy is $H(X) = 1.3527$ bit. The gap between the obtained codeword length and the entropy is 0.05 bit/symbol. Even though this is only 3.5% of the entropy, it is an unnecessary gap for optimal coding. So, instead of constructing the code symbolwise, group pairs of symbols and construct a Huffman code for this case.

(x_1x_2)	$p(x_1x_2)$	$y(x_1x_2)$	$\ell_{x_1x_2}$
11	0.36	1	1
12	0.15	010	3
13	0.09	0001	4
21	0.15	011	3
22	0.0625	0011	4
23	0.0375	00001	5
31	0.090	0010	4
32	0.0375	000000	6
33	0.0225	000001	6

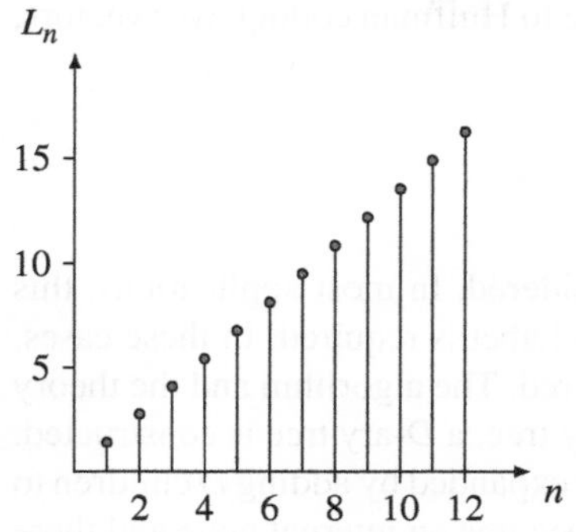

Figure 4.15 Average codeword length for Huffman coding by vectors of length up to $n = 12$.

The average codeword length of this code is $L_2 = 2.7775$ bit, which gives $L_2^{(1)} = L_2/2 = 1.3887$ bit/symbol. Even more improvements can be obtained if vectors of length three, $(x_1x_2x_3)$, are used in the Huffman construction. Then the average codeword length is $L_3 = 4.1044$ bits, and the average codeword length per symbol is $L_3^{(1)} = L_3/3 = 1.3681$ bit/symbol. As seen, the efficiency of the code increases and the codeword length per symbol is getting closer to the entropy. In Figure 4.15, the average codeword lengths are shown when vectors of lengths $n = 1, 2, \ldots, 12$ are used in the Huffman code. In Figure 4.16, the corresponding average codeword length per symbol is shown. Here the upper and lower bounds are also shown as

$$H(X) \leq L_n^{(1)} \leq H(X) + \frac{1}{n} \tag{4.53}$$

Clearly, there is an improvement in terms of codeword lengths per symbol, and the larger n becomes the closer to the entropy it can get. However, the price paid here is the extra complexity for building the code tree for vectors instead of symbols. The number of vectors will grow exponentially with n, and the tree for $n = 12$ in this example is built with $3^{12} = 531\,441$ leaves. In the Section 4.5, the principles of Arithmetic coding will be described. The idea is to perform encoding over vectors in a

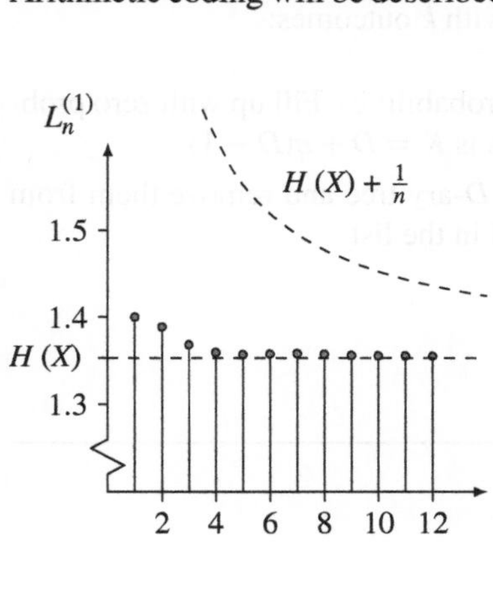

Figure 4.16 Average codeword length per symbol for Huffman coding by vectors of length up to $n = 12$.

slightly suboptimal way. The results are comparable to Huffman coding over vectors, but without the exponential complexity growth.

4.4.1 Nonbinary Huffman Codes

So far only binary Huffman codes have been considered. In most applications, this is enough but there are also cases when a larger alphabet is required. In these cases, nonbinary, or D-ary, Huffman codes can be considered. The algorithm and the theory are in many aspects very similar. Instead of a binary tree, a D-ary tree is constructed. Such a tree with depth 1 has D leaves. Then it can be expanded by adding D children to one of the leaves. In the original tree, one leaf has become an internal node and there are D new leaves, in total there are now $D-1$ additional leaves. Every following such expansion will also give $D-1$ additional leaves, which gives the following lemma.

Lemma 4.2 The number of leaves in a D-ary tree is

$$D + q(D-1) \tag{4.54}$$

for some nonnegative integer q. □

This means there can be unused leaves in the tree, but for an optimal code there must be at most $D-2$. Furthermore, these unused leaves must be located at the same depth and it is possible to rearrange such that they stem from the same parent node. This observation corresponds to the binary case that there are no unused leaves.

The code construction for the D-ary case can then be shown in a similar way as for the binary case. Although not shown here, the constructed code is optimal. The algorithm for constructing a D-ary Huffman code is given below.

Algorithm 4.2 (D-ary Huffman code)
To construct the code tree for a random variable with k outcomes:

1. Sort the source symbols according to their probabilities. Fill up with zero probable nodes so that the total number of nodes is $K = D + q(D-1)$.
2. Connect the D least probable symbols in a D-ary tree and remove them from the list. Add the root of the tree as a symbol in the list.
3. If only one symbol left in the list
   STOP
   Else
   GOTO 2

The procedure is shown with an example.

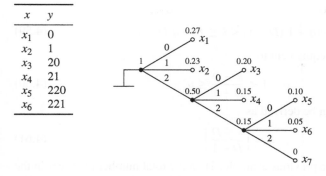

x	y
x_1	0
x_2	1
x_3	20
x_4	21
x_5	220
x_6	221

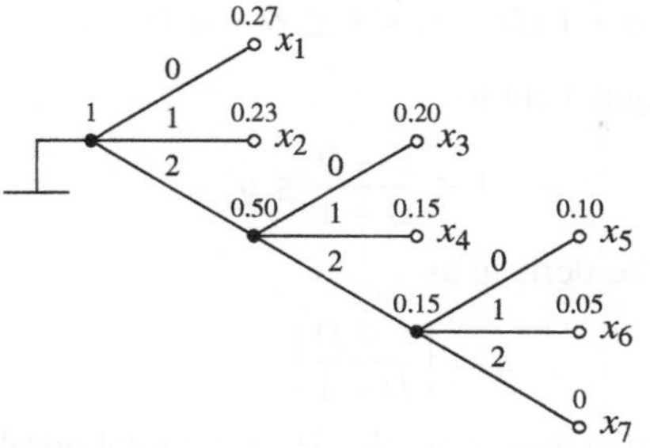

Figure 4.17 A 3-ary tree for a Huffman code.

Example 4.8 Construct an optimal code with $D = 3$ for the random variable X with $k = 6$ outcomes and probabilities

x :	x_1	x_2	x_3	x_4	x_5	x_6
$p(x)$:	0.27	0.23	0.20	0.15	0.10	0.05

First, find the number of unused leaves in the tree. There will be $K = D + q(D - 1)$ leaves in the tree where K is the least integer such that $K \geq k$ and q integer. That is,

$$q = \left\lceil \frac{k - D}{D - 1} \right\rceil = \left\lceil \frac{6 - 3}{3 - 1} \right\rceil = \left\lceil \frac{3}{2} \right\rceil = 2 \tag{4.55}$$

and the number of leaves in the 3-ary tree is

$$K = 3 + 2(3 - 1) = 7 \tag{4.56}$$

Since there are only six codewords used, there will be one unused leaf in the tree. To incorporate this in the algorithm add one symbol, x_7, with probability 0. Then the code tree can be constructed as in Figure 4.17. The branches are labeled with the code alphabet $\{0, 1, 2\}$. In the figure, the tree representation is also translated into a code table.

The average length can as before be calculated with the path length lemma,

$$L = 1 + 0.5 + 0.15 = 1.65 \tag{4.57}$$

As a comparison, the lower bound of the length is the 3-ary entropy,

$$H_3(X) = \frac{H(X)}{\log 3} \approx \frac{2.42}{1.59} = 1.52 \tag{4.58}$$

Even though the lower bound is not reached, this is an optimal code and it is not possible to find a prefix code with less average length.

To start the algorithm, the number of unused leaves in the tree must be found. The relation between the number of source symbol alternatives k and the number of leaves in the tree is

$$D + (q-1)(D-1) < k \leq D + q(D-1) \tag{4.59}$$

By rearrangement, this is equivalent to

$$q - 1 < \frac{k-D}{D-1} \leq q \tag{4.60}$$

Since q is an integer, it can be derived as

$$q = \left\lceil \frac{k-D}{D-1} \right\rceil \tag{4.61}$$

which was also used in the previous example. Then the total number of leaves in the tree is $K = D + q(D-1)$ and the number of unused leaves becomes $m = K - k = D - q(D-1) - k$. Assuming that all unused leaves are located in the same subtree, i.e. having the same parent node, the corresponding number of used leaves in that subtree is $N = D - m$. Then, the total number of used leaves in the tree is the same as the total number of symbols, and

$$k = q(D-1) + N \tag{4.62}$$

In an optimal code, there must be at least two used leaves in the subtree with the unused leaves, i.e. $2 \leq N \leq D$, or equivalently, $0 \leq N - 2 < D - 1$. Subtracting two from the above equation yields

$$k - 2 = q(D-1) + N - 2 \tag{4.63}$$

where it is assumed that $k \geq 2$. From Euclid's division theorem,[3] the following can be derived.

Theorem 4.8 The number of unused leaves in the tree for an optimal D-ary prefix code with k codewords is

$$m = D - N \tag{4.64}$$

where

$$N = R_{D-1}(k-2) + 2 \tag{4.65}$$

is the number of used variables in the first iteration of Huffman's algorithm. □

With this result, the algorithm for constructing a D-ary Huffman code can be slightly changed. Instead of filling up with m zero probable dummy symbols in the first step, the number of merged nodes in the first iteration is N, i.e. the number of used leaves

[3] **Euclid's division theorem:** Let a and b be two positive integers. Then there exist two unique integers q and r such that

$$a = q \cdot b + r, \quad 0 \leq r < b$$

where $r = R_b(a)$ is the reminder and $q = \lfloor \frac{a}{b} \rfloor$ the quotient. An alternative notation of the reminder is the modulo operator, $a \equiv r \pmod{b}$.

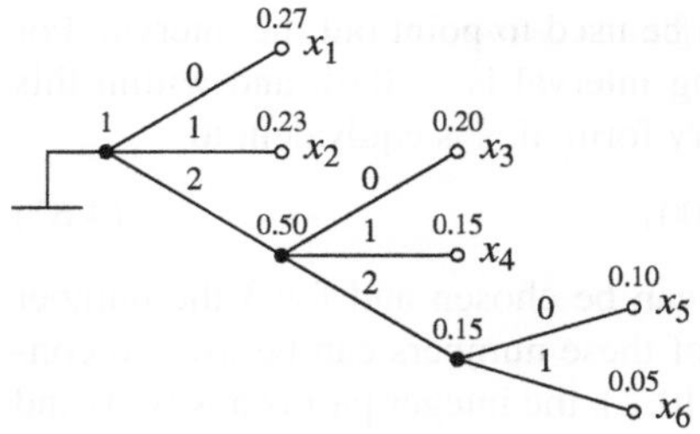

Figure 4.18 A 3-ary code tree for the Huffman code in Example 4.8.

in the subtree with the unused leaves. For Example 4.8, it means $N = R_2(4) + 2 = 2$ nodes should be merged in the first iteration. The result is then a tree where one of the subtrees might not be a full tree. For the case of Example 4.8, the tree is shown in Figure 4.18. The result is of course the same as in Figure 4.17 with the dummy node absent.

4.5 ARITHMETIC CODING

In Theorem 4.6, it is stated that the average codeword length of an optimal code is bounded by

$$H(X) \leq L < H(X) + \frac{1}{n} \tag{4.66}$$

if the encoding is done over vectors of length n. This is exemplified in Example 4.7 where a source of i.i.d. symbols in $\{1, 2, 3\}$ is encoded using vectors of lengths up to $n = 12$. There can clearly be a gain in Huffman coding by extending the code over vectors, but at the cost of an exponential growth of the number of codewords, and thus the complexity in the encoder and decoder. The idea of arithmetic coding is to consider the probabilities for vectors and to build a code with codeword length that grows as the logarithm of the probability for the vector, but without the exponential growth in complexity. Asymptotically, that should give a similar bound on the average codeword length as in Theorem 4.6.

Consider the same distribution as in Example 4.7, i.e.

x	$p(x)$	$F(x)$
1	0.6	0.6
2	0.25	0.85
3	0.15	1.0

Here the probability function has been complemented with the distribution function, $F(x) = \sum_{i=1}^{x} p(x)$. Then the interval[4] $\big[F(x-1), F(x)\big)$ can be used to point out symbol

[4] The interval $x \in [a, b)$ means the numbers $a \leq x < b$.

x, and any number within $\big[F(x-1), F(x)\big)$ can be used to point out the interval. For example, to point out $x = 1$, the corresponding interval is $[0, 0.6)$, and within this interval the number 0.5 can be chosen. In binary form, this is equivalent to

$$0.5 = 0.10000_2 \tag{4.67}$$

Similarly, for 2 the number $0.75 = 0.11000_2$ can be chosen and for 3 the number $0.875 = 0.11100_2$. The binary representation of these numbers can be used to construct the codewords. Since all numbers are in $[0, 1)$, the integer part is always 0 and can be omitted. A prefix code can in this case be constructed by truncating the binary fractional part after the first zero,

x	$p(x)$	$F(x)$	Num	Bin	y
1	0.6	0.6	0.5	0.10000	10
2	0.25	0.85	0.75	0.11000	110
3	0.15	1.0	0.875	0.11100	1110

Clearly, in this case the code is not very impressive compared to the Huffman code, but when considering vectors the same principle can give a good code at moderate complexity. However, before that there are three main obstacles to overcome:

1. How can $F(x_1x_2 \dots x_n)$ be derived in an efficient way?
2. How should the number in the interval be chosen?
3. For long vectors, the intervals will be very small and numerical problems must be addressed.

For the first part, finding $F(x_1x_2 \dots x_n)$, consider the sequence $x_1x_2x_3 = 231$. The interval is then $\big[F(223), F(231)\big)$ since (223) is the vector ordered before (231). The first symbol $x_1 = 2$ points out the interval $\big[F(1), F(2)\big) = [0.6, 0.85]$, as seen before. The sought interval needs to be a subinterval to this, and to get numbers a sequence of two symbols divide it proportional to the probabilities. In Figure 4.19, the numbers for $F(x_1x_2x_3)$ are derived. At the second level, the numbers are

$$F(13) = F(1) = 0.6 \tag{4.68}$$
$$F(21) = F(1) + p(2)F(1) = 0.6 + 0.25 \cdot 0.6 = 0.75 \tag{4.69}$$
$$F(22) = F(1) + p(2)F(2) = 0.6 + 0.25 \cdot 0.85 = 0.8125 \tag{4.70}$$
$$F(23) = F(1) + p(2)F(3) = F(2) = 0.85 \tag{4.71}$$

Thus, the vector $x_1x_2 = 23$ points out the interval $\big[F(22), F(23)\big) = [0.8125, 0.85)$. Similarly, to get the interval for $x_1x_2x_3 = 231$ the interval for $x_1x_2 = 23$ should be split according to the probabilities in the same way,

$$F(223) = F(22) = 0.8125 \tag{4.72}$$
$$F(231) = F(22) + p(23)F(1) = 0.8125 + 0.25 \cdot 0.15 \cdot 0.6 = 0.8350 \tag{4.73}$$

and the interval is $\big[F(223), F(231)\big) = [0.8125, 0.8350]$.

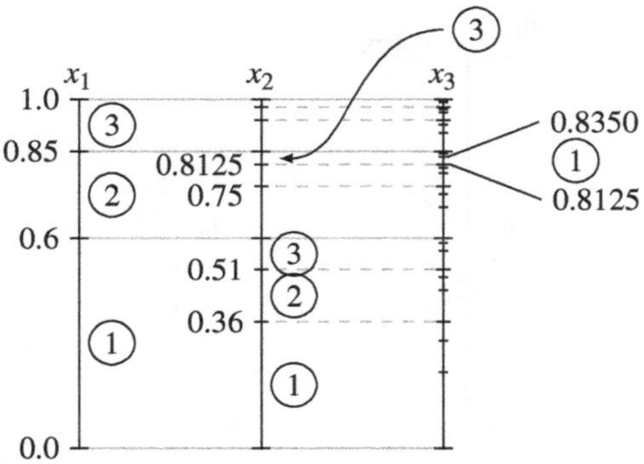

Figure 4.19 Derivation of $F(x_1x_2x_3)$ for a vector of length three.

When continuing the derivations for longer vectors, the interval becomes more and more narrow. For a graphical view like in Figure 4.19, it is more illustrative to magnify the interval in between each step. In Figure 4.20, an illustration is given for the interval corresponding to the vector $\boldsymbol{x} = x_1x_2x_3x_4x_5x_6x_7 = 2312112$.

In each step, the interval is split according to the probabilities and the subinterval according to the present symbol is chosen. In that way, the interval for $\boldsymbol{x} =$ 2312112 is found as

$$\big[F(2312111), F(2312112)\big) = [0.82722, 0.82772) \tag{4.74}$$

Notice that from the construction, the width of the interval is the probability of the vector, so $p(2312112) = F(2312112) - F(2312111) \approx 0.00050625$. As long as the accuracy of the interval limits is adequate, this procedure can be used for arbitrarily long sequences at a constant complexity per added symbol. More generally, the update process for symbol x_k starts with the interval from the previous step, $\big[a_{k-1}, b_{k-1}\big)$ (see Figure 4.21). The new interval boundaries for the symbol are derived as

$$a_k = a_{k-1} + (b_{k-1} - a_{k-1})F(x_k - 1) \tag{4.75}$$

$$b_k = a_{k-1} + (b_{k-1} - a_{k-1})F(x_k) \tag{4.76}$$

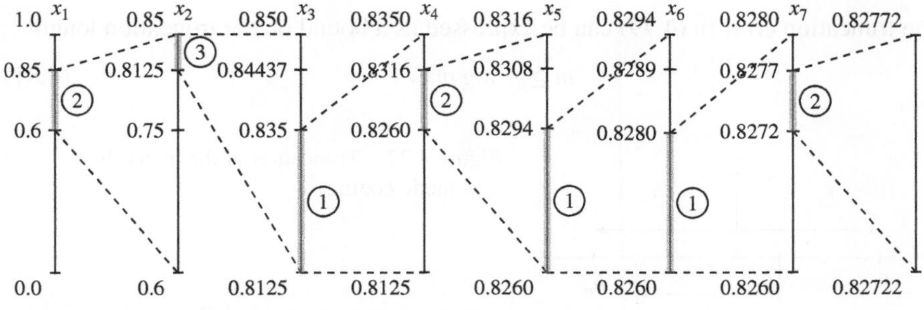

Figure 4.20 Derivation of the interval for $\boldsymbol{x} = 2312112$.

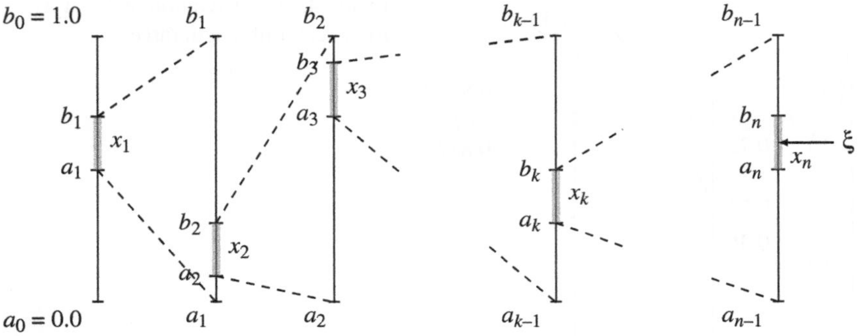

Figure 4.21 General update of the intervals along the encoded sequence.

where $F(x_k - 1) = 0$ if x_k is the lowest ordered symbol, e.g., $x_k = 1$ in the above example.

After the nth symbol, a number inside the interval should be chosen to represent the vector, in Figure 4.21 denoted by ξ. To get a codeword for the sequence, the binary representation of ξ is truncated such that all possible continuations of it lie inside the interval. To get this value, start by assigning ξ as the mid of the interval,

$$\xi = \frac{b_n + a_n}{2} \tag{4.77}$$

(see Figure 4.22). Expressed in a binary form, the value is

$$\xi = \{0.y_1 y_2 \dots y_m y_{m-1} \cdots\}_2 \tag{4.78}$$

The codeword is a truncated version of the decimal part of the number, $y = y_1 y_2 \dots y_m$. The truncation error is at most $\frac{1}{2^m}$, and to guarantee a uniquely decodable code this interval around ξ must be enclosed at the interval $[a_n, b_n)$, or in other words

$$\frac{b_n - a_n}{2} \geq \frac{1}{2^m} \tag{4.79}$$

Since the width of the interval is the probability of the vector,

$$b_n - a_n = p(x_1)p(x_2) \cdots p(x_n) = p(x_1 x_2 \dots x_n) = p(\boldsymbol{x}) \tag{4.80}$$

the truncation error in (4.79) can be expressed as a bound on the truncation length

$$m \geq -\log p(\boldsymbol{x}) + 1 \tag{4.81}$$

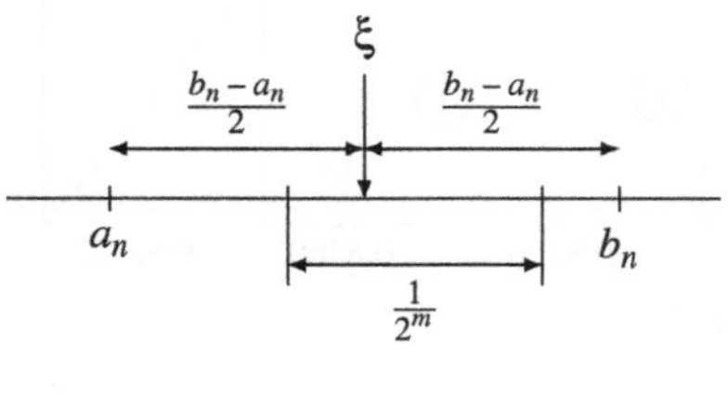

Figure 4.22 Truncation of the intervals in arithmetic coding.

and the codeword length can be chosen as

$$\ell(x) = \left\lceil -\log p(x) \right\rceil + 1 \tag{4.82}$$

For the example given above, the midvalue is

$$\xi = \frac{F(2312112) + F(2312111)}{2} = 0.82747 \tag{4.83}$$

$$= \{0.110100111101010\ldots\}_2 \tag{4.84}$$

The codeword length should be $\ell(2312112) = \lceil -\log(0.00050625) \rceil + 1 = 12$, so the codeword is formed by the first 12 bits of the binary representation of ξ,

$$y = 110100111101 \tag{4.85}$$

The interval where all numbers starting with $\{0.110100111101\ldots\}_2$ is given by $[0.82739, 0.82764)$, which is a true subset of $[0.82722, 0.82772)$, and there cannot be any other input vector giving the same codeword.

Decoding the vector means constructing the same intervals as for the encoding. The codeword directly translates to $\tilde{\xi} = 0.y_1y_2\ldots y_m$, that is a truncated version of ξ known to lie in the same interval as ξ. For each step, the calculation of subinterval containing $\tilde{\xi}$ is chosen and expanded. After n steps, the correct subinterval has been chosen, and this directly gives the encoded vector.

From (4.82), the average codeword length for encoding of a vector of length n can be derived as

$$L_n = E\big[\ell(X)\big] = \sum_x p(x)\left\lceil -\log p(x) \right\rceil + 1$$

$$< -\sum_x p(x)\log p(x) + 2 = nH(X) + 2 \tag{4.86}$$

where the last equality follows from the assumption of i.i.d. symbols. Hence, the average codeword length per symbol can be bounded by $L < H(X) + \frac{2}{n}$. This, together with the fact that the codeword length is lower bounded by the entropy, can be expressed as a theorem.

Theorem 4.9 The average codeword length per symbol for an arithmetic source encoder is bounded by

$$H(X) \leq L < H(X) + \tfrac{2}{n} \tag{4.87}$$

□

Clearly, as the length of the encoded vector is increased, the codeword length will approach the entropy almost as fast as for Huffman coding over vectors, but without the exponential complexity growth.

However, for an implementation of the algorithm as described above, at some point there will be numerical problems since the intervals are getting smaller and smaller. To circumvent this obstacle, it is possible to derive the codeword along with

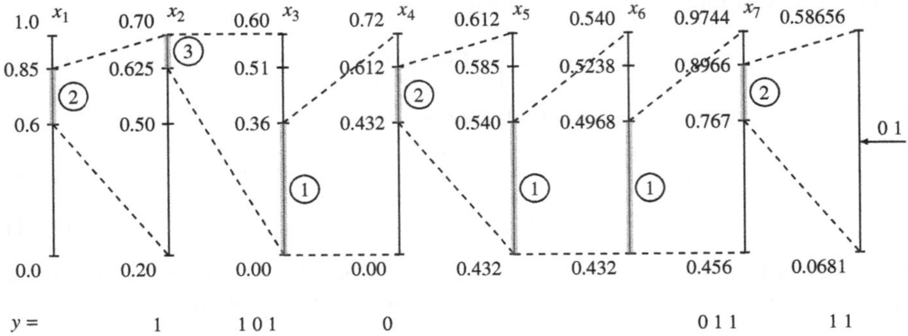

Figure 4.23 Arithmetic coding with scaled scales.

the interval expansions. Consider again the encoding of $\boldsymbol{x} = 2312112$ in Figure 4.20. After the first input symbol $x_1 = 2$, the interval is $[0.6, 0.85)$. Since the entire interval is greater than 0.5, all numbers in it will have the binary representation starting with $0.1\ldots$. Since all the forthcoming intervals are true subintervals, this is also the start of the representing value ξ, and the first bit of the codeword must be 1. Therefore, this bit can be removed from the derivations by multiplying with two and subtraction of one for all the numbers in the interval (see Figure 4.23). The multiplication of two means moving the decimal point one step to the right, and the subtraction of one means removing the 1 that was decided for the codeword.

In the next step $x_2 = 3$, which gives the interval $[0.625, 0.700)$, which is also entirely above 0.5. The same argument as above gives a 1 to the codeword, and the interval should be doubled and subtracted by one, giving $[0.25, 0.40)$. This new interval is entirely below 0.5, meaning all numbers in it will have a binary representation starting with $0.0\ldots$, and the next digit of the codeword has to be a 0. To move the decimal point one step to the right, the interval should be multiplied with two to get $[0.5, 0.8)$. Again the interval is entirely above 0.5 and the next bit in the codeword is 1, while the scaled interval becomes $[0.0, 0.6)$. Now, this interval is not only on the one side of 0.5 and cannot be scaled anymore. So, for the second step in the algorithm the digits 1 0 1 was added to the codeword and the interval scaled to $[0.0, 0.6)$. In Figure 4.23, the remaining symbols from x_3 to x_7 are shown together with the scaling and the concluded bits in the codeword.

After the last symbol x_7, the remaining of the codeword can be derived in the same way as before. For this example, the midvalue of the last interval is $\xi = 0.2592 = \{0.0100\ldots\}_2$. The remaining codeword length is

$$\lceil -(0.58656 - 0.06816) \rceil + 1 = 2 \tag{4.88}$$

Hence, the last part of the codeword is 1 0, and the complete codeword is $y = 110100111101$, which of course is the same as before.

PROBLEMS

4.1 Consider the code $\{0, 01, 10\}$.

(a) Is it nonsingular?

(b) Is it uniquely decodable?

(c) Is it a prefix code?

4.2 For each of the following sets of codeword lengths, decide whether there exists a binary prefix code. If it exists, construct a code.

(a) $\{1, 2, 3, 4, 5\}$

(b) $\{2, 2, 3, 3, 4, 4, 5, 5\}$

(c) $\{2, 2, 2, 3, 4, 4, 4, 5\}$

(d) $\{2, 3, 3, 3, 4, 5, 5, 5\}$

4.3 In the following tables, the probabilities for a random variable X are given together with a binary prefix code.

x	$p(x)$	y
x_1	0.2	0
x_2	0.2	100
x_3	0.2	101

x	$p(x)$	y
x_4	0.2	110
x_4	0.1	1110
x_6	0.1	1111

(a) Draw a binary tree that represents the code.

(b) Derive the entropy $H(X)$ and the average codeword length $L = E[\ell]$.

(c) Is it possible that the code is optimal for X?

(d) Is the code optimal for X?

4.4 Show, by induction, that a binary tree with k leaves has $k - 1$ inner nodes.

4.5 For a given k-ary source, the probability function is estimated to be $q(x)$, $x \in \{0, 1, \dots, k - 1\}$. Consider the case when an optimal source code for this source is chosen according to this distribution (neglect the truncation error in the logarithm). Let the true distribution for the symbols be $p(x)$, $x \in \{0, 1, \dots, N - 1\}$. Show that the relative entropy is the penalty in bits per source symbol due to the estimation error.

4.6 Construct optimal codes for the random variables with the following distributions. Also derive the average codeword lengths and the entropies.

(a)

x :	x_1	x_2	x_3	x_4
$p(x)$:	1/4	1/4	3/8	1/8

(b)

x :	x_1	x_2	x_3	x_4	x_5
$p(x)$:	1/3	1/6	1/6	2/9	1/9

(c)

x :	x_1	x_2	x_3	x_4	x_5	x_6	x_7
$p(x)$:	0.3	0.2	0.2	0.1	0.1	0.06	0.04

4.7 A binary information source with probabilities $P(X = 0) = \frac{3}{5}$ and $P(X = 1) = \frac{2}{5}$ produces a sequence. The sequence is split into blocks of 3 bits each that should be encoded separately. In the table below, such a code is given. Is the suggested code optimal? If not, construct one and derive the average gain compared to the uncoded case.

x	y
000	0
001	100
010	110
011	11100

x	y
100	101
101	11101
110	11110
111	11111

4.8 The first-order statistics of the letters in the English alphabet is given below. It lists, for each letter, the total average number of occurrences out of a 1000 letter text.

x	Nbr
A	73
B	9
C	30
D	44
E	130

x	Nbr
F	28
G	16
H	35
I	74
J	2

x	Nbr
K	3
L	35
M	25
N	78
O	74

x	Nbr
P	27
Q	3
R	77
S	63
T	93

x	Nbr
U	27
V	13
W	16
X	5
Y	19
Z	1

Construct a binary Huffman code for the letters. What is the average codewords length? How does this compare to the entropy of the letters? How does it compare to a coding where the statistics is not taken into consideration?

4.9 Assume a binary random variable X with $P(X = 0) = 0.1$ and $P(X = 1) = 0.9$.

(a) Find the average codeword length of an optimal source code for X.

(b) Consider vectors of n i.i.d. symbols, $X^{(n)} = X_1X_2 \dots X_n$. Construct optimal source codes for the cases $n = 2, 3, 4$. What is the average codeword lengths per binary source symbol?

(c) Compare the above results with the entropy of X.

4.10 A *unary* source code is a mapping from the positive integers to binary vectors such that the codeword for the integer n consists of $n - 1$ zeros followed by a one, i.e.

n	y
1	1
2	01
3	001
4	0001
5	00001
⋮	⋮

This is clearly a prefix code since there is only one 1, located at the end of each codeword.

On Wikipedia (https://en.wikipedia.org/wiki/Unary_coding),[5] it is stated that the unary code is optimal for the probability function

$$P(n) = (k-1)k^{-n}, \quad n = 1, 2, \dots, \text{ and } k \geq 1.61803\dots$$

The aim of this problem is to clarify if this is correct. However, we will only consider probability functions where k is an integer, i.e. $k = 2, 3, 4, \dots$ is considered.

(a) Is it true that for any integer $k \geq 2$ the function $P(n) = (k-1)k^{-n}$, $n = 1, 2, 3, \dots$ is a probability function?

(b) Is it true that for $k = 2$ the unary code is optimal? What is the average codeword length?

(c) Is it true that for any integer $k = 2, 3, 4, \dots$ the unary code is optimal? What is the average codeword length?

4.11 A random variable X has k outcomes and probabilities $p_i = P(X = x_i)$. To construct a source code for the variable, order the outcomes according to the probabilities, with the highest to the left, i.e. the vector becomes $x_1, x_2, \dots, x_k$, where $p_1 \geq p_2 \geq \cdots \geq p_k$. Split the vector in two parts such that their sums are as close as possible. That is, find an index q such that

$$\left| \sum_{i=1}^{q} p_i - \sum_{j=q+1}^{k} p_j \right|$$

is as small as possible. Label the left part with 0 and the right part with 1. Continue the procedure iteratively in the same way with each of the parts until there are only two outcomes in the vector. The labeling constitute the codeword for each outcome.

The described method is often referred to as Fano coding. For each of the probability vectors below find the Fano code and state if it is also a Huffman code and/or if it is optimal.

$$P_a = \left(\frac{4}{10}, \frac{3}{10}, \frac{2}{10}, \frac{1}{10}\right)$$

$$P_b = \left(\frac{6}{21}, \frac{5}{21}, \frac{4}{21}, \frac{3}{21}, \frac{2}{21}, \frac{1}{21}\right)$$

$$P_c = \left(\frac{15}{43}, \frac{7}{43}, \frac{7}{43}, \frac{7}{43}, \frac{7}{43}\right)$$

4.12 Use arithmetic coding to compress the string $\boldsymbol{x} = \text{d a c b d a a a d a}$, if the distribution is given by

x:	a	b	c	d
$p(x)$:	0.5	0.3	0.1	0.1

4.13 Use the same distribution as in Problem 4.12 and decode the following codeword from an arithmetic encoder.

$$y = 01111000111011100$$

[5] You can of course not use any formulas found on Wikipedia, or elsewhere on Internet, without verifying it.

4.14 Construct a binary random vector $x = x_1 x_2 \ldots x_n$ of i.i.d. symbols with the distribution

x:	0	1
$p(x)$:	1/4	3/4

Compress the vector using arithmetic coding for $n = 10\,000$. What is the average codeword length per symbol? Compare with the entropy.

Note that this problem is best suited for computer implementation and that the intervals need to be scaled to avoid numerical problems.

CHAPTER 5

ADAPTIVE SOURCE CODING

IN THE PREVIOUS chapter, it was shown by use of Kraft inequality that the optimal average codeword length for a source code equals the entropy, i.e., the uncertainty, of the source. With Huffman's algorithm, it is possible to find an optimal code, symbol by symbol, which is no more than 1 bit far from the entropy. However, in this analysis there were some assumptions. First, the source symbols are assumed to be independent and identically distributed (i.i.d.). Considering, for example normal text or images, this assumption yields individual letters or pixels to be independent of their neighbors. In reality, most sources are much more complex than this. A second assumption is that the statistics of the source is known. Again, in practical examples like texts or images, the distribution of letters or pixels is dependent on the language or content. So, it turns out that none of these assumptions are very realistic. In this chapter, algorithms that can perform compression for sources when the statistics is unknown are considered. These types of algorithms are often called universal source coding algorithms.

5.1 THE PROBLEM WITH UNKNOWN SOURCE STATISTICS

For most sources, e.g., text, images, or audio, one problem to use Huffman codes is that the source statistics is not known. Even though the statistics is unknown, there are often some presumptions about it, e.g., for a normal English text the frequency for each letter is relatively known. So, one obvious way to construct a coding scheme would be to use a predefined code for this content type. This will work quite good for English texts. However, if a text in another language is applied, the algorithm will not work as good any more without changing the code. It gets even worse if it is used on, e.g., a list of names, a program code, or even an image, which has a completely different distribution of the characters.

The compression loss made by assuming the wrong distribution can be derived by assuming that $p(x)$ is the true distribution for the random variable X and $q(x)$ the estimated. The optimal codeword length for the source is $L^{(\text{opt})} = E_p[-\log p(x)] = H(X)$ and when using $q(x)$ for code construction the average length becomes

Information and Communication Theory, First Edition. Stefan Höst.

$L_q = E_p[-\log q(x)]$. The loss between the optimal length and the achieved is

$$\begin{aligned} L_q - L^{(\text{opt})} &= -\sum_x p(x) \log q(x) + \sum_x p(x) \log p(x) \\ &= \sum_x p(x) \log \frac{p(x)}{q(x)} = D\big(p(X) || q(X)\big) \end{aligned} \tag{5.1}$$

Hence, the relative entropy comes back here as the penalty for having mismatch in the distribution of the code and the source.

In many cases when the source statistics is unknown, it is possible to use adaptive versions of the source coding algorithms. That is, instead of defining the code for a specific, predefined distribution, the algorithm estimates the statistics of the source.

5.2 ADAPTIVE HUFFMAN CODING

A straightforward way to construct an adaptive version of the Huffman algorithm is to have a two-sweep algorithm. In the first sweep, the complete source sequence is processed and the distribution of the letters is estimated. Then, in the second sweep, a Huffman code for this distribution can be used to generate the code sequence. This requires the source text to be finite since the first pass must finish. The procedure can give a fairly good compression rate but has the disadvantage that the code table must be submitted along with the code string, which of course increases the average codeword length, especially for short sequences.

In this section, an adaptive, or dynamic, version of Huffman coding will be described where the statistics for the current symbol is estimated from the past symbols of the sequence. That means the code tree is reshaped along with the encoding process. It is also possible for the receiver to build the same code table, and, hence, it must not be transmitted.

In 1973, Faller [21] published a one-sweep adaptive Huffman algorithm. The idea is to use the past symbols of the sequence to estimate the distribution for the code construction. One way to do this is to rebuild the code tree from scratch after each new symbol, but this would be a waste of computational resources. Faller found a way to update the distribution and the code tree without starting all over. The algorithm was improved by Gallager in 1978 [22], introducing the sibling property, and by Knuth in 1985 [23], who also considered zero weight symbols in the tree. Knuth also realized that as well as adding one symbol to the tree, one symbol can be deleted, opening for a windowed version of the algorithm. In 1987, Vitter [24] made further improvements of the algorithm, rearranging the update procedure and thereby decreasing the number of computations. The description here follows Knuth [23].

Assuming a very long sequence of i.i.d. symbols as a source, there might be problems with both delay and buffering if using a two-sweep algorithm. Instead, for each step an estimate of the source distribution is derived based on the previous symbols in the sequence. At the beginning of the sequence, the estimate will be rough, but quite soon it will converge and give a good estimate, as long as the source statistics does not change along the sequence. The key component for constructing an adaptive version of the Huffman code is to update the tree according to the changes in the estimated distribution.

A code tree for a binary prefix code for an alphabet of length k can be represented by a data structure in a vector $\boldsymbol{q} = (q_1, q_2, \ldots, q_m)$. If the code tree contains no empty leaves, the length of the vector is $m = 2k - 1$, i.e., k leaves and $k - 1$ inner nodes. Each element in the vector contains information of weight w_i, denoting the number of occurrences of the node letter, and pointers to the elements, representing parent and children nodes.[1] The weight of the root node w_1 is the sum of all the letters in the source sequence seen so far, which is also the sum of the weights for the leaves in the tree. Hence, an estimation of the probability for each node in the tree is found by $\hat{p}_i = w_i / w_1$. Since the probability of the root node equals the sum of the leaf probabilities, the average codeword length can be found from the path length lemma as the sum of all probabilities in the tree, excluding the root weight,

$$E[\ell] = \sum_{i=2}^{2k-1} \frac{w_i}{w_1} \tag{5.2}$$

The *sibling property* refers to the case when the vector is ordered according to the weights of the elements, and that each sibling pair in the tree is located adjacently in the vector. The ordering in this text will be made with the root node represented by q_1. The sibling property is closely related to the construction of a Huffman code. To obtain a vector with the sibling property, start with q_1 representing the root in the Huffman tree. Then, add sibling pairs to the vector in a reverse order as they are merged in the code tree construction. That means all code trees constructed with the Huffman algorithm can be represented by a vector with the sibling property. Conversely, all vectors with the sibling property can be represented by a code tree according to the Huffman algorithm. This is stated in the following theorem from Gallager [22].

Theorem 5.1 A code tree can be represented by a vector with the sibling property if and only if it represents a Huffman code. □

The property is illustrated by the following example.

Example 5.1 Given an alphabet of five letters $\{a, b, c, d, e\}$ with the weights $\{9, 10, 3, 8, 12\}$. The code tree in Figure 5.1 is built according to the Huffman algorithm.

In the tree, nodes are labeled q_i starting at the root with q_1 and ending in the first merged subtree with q_8 and q_9. The numbering is performed in an opposite order

[1] To represent the tree, it suffice to have information about either children or parent nodes. However, since the tree will be traversed both from the root to the leaves and vice versa it is convenient to store both directions.

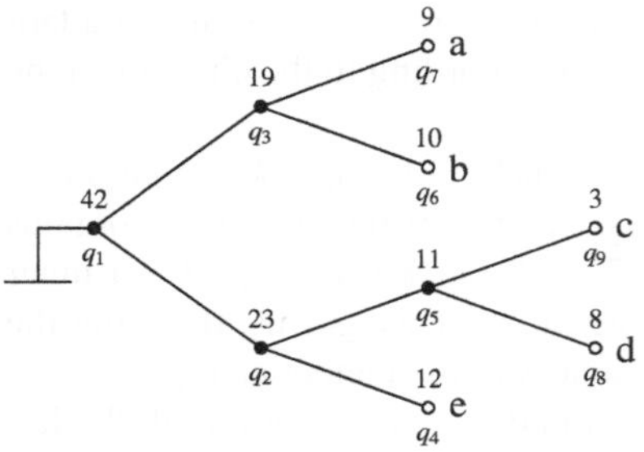

Figure 5.1 The code tree for the five-letter code.

as the merging in the Huffman algorithm. Hence, by writing the labels as a vector will satisfy the sibling property,

$$\begin{array}{ccccccccc} & & & e & & b & a & d & c \\ q_1 & q_2 & q_3 & q_4 & q_5 & q_6 & q_7 & q_8 & q_9 \\ 42 & 23 & 19 & 12 & 11 & 10 & 9 & 8 & 3 \end{array} \tag{5.3}$$

In (5.3), the weights of the nodes are written below the node label and for the leaves the letter symbol is written above. In this example, as well as the continuation of this section, the information about children and parent nodes is omitted from the vectors. This can instead be read from the tree picture. Naturally, in a computer implementation they should be included in the vector as indices or pointers.

The weight of the inner nodes is the sum of the weights of their child nodes, i.e. the number of letter occurrences for the leaves in the subtree. The total number of letters in the sequence is thus the weight of the root node, and normalizing all weights with this yields the estimated probabilities. Reading one more letter from the source sequence, the weights in the path from the letter back to the root should be incremented by one. As long as this preserves the Huffman property of the tree, this can be done. Translated to the sibling property, this means the weights can be incremented as long as for each i in the path the weights have a superior neighbor, i.e.

$$w_i < w_{i-1}, \quad \forall i \in \mathcal{P} \tag{5.4}$$

where $\mathcal{P}$ denotes the set of indices in the path from the leaf of the letter to the root. The procedure is shown in the following example.

Example 5.2 In the tree from Example 5.1, assume the next symbol in the source sequence is "c," so the tree should be updated. In the vector representation, marking all the entries that should be incremented bold, gives

$$\begin{array}{ccccccccc} & & & e & & b & a & d & \boldsymbol{c} \\ \boldsymbol{q_1} & \boldsymbol{q_2} & q_3 & q_4 & \boldsymbol{q_5} & q_6 & q_7 & q_8 & \boldsymbol{q_9} \\ \mathbf{42} & \mathbf{23} & 19 & 12 & \mathbf{11} & 10 & 9 & 8 & \mathbf{3} \end{array} \tag{5.5}$$

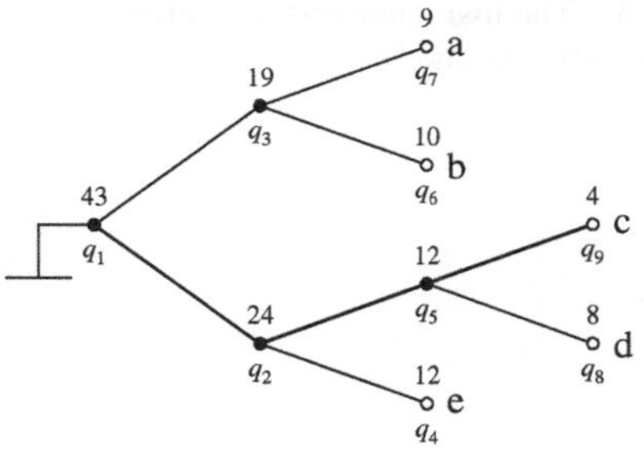

Figure 5.2 The five-letter code tree updated with a "c."

The nodes included in the path are thus $\{q_1, q_2, q_5, q_9\}$. Since $q_2 < q_1$, $q_5 < q_4$ and $q_9 < q_8$, the requirement in (5.4) is fulfilled and all nodes in the path can be incremented to get

$$\begin{array}{ccccccccc} & & & e & & b & a & d & c \\ \mathbf{q_1} & \mathbf{q_2} & q_3 & q_4 & \mathbf{q_5} & q_6 & q_7 & q_8 & \mathbf{q_9} \\ \mathbf{43} & \mathbf{24} & 19 & 12 & \mathbf{12} & 10 & 9 & 8 & \mathbf{4} \end{array} \tag{5.6}$$

The updated code tree is shown in Figure 5.2.

If at some point along the path from the leaf to the root, it occurs that (5.4) is not fulfilled, there has to be one or several nodes with the same weight. Obviously, in a Huffman tree it is possible to swap subtrees with equal weight without changing the average codeword length. Consequently, when adding one occurrence of a letter, first find the corresponding index in the vector $\boldsymbol{q}$, denoted here as α. Then, following the path from the α to the root, for each node i check if $w_i < w_{i-1}$. If this is not fulfilled, find the index j such that $w_j = w_i$ and $w_j < w_{j-1}$, and swap the children of node i with the children of node j. After this operation, all the nodes in the path from α to the root satisfy (5.4) and the path can be updated.

Example 5.3 Continuing from the resulting tree in Example 5.1 and adding one more "c" to the tree. The vector is

$$\begin{array}{ccccccccc} & & & e & & b & a & d & c \\ \mathbf{q_1} & \mathbf{q_2} & q_3 & q_4 & \mathbf{q_5} & q_6 & q_7 & q_8 & \mathbf{q_9} \\ \mathbf{43} & \mathbf{24} & 19 & 12 & \mathbf{12} & 10 & 9 & 8 & \mathbf{4} \end{array} \tag{5.7}$$

Following the path from the leaf gives $w_9 < w_8$, so the first branch satisfies the condition. But $w_5 \not< w_4$ so here the tree must be rearranged. The lowest ordered node with the same weight as q_5 is q_4, thus their children should be swapped. Then q_5 becomes a leaf with the letter "e", and q_4 gets two children, q_8 and q_9, resulting in the vector

$$\begin{array}{ccccccccc} & & & & e & b & a & d & c \\ \mathbf{q_1} & \mathbf{q_2} & q_3 & \mathbf{q_4} & q_5 & q_6 & q_7 & q_8 & \mathbf{q_9} \\ \mathbf{43} & \mathbf{24} & 19 & \mathbf{12} & 12 & 10 & 9 & 8 & \mathbf{4} \end{array} \tag{5.8}$$

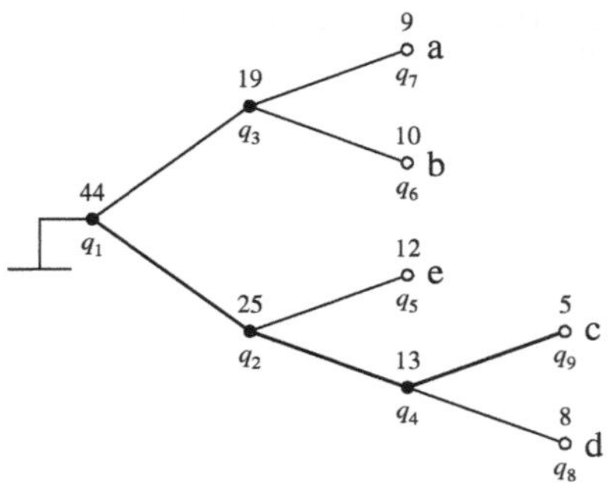

Figure 5.3 The five-letter code tree after updating with a second "c".

Continuing the check of the condition gives $w_4 < w_3$ and $w_2 < w_1$, and the node weights along the path can be incremented. The resulting vector is

$$\begin{array}{ccccccccc} & & & & e & b & a & d & c \\ \mathbf{q_1} & \mathbf{q_2} & q_3 & \mathbf{q_4} & q_5 & q_6 & q_7 & q_8 & \mathbf{q_9} \\ \mathbf{44} & \mathbf{25} & 19 & \mathbf{13} & 12 & 10 & 9 & 8 & \mathbf{5} \end{array} \tag{5.9}$$

The updated tree is shown in Figure 5.3.

The procedure for updating the tree as described above has one drawback; all nodes must be represented in the tree from the beginning. The parent node of two zero weight nodes will also have zero weight, and thus in the update procedure the subtree root will be swapped with one of the leaves, collapsing that part of the tree. To circumvent this flaw, all unused letters, i.e. zero weight leaves, are collected in one subtree represented as one single node in the algorithm. This single zero weight node is often referred to as *not yet transmitted*, abbreviated as NYT. The codeword for a NYT node is then the codeword for the NYT tree root concatenated with the codeword for the path in the NYT tree. Since all the NYT nodes have zero weight, the best code is to have equal length codewords. In general, the number of NYT nodes is not a power of two. Thus, an optimal construction for the NYT tree with r symbols has leaves at depth $\lfloor \log r \rfloor$ and $\lfloor \log r \rfloor + 1$, as shown in Figure 5.4.

In the figure, there are s leaves at depth $\lfloor \log r \rfloor$ and t nodes that are further expanded by a depth one subtree, which gives $2t$ leaves at depth $\lfloor \log r \rfloor + 1$. Since $t + s = 2^{\lfloor \log r \rfloor}$ and $2^{\lfloor \log r \rfloor} \leq r < 2^{\lfloor \log r \rfloor + 1}$, the number of expanded nodes at level $\lfloor \log r \rfloor$ is $t = r - 2^{\lfloor \log r \rfloor}$. Hence, the number of leaves at depth $\lfloor \log r \rfloor + 1$ is $2t = 2r - 2^{\lfloor \log r \rfloor + 1}$ and the number of leaves at depth $\lfloor \log r \rfloor$ is $s = 2^{\lfloor \log r \rfloor} - t = 2^{\lfloor \log r \rfloor + 1} - r$.

Ordering the NYT symbols in a predetermined order in the tree, e.g., alphabetical, the NYT codeword can be determined. When a NYT symbol is used, it is removed from the NYT list and the size of the tree decreased accordingly. When adding the new symbol to the tree, one has to take care of the swapping. The new reduced NYT node will be positioned as a sibling to the newly added node (see Figure 5.5).

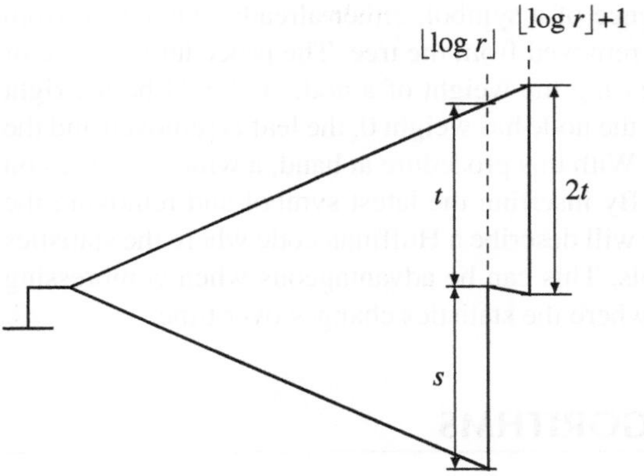

Figure 5.4 The subtree for the NYT nodes.

That is, to add a symbol γ from the NYT node to the tree, first add a parent node and then attach γ and the NYT node as siblings to this, both with zero weight. In Figure 5.5a, the tree before adding the new node γ is shown. Here, the sibling leaf to the NYT node is x with weight w_l. If x is located at position l in the vector, then the adjacent cell q_{l+1} is the NYT node. Figure 5.5b shows the addition of a parent node, replacing the NYT node at q_{l+1}, and node q_{l+2} containing γ and the NYT node where s is excluded.

If the incremental procedure in the vector is started from the new node q_{l+2} in the tree, the algorithm will start by placing it at the lowest index of all nodes with the same weight, i.e. it will swap q_{l+1} with its parent q_{l+1}. To circumvent this, the procedure is instead to increment the newly added node q_{l+2}, as shown in Figure 5.5c, and then start to increment the path from its parent q_{l+1} to the root (Figure 5.5d).

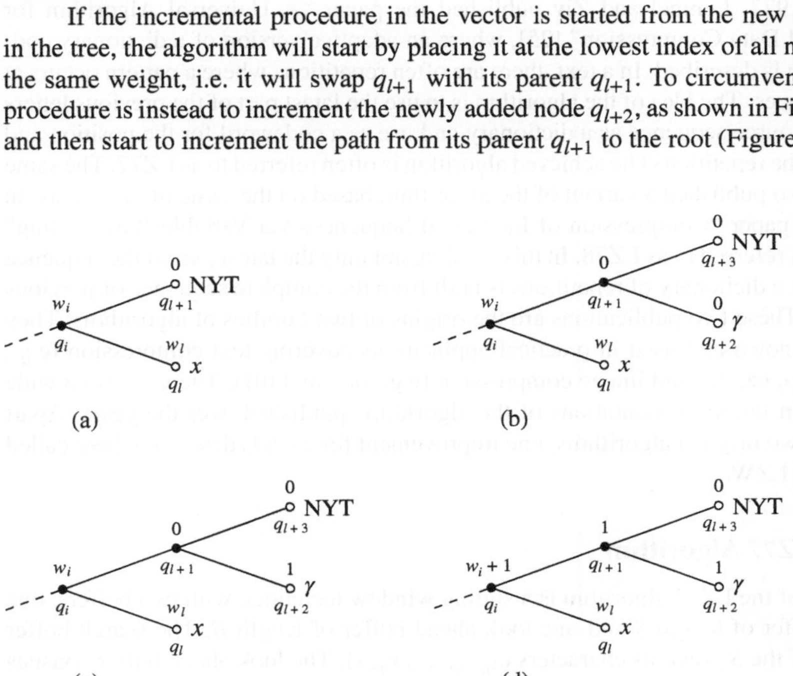

Figure 5.5 Adding one symbol from the NYT node to the tree.

Similar to adding an occurrence of a symbol, either already existing or from the NYT node, one symbol can be removed from the tree. The procedure is more or less the opposite, and before decreasing the weight of a node it should be the right most of all with the same weight. If the node has weight 0, the leaf is removed and the symbol is placed in the NYT node. With this procedure at hand, a windowed version of the algorithm can be deployed. By inserting the latest symbol and removing the symbol M steps in the past, the tree will describe a Huffman code where the statistics is based on the previous M symbols. This can be advantageous when compressing very long sequences or sequences where the statistics changes over time.

5.3 THE LEMPEL–ZIV ALGORITHMS

An alternative approach to solve the problem of source coding is to use a dictionary with the most frequent patterns. Then the codeword would be the index of the dictionary where the pattern is found. For example, considering a text where the statistics is relatively known it is possible to construct a code by finding the most probable sequence and assigning them short codewords.. In cases when the source content is specific and statistics known, the method can give a good compression, but in most cases some sort of adaptive dictionary is needed. In this section, a family of algorithms is described, where the dictionary is built from the past compressed sequence.

In 1977, Lempel and Ziv published the paper "A Universal Algorithm for Sequential Data Compression" [25], where an adaptive version of a dictionary coding system is described. In a text, there are often repetitions where a pattern occurs at several places. The idea of the algorithm is to use the latest part of the previous letters from the source sequence as a dictionary and use as a codeword for the position and length of the repetition. The achieved algorithm is often referred to as LZ77. The same authors also published a variant of the algorithm, based on the same observations, in 1978 in a paper "Compression of Individual Sequences via Variable-Rate Coding" [26], often referred to as LZ78. In this version, not only the latest part of the sequence is used but a dictionary of repetitions is built from the complete sequence of previous symbols. These two publications are the origins of two families of algorithms. They are well-known and used in practical applications covering text compression (e.g., deflate, zip, gz, 7z) and image compression (e.g., png and tiff). There is also a wide variation in the implementations of the algorithms published over the years. Apart from the two original algorithms, one improvement for each is described, here called LZSS and LZW.

5.3.1 LZ77 Algorithm

The base of the LZ77 algorithm is a sliding window technique with two buffers, one search buffer of length S and one look ahead buffer of length B. The search buffer consists of the S previous characters $(x_{n-S}, \dots, x_{n-1})$. The look ahead buffer consists of the next B characters $(x_n, \dots, x_{n+B-1})$. In Figure 5.6, a schematic view of the buffers is shown. If the character x_n at time n should be encoded, it should be a not too low probability that x_n and a couple of its successors can also be found among the S

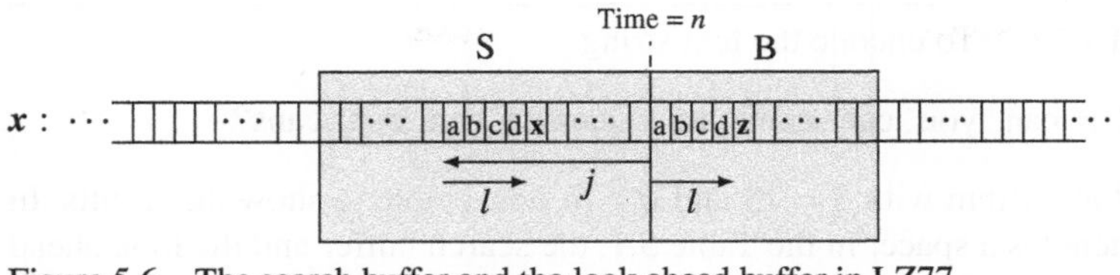

Figure 5.6 The search buffer and the look ahead buffer in LZ77.

previous characters. In this way, the search buffer is used as an adaptive dictionary. The algorithm searches the dictionary for the longest match of x_n and its successors in the look ahead buffer. If there is such a match, it will give an index j to the start of the match and a length l of the match. The index j is the offset between the match in the search buffer and in the look ahead buffer, i.e. the number of positions from x_n to the start of the match. In the case when there is no match, set $j, l = 0, 0$. To be able to continue after a missing match, the next character in the string should also be attached to the codeword. Hence, the codeword for the match is (j, l, c), where c is the character following the match in the sequence. In Algorithm 5.1, the procedure is described.

Algorithm 5.1 (LZ77)

Initialization:

  Initialize the search buffer with the first S characters of the text,

  Set $n = S + 1$.

Continue until n at end of text

  Find all offsets of x_n in the search buffer.

  If number offsets > 0 [There is at least one match]

    Find the offset with the longest match. This gives index j and length l.

    Set the codeword to (j, l, c), where $c = x_{n+l}$ is the character after the match.

    Set $n \leftarrow n + l + 1$.

  Else [There are no matches]

    Set the codeword to $(0, 0, c)$, where $c = x_n$.

    Set $n \leftarrow n + 1$.

  Update search buffer and look ahead buffer

The two numbers j and l are bounded by the buffer lengths, $j \leq S$ and $l \leq B$. So, to express these in a binary form it is needed $\lceil \log(S + 1) \rceil$ and $\lceil \log(B + 1) \rceil$ bits, respectively. Here, it is assumed that number zero should also be possible to set in the codeword, which is the reason for the argument $S + 1$ and $B + 1$. If the normal ASCII table is used to express the characters in a binary form, each character is represented by 8 bits. Hence, a binary representation of a codeword requires

$$\ell\big((j, l, c)\big) = \lceil \log(S + 1) \rceil + \lceil \log(B + 1) \rceil + 8 \tag{5.10}$$

The next example will further explain the flow of the algorithm.

Example 5.4 (LZ77) To encode the text string

$$T = \text{can␣you␣can␣cans␣as␣a␣canner␣can␣can␣cans?}$$

using the LZ77 algorithm with $S = 15$ and $B = 7$, use a table to show the results. In the string "␣" denotes a space. In the Table 5.1, the search buffer and the look ahead buffer are shown for each codeword generated.

In the table, a match of the first letter from the B buffer is denoted by a ∘ and underlined the match length for each match. The longest match is also underlined in the B buffer. The initialization of the S buffer is done with the 16 first letters of the source text. The steps in the algorithm are as follows:

1. To start, the first character of the B buffer is "s." This does not exist in the S buffer so the codeword becomes (0,0,s).
2. In the second step, there are three matches of the first symbol, "␣", in the B buffer. All three of them give a match of length 1, so any of the codewords (5,1,a), (9,1,a), or (13,1,a) can be used. In our case, the first match from the right is chosen.
3. There is only one match of "s". This gives a match length of 3, and the codeword is (3,3,␣).
4. There are two "c"in the S buffer, on position 10 and 14. Both have a match length of 3. Again chosen the codeword with the shortest back track, (10,3,n), is chosen.
5. "e"cannot be found in the S buffer, giving the codeword (0,0,e).
6. "r"cannot be found in the S buffer, giving the codeword (0,0,r).
7. There are three "␣"in the S buffer, on 7, 9, and 12. The first has the longest match, four, giving the codeword (7,4,␣).
8. In the S buffer, there are two "c", at 4 and 11. The match at 11 has length 3, and the match at 4 has length 7; hence the codeword is (4,7,s). Notice that the

TABLE 5.1 Encoding table of Example 5.4.

Step	S buffer	B buffer	Codeword
1	can␣you␣can␣can	s␣as␣a␣	(0,0,s)
2	an␣you␣can␣cans	␣as␣a␣c	(5,1,a)
3	␣you␣can␣cans␣a	s␣a␣can	(3,3,␣)
4	␣can␣cans␣as␣a␣	canner␣	(10,3,n)
5	␣cans␣as␣a␣cann	er␣can␣	(0,0,e)
6	cans␣as␣a␣canne	r␣can␣c	(0,0,r)
7	ans␣as␣a␣canner	␣can␣ca	(7,4,␣)
8	s␣a␣canner␣can␣	can␣cans	(4,7,s)
9	er␣can␣can␣cans	?	(0,0,?)

match continues into the B buffer. This is not a problem since at decoding the S buffer is known at each step, and the match is built up from left to right.

9. There is no "?" in the S buffer so the codeword is (0,0,?).

Summarizing, the codewords are

$$C = (0,0,\text{s})(5,1,\text{a})(3,3,\text{␣})(10,3,\text{n})(0,0,\text{e})(0,0,\text{r})(7,4,\text{␣})(4,7,\text{s})(0,0,?)$$

In the code sequence, the initialization of the S buffer is not shown, but these 15 characters must also be included. Assuming the characters are encoded by the ASCII table using 8 bits each, the uncoded string requires $42 \cdot 8 = 336$ bits. Similarly, initialization of the buffer requires $15 \cdot 8 = 120$ bits. Each codeword needs $\lceil \log 16 \rceil + \lceil \log 8 \rceil + 8 = 15$ bits, and there are nine codewords. In total, the encoded sequence can be represented with $120 + 9 \cdot 15 = 255$ bits. Even for this short string, with relatively small buffers, there is a compression gain and the compression ratio is $R = 1.32$.

At the receiver side, the S buffer at the decoder is initialized with the same initialization sequence, included in the code sequence. In Table 5.2, the procedure is shown, and the steps explained below.

1. The codeword indicates there is no match in the S buffer, and the decoded symbol is just the character from the codeword, "s".
2. The match is five steps into the S buffer, but it is only one long giving the decoded symbol "␣". Then, with the extra symbol "a"from the codeword appended, gives the source string "␣a".
3. The match is at position 3 with a length of 3, which gives "s␣a" and then the codeword symbol "␣" is appended to get "s␣a␣".
4. At position 10 and length 3, there is the match "can", and "n"is added from the codeword.
5. There is no match, and the codeword gives "e".
6. There is no match, and the codeword gives "r".

TABLE 5.2 The decoding procedure in Example 5.4.

Step	Codeword	S buffer	New symbols
1	(0,0,s)	can␣you␣can␣can	s
2	(5,1,a)	an␣you␣can␣cans	␣a
3	(3,3,␣)	␣you␣can␣cans␣a	s␣a␣
4	(10,3,n)	␣can␣cans␣as␣a␣	cann
5	(0,0,e)	␣cans␣as␣a␣cann	e
6	(0,0,r)	cans␣as␣a␣canne	r
7	(7,4,␣)	ans␣as␣a␣canner	␣can␣
8	(4,7,s)	s␣a␣canner␣can␣	can␣cans
9	(0,0,?)	er␣can␣can␣cans	?

7. The match is "␣can", and the codeword symbol "␣".
8. At position 4, there is a match of length 7. Starting to write the decoded symbols, these are located in the sequence directly following the S buffer. The first four symbols are given by the S buffer as "can␣", but then there should be three more symbols. Since, at this point "can␣" is found as the four first characters in the B buffer, the remaining symbols are "can". Hence, together with the codeword symbol the decoded string is "can␣cans".
9. There is no match, and the codeword gives "?".

5.3.2 LZSS Algorithm

There are a variety of improvements that can be applied for the algorithm to give a better compression ratio. Here one such improvement will be described, published in 1982 [27]. Often, this version of the algorithm is called LZSS, after the inventors James Storer and Thomas Szymanski. It is based on the observation that in each codeword there is also an uncoded character. This is only necessary in the case when there is no match. In the case when there is a match, it is a waste to transmit an uncoded symbol. However, just omitting it will give unequal lengths of the codewords for the two cases. To let the receiver know it the codeword corresponds to a match or not, each codeword is preceded by a binary prefix. The modified algorithm is described in Algorithm 5.2.

Algorithm 5.2 (LZSS)

Initialization:
- Initialize the search buffer with the first S characters of the text, and set $n = S + 1$.

Continue until end of text
- Find all offsets of x_n in the search buffer.
- If number offsets > 0 [There is at least one match]
  - Find the offset with the longest match. This gives an index j and a length l.
  - Set the codeword to $(0, j, l)$
  - Set $n \leftarrow n + l$.
- Else [There are no matches]
  - Set the codeword to $(1, c)$ where $c = x_n$.
  - Set $n \leftarrow n + 1$.
- Update search buffer and look ahead buffer

The next example illustrates the algorithm using the same text as in the previous example.

TABLE 5.3 Encoding table of Example 5.5.

Step	*S* buffer	*B* buffer	Codeword
1	can␣you␣can␣can	s␣as␣a␣	(1,s)
2	an␣you␣can␣cans	␣as␣a␣c	(0,5,1)
3	n␣you␣can␣cans␣	as␣a␣ca	(0,4,1)
4	␣you␣can␣cans␣a	s␣a␣can	(0,3,3)
5	u␣can␣cans␣as␣a	␣canner	(0,10,4)
6	n␣cans␣as␣a␣can	ner␣can	(0,1,1)
7	␣cans␣as␣a␣cann	er␣can␣	(1,e)
8	cans␣as␣a␣canne	r␣can␣c	(1,r)
9	ans␣as␣a␣canner	␣can␣ca	(0,7,4)
10	as␣a␣canner␣can	␣can␣ca	(0,4,7)
11	nner␣can␣can␣ca	ns	(0,4,1)
12	ner␣can␣can␣can	s	(1,s)
13	er␣can␣can␣cans	?	(1,?)

Example 5.5 (LZSS) To encode the text string

$$T = \text{can␣you␣can␣cans␣as␣a␣canner␣can␣can␣cans?}$$

using the LZSS algorithm with $S = 15$ and $B = 7$, a table is used to show the results. In the Table 5.3, the search buffer and the look ahead buffer are shown for each codeword generated.

The encoding follows the same principles as for LZ77. A match of the first letter in the *B* buffer is marked in the *S* buffer with $\circ$ and the longest match is underlined. In step 10 it is seen that the match is of length 7, but it is limited by the length of the look ahead buffer. If the buffer would be longer, so would the match. The choice of buffer lengths becomes a trade-off between having enough matches in the dictionary and lengths in the match, and the number of bits used to describe the codewords.

The codeword for the case when there is a match can now be represented with $1 + \lceil \log 16 \rceil + \lceil \log 8 \rceil = 8$ bits. In the case when there is no match, the codeword length $1 + 8 = 9$ bits. The code sequence consists of eight codewords of length 8 and five of length 9, resulting in totally $15 \cdot 8 + 8 \cdot 8 + 5 \cdot 9 = 229$ bits including the initialization of the *S* buffer. The uncoded sequence gives, as before, 336 bits, and the compression ratio is increased to $R = 1.47$.

In Table 5.4, the decoding of the sequence is shown.

There are other improvements published for the algorithm, but they will not be described in this text. However, there is also another type of improvement that is common to use. It is based on using a two-pass procedure, where the first processing consists of LZ77. Then the tuples in the codeword can be viewed as random variables and encoded by a variable length code, e.g., a Huffman code, which will

TABLE 5.4 Decoding table of Example 5.5.

Step	Codeword	*S* buffer	New symbols
1	(1,s)	can␣you␣can␣can	s
2	(0,5,1)	an␣you␣can␣cans	␣
3	(0,4,1)	n␣you␣can␣cans␣	a
4	(0,3,3)	␣you␣can␣cans␣a	s␣a
5	(0,10,4)	u␣can␣cans␣as␣a	␣can
6	(0,1,1)	n␣cans␣as␣a␣can	n
7	(1,e)	␣cans␣as␣a␣cann	e
8	(1,r)	cans␣as␣a␣canne	r
9	(0,7,4)	ans␣as␣a␣canner	␣can
10	(0,4,7)	as␣a␣canner␣can	␣can␣ca
11	(0,4,1)	nner␣can␣can␣ca	n
12	(1,s)	ner␣can␣can␣can	s
13	(1,?)	er␣can␣can␣cans	?

further improve the compression rate. The widely used text compression ZIP and the image format PNG follow this principle.

In the description presented above, the algorithm has, for simplicity, been initiated with the first *S* symbols of the source sequence. A disadvantage with this is that the buffer has also to be transmitted along with the code sequence. Alternatively, the algorithm can be initialized with a known sequence, e.g., a sequence of *S* blanks. There will still be a start-up phase since there are probably not many matches in the beginning, but hopefully this will give a shorter code sequence than the uncoded buffer. How the initialization is done must be agreed between the transmitter and receiver. This is typically a part of the standardization of the compression protocol.

5.3.3 LZ78 Algorithm

In 1978, Lempel and Ziv published a follow-up article on the previous one [26]. In this, they gave an alternative version of a compression algorithm, often called LZ78. It is based on the same idea that a dictionary is built during a one-pass sweep of the text, and it is based on the previously encoded part of the source sequence. The first algorithm, LZ77, can get into problems when the repetitions in the text are longer than the search buffer. Then it might actually lead to an expansion of the text. The problem is that the dictionary is built with a sliding window technique, and not the complete previous text is represented. In LZ78, a dictionary is built from all the past repetitions in the encoded sequence. The dictionary can be represented in a tree structure, making it memory efficient and easy to search. When a new match is found, the dictionary is expanded by this matching string concatenated with the following symbol. In that way, a dictionary is built with matches of different lengths. To see how this works, first set up the algorithm in Algorithm 5.3.

Algorithm 5.3 (LZ78)

Initialize

  The dictionary contains the empty symbol with index 0
  $n = 1$ [The first symbol]
  Ind = 1 [Next index in the dictionary]

Continue until n at end of text
  Find x_n in the dictionary.
  If there is a match in the dictionary
    Find the longest match $x_n \dots x_{n+l-1}$ in the dictionary.
    Set the codeword (Ind_m, x_{n+l}), where Ind_m is the index of the match
    Add $x_n \dots x_{n+l-1}x_{n+l}$ with index *Ind* to the dictionary
    Set $n \leftarrow n + l + 1$.
  Else [x_n not in dictionary]
    Set the codeword to $(0, x_n)$
    Add x_n with index *Ind* to the dictionary
  Ind ← Ind + 1

At step Ind, the match index of the codeword can be in the interval [0, Ind − 1], and it is needed $\lceil \log \text{Ind} \rceil$ bits to describe it. If the alphabet size is k, it is needed in total $L_{\text{Ind}} = \lceil \log \text{Ind} \rceil + \lceil \log k \rceil$ bits to describe the codeword at step Ind.

In the next example, the same text string as before is used to show the encoding and the decoding with the LZ78 algorithm.

Example 5.6 Once again, use the same text string,

$$T = \text{can␣you␣can␣cans␣as␣a␣canner␣can␣can␣cans?}$$

and use the LZ78 algorithm for compression. First the text is shown again with the encoded parts underlined and marked with an index of the codeword. Then, in Table 5.5, the corresponding codewords and the dictionary are shown. Also, the binary representation of the codeword is shown to the right-hand side in the table. For the binary representation of the uncoded letters, assume the ASCII table with 8 bits for each character. The following shows the codeword representation for the parts in the source text.

$T=$	c	a	n	␣	y	o	u	␣c	an	␣ca	ns	␣a	s	␣a␣	ca	nn	e	r	␣can	␣can␣	can	s?
	1	2	3	4	5	6	7	8	9	10	11	12	13	14	15	16	17	18	19	20	21	22

In total, 255 bits is needed to represent the codewords and the uncoded sequence needs 336 bits, which gives a compression ratio of $R = 1.32$. The representation of the dictionary is better done in a tree structure. For this example, it is shown in Figure 5.7. The decoding is done straightforward following the same table. The dictionary can be built in the same way at the decoder side since everything about the sequence up to the current codeword is known.

TABLE 5.5 The encoding table of Example 5.6.

Ind	Codeword	Dictionary	Bits	Binary
Init 0		–		
1	0,c	c	0	01100011
2	0,a	a	1	0 01100001
3	0,n	n	2	00 01101110
4	0,␣	␣	2	00 00100000
5	0,y	y	3	000 01111001
6	0,o	o	3	000 01101111
7	0,u	u	3	000 01110101
8	4,c	␣c	3	100 01100011
9	2,n	an	4	0010 01101110
10	8,a	␣ca	4	1000 01100001
11	3,s	ns	4	0011 01110011
12	4,a	␣a	4	0100 01100001
13	0,s	s	4	0000 01110011
14	12,␣	␣a␣	4	1100 00100000
15	1,a	ca	4	0001 01100001
16	3,n	nn	4	0011 01101110
17	0,e	e	5	00000 01100101
18	0,r	r	5	00000 01110010
19	10,n	␣can	5	01010 01101110
20	19,␣	␣can␣	5	10011 00100000
21	15,n	can	5	01111 01101110
22	13,?	s?	5	01101 00111111

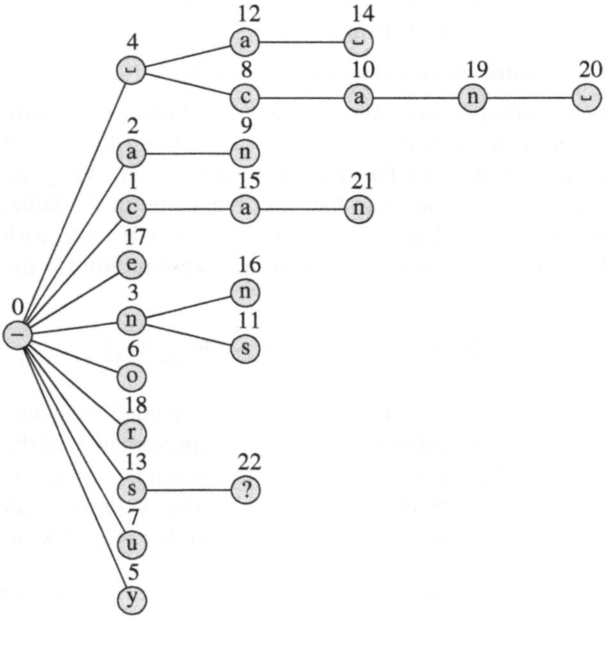

Figure 5.7 The dictionary in a tree form for example 5.6.

As with the LZ77 algorithm, there are several possible improvements for this algorithm. The most well known was published by Welsh in 1984 [28]. It is usually named the LZW algorithm and is described next.

5.3.4 LZW Algorithm

In LZ78, there is an uncoded character in each codeword. The only reason for having this is to cope with the case when the first encoded character is not in the dictionary. This is the same problem that in LZ77 was solved with two different codewords. Here, the dictionary can instead be initialized with the complete alphabet, which is known both at the encoder and decoder side. In this way, it is no longer needed to have the extra character in the codeword and it can be just the index of the match. It is still possible on the receiver side to extend the dictionary with the match concatenated with the following character.

Algorithm 5.4 (LZW)

Initialize

  Initialize the dictionary with the complete alphabet $\mathcal{A}$.
  $n = 1$ [The first symbol in the sequence]
  $\text{Ind} = |\mathcal{A}| + 1$ [Next index in the dictionary]

Continue until n at end of text

  Find x_n in the dictionary.
  Find the longest match $x_n \dots x_{n+l-1}$ in the dictionary.
  Set the codeword (Ind_m), where Ind_m is the index of the match
  Add $x_n \dots x_{n+l-1}x_{n+l}$ with index Ind to the dictionary
  Set $n \leftarrow n + l$.
  $\text{Ind} \leftarrow \text{Ind} + 1$

As in LZ78, the number of bits to describe the index of the codeword depends on the length of the dictionary.

Once again, the same sequence will be used to show encoding and decoding using the LZW algorithm.

Example 5.7 What first needs to be decided is what alphabet to use in the example. Of course, it can be viewed as the alphabet consists of the present letters, $\mathcal{A}_1 = \{\text{␣}, ?, a, c, e, n, o, r, s, u, y\}$. That would give a very unfair estimate of the compression rate since it is compared with the uncoded case where the characters are taken from the ASCII table with 8 bits. Therefore, here the complete ASCII table with 256 characters is used as the alphabet. To shorten the table of the encoding, only the entrance of the alphabet that is actually used in the text is shown.

First, the text, again with the parts that are encoded underlined and marked with the corresponding index (codeword), is shown:

$T=$ c a n ␣ y o u ␣ ca n␣ can s ␣a s␣ a ␣c an n e r ␣ca n␣c an ␣can s ?
1 2 3 4 5 6 7 8 9 10 11 12 13 14 15 16 17 18 19 20 21 22 23 24 25 26 27

TABLE 5.6 The encoding table of Example 5.7.

n	Codeword	Ind	Dictionary	Bits	Binary
		32	␣		
		63	?		
		97	a		
		99	c		
		101	e		
	Init	110	n		
		111	o		
		114	r		
		115	s		
		117	u		
		121	y		
1	99	256	ca	8	01100011
2	97	257	an	9	001100001
3	110	258	n␣	9	001101110
4	32	259	␣y	9	000100000
5	121	260	yo	9	001111001
6	111	261	ou	9	001101111
7	117	262	u␣	9	001110101
8	32	263	␣c	9	000100000
9	256	264	can	9	100000000
10	258	265	n␣c	9	100000010
11	264	266	cans	9	100001000
12	115	267	s␣	9	001110011
13	32	268	␣a	9	000100000
14	97	269	as	9	001100001
15	267	270	s␣a	9	100001011
16	97	271	a␣	9	001100001
17	263	272	␣ca	9	100000111
18	257	273	ann	9	100000001
19	110	274	ne	9	001101110
20	101	275	er	9	001100101
21	114	276	r␣	9	001110010
22	272	277	␣can	9	100010000
23	265	278	n␣ca	9	100001001
24	257	279	an␣	9	100000001
25	277	280	␣cans	9	100010101
26	115	281	s?	9	001110011
27	63	282	?☒	9	000111111

In Table 5.6, it is shown how the codewords are formed and the dictionary is built. In total, 242 bits are needed to represent the codewords, giving a compression ratio of $R = 1.39$, which is a slight improvement from the LZ78 algorithm. Similar to LZ78, the dictionary can be represented in a tree structure, as in Figure 5.8. As with the table, the letters of the initialized alphabet that are not used are omitted in the tree representation.

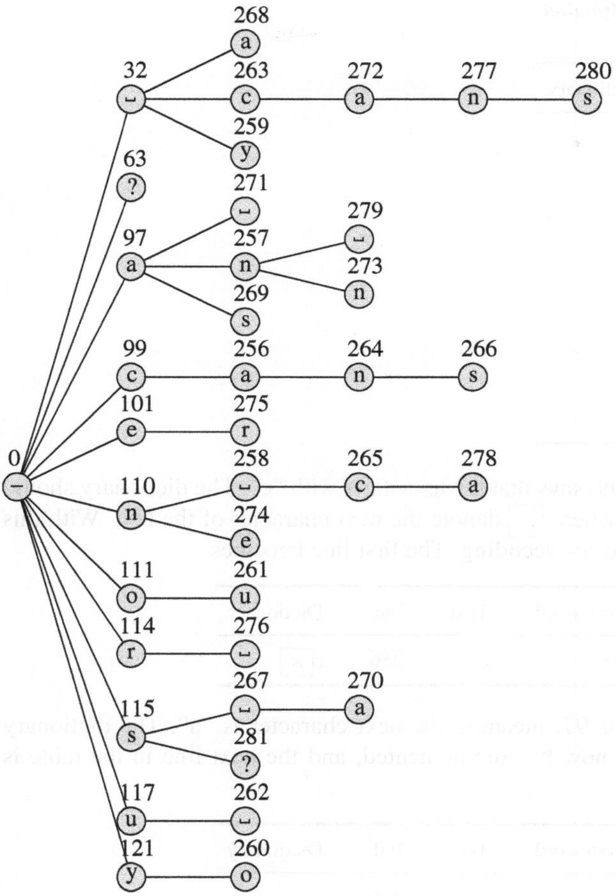

Figure 5.8 The dictionary in a tree form for Example 5.7.

What is left to show is how the same dictionary can be built at the receiver side. The codeword specifies the index of the dictionary, and that gives directly the corresponding text. But then the new entrance in the dictionary consists of this text concatenated with the next character of the text. At the time of decoding the codeword, the next symbol is not known. However, it will be as soon as the next codeword is decoded. Hence, at that point it is always possible to complement the new entrance in the dictionary and set it in the tree. In the next example, the first couple of steps in decoding of the previously encoded sequence is shown.

Example 5.8 Initialize the decoding of LZW with the ASCII table and decode

$$C = (99, 97, 110, 32, 121, 111, 117, 32, 256, 258, 264, 115, 32, 97, 267, 97, 263, 257, 110, 101, 114, 272, 265, 257, 277, 115, 63)$$

The initialization is of course done with the whole ASCII table as for the encoding, but only the part that is used is shown in Table 5.7.

TABLE 5.7 Initialization of alphabet in Example 5.8.

Ind	Dictionary
32	␣
63	?
97	a
99	c
101	e
110	n
111	o
114	r
115	s
117	u
121	y

The first codeword, 99, says that the text starts with "c". The dictionary should be expanded with "c[×]", where [×] denote the next character of the text. With this as a start, fill in a table used for decoding. The first line becomes

n	Codeword	Text	Ind	Dictionary
1	99	c	256	c[×]

The next codeword is 97, meaning the next character is "a". The dictionary entrance at index 256 can now be complemented, and the next line in the table is processed.

n	Codeword	Text	Ind	Dictionary
1	99	c	256	ca
2	97	a	257	a[×]

Similarly, the dictionary entrance at index 257 can be completed when the next codeword is decoded. This is 110, which gives an "n". The table becomes

n	Codeword	Text	Ind	Dictionary
1	99	c	256	ca
2	97	a	257	an
3	110	n	258	n[×]

In this way, it is always possible to fill in all entrances of the dictionary. To complete the dictionary entrance, the first character of the next decoded pattern is appended. Since the next pattern always consists of at least one character, this procedure will always work. The complete decoding is shown in Table 5.8.

TABLE 5.8 The decoding table of Example 5.8.

n	Codeword	Text	Ind	Dictionary
1	99	c	256	ca
2	97	a	257	an
3	110	n	258	n␣
4	32	␣	259	␣y
5	121	y	260	yo
6	111	o	261	ou
7	117	u	262	u␣
8	32	␣	263	␣c
9	256	ca	264	can
10	258	n␣	265	n␣c
11	264	can	266	cans
12	115	s	267	s␣
13	32	␣	268	␣a
14	97	a	269	as
15	267	s␣	270	s␣a
16	97	a	271	a␣
17	263	␣c	272	␣ca
18	257	an	273	ann
19	110	n	274	ne
20	101	e	275	er
21	114	r	276	r␣
22	272	␣ca	277	␣can
23	265	n␣c	278	n␣ca
24	257	an	279	an␣
25	277	␣can	280	␣cans
26	115	s	281	s?
27	63	?	282	?☒

5.4 APPLICATIONS OF SOURCE CODING

The LZ algorithms have over the years become very popular in a variety of applications, both text coding and image coding. In text compression, there exist several widely used algorithms, like ZIP and GZIP. The compression engine in ZIP and GZIP is virtually the same, viz *deflate* [29], which is based on a combination of LZ77 followed by Huffman compression. In GZIP, the deflate algorithm is wrapped in a header, with information like time and file system, and a tailing cyclic redundancy check (CRC). The ZIP format is a bit more general and can be a container for multiple files. In that case, each file is compressed separately using the deflate algorithm. Consequently, each file can be extracted separately, and it is also possible to append more files without decompressing the whole file.

A similar scheme, an LZ algorithm followed by a Huffman compression, can be found in the image compression formats GIF and PNG. At the beginning of

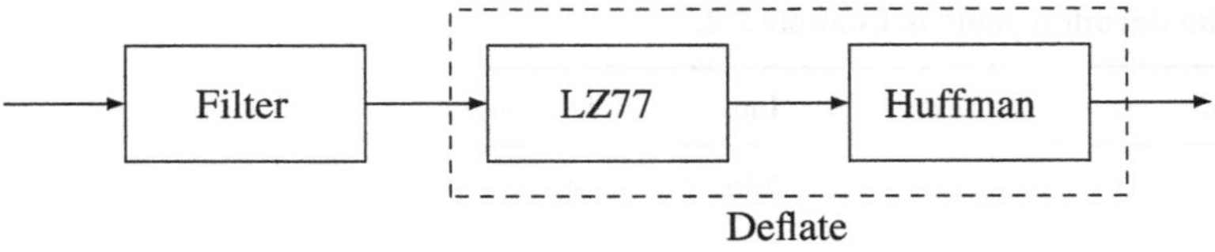

Figure 5.9 The principles of compression in PNG.

the Internet, or at least the Internet viewed through a web browser, there were two dominating image formats, JPEG and GIF. JPEG is based on transform coding and is described in Chapter 11, but GIF is built with an LZW algorithm followed by a Huffman code. When the first widely spread web browser, MOSAIC, was launched in 1993 it also meant an increased use of images. In December 1994, Unisys, who had patent rights for LZW, declared that they were going to start claiming license fees for all programs using it, including all programs using GIF [31]. The Internet community gave a swift answer, and shortly after the PNG group was formed. Already in March 1995, the ninth draft of PNG (Portable Network Graphics) was published, and PNG reference images posted [30, 31]. By using LZ77 instead of LZW, more specifically deflate, it is guaranteed to be a format that can be freely used without any patent issues. Compared to GIF, it has the advantage to support true color instead of only a palette as for GIF. It also gives a superior compression for most images compared to GIF.

In Figure 5.9, a simple block diagram is given for compression scheme used in PNG. Since images typically do not give enough recursions in the pixels to use an LZ algorithm efficiently, the pixel stream is first filtered by a relatively simple prediction filter. There are five different settings for the filtering, as given in Figure 5.10. In the figure, the current pixel is denoted by x, and for filter alternative 0 this is used directly as an input to the deflate compression. The two next settings use the pixel directly to left of the current pixel, alternatively directly above the current pixel, as a

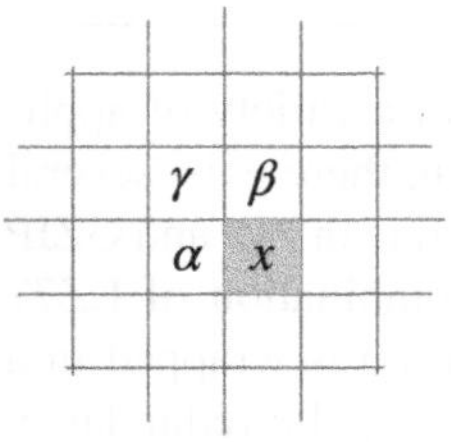

Index	Filter specification
0:	x
1:	$x - \alpha$
2:	$x - \beta$
3:	$x - \frac{\alpha + \beta}{2}$
4:	$x - a$, where $a = \underset{p \in \{\alpha, \beta, \gamma\}}{\operatorname{argmin}} \{\lvert p - (\alpha + \beta - \gamma) \rvert\}$

Figure 5.10 Prefilter in PNG. In the left picture, the currently encoded pixel is denoted by x and its neighboring pixels α, β, and γ. The pixel sequence of the image is read from left to right and top to bottom, so these neighboring pixel values are available both in the transmitter and receiver.

prediction. These pixel values are called α and β, respectively, in the figure. Then the compressed value is the difference between the current pixel and the prediction. The fourth filter is a combination where the filtered value is the difference between x and the average of the left and above pixels α and β. The fifth version of the prediction means considering the three pixels, α, β, and γ and uses the one of the closest to the value $\alpha + \beta - \gamma$.

The idea with the filtering is that in most images, the neighboring pixels are correlated and have values close to each other. Then, on average, their differences should be relatively close to zero, and the distribution becomes more narrow. That also means the probability of repetitions in the pixel sequence becomes higher. The resulting filtered sequence is then compressed by the deflate algorithm.

Deflate [29] was originally designed as a part of ZIP and combines LZ77 with Huffman coding as depicted in Figure 5.9. It is by design a general compression specification, independent of platform and type of data. It is widely known that deflate can be implemented by use of nonpatented algorithms, meaning it can be freely used. The Huffman code can be given as a predefined tree, or by including the tree in the file. When including it, the tree is arranged such that it can be described by the list of codeword lengths.

The input stream to deflate is interpreted as a stream of bytes, i.e. the symbols are of the values 0–255. For PNG, that means values of the filter are interpreted as bytes. Then, the first step in deflate is the LZ77 encoder. It will look for matches of lengths in 3–258, at a backsearch depth of 1–32768. If it cannot find a match of this length, it should transmit the byte itself. The byte value and the match length are treated as the first symbol, encoded as an index value between 0 and 285. If the value is in 0–255 range, it represents a byte value and values between 257 and 285 represent match lengths according to Table 5.9.

The index value, in Table 5.9 denoted Ind, between 0 and 285 is encoded by a Huffman code. As seen in the table, some code values for match lengths represent multiple lengths. In those cases, the code value for the index is appended with a binary

TABLE 5.9 Representations of match length in deflate.

Ind	Bits	Length	Ind	Bits	Length	Ind	Bits	Length
257	0	3	267	1	15,16	277	4	67–82
258	0	4	268	1	17,18	278	4	83–98
259	0	5	269	2	19–22	279	4	99–114
260	0	6	270	2	23–26	280	4	115–130
261	0	7	271	2	27–30	281	5	131–162
262	0	8	272	2	31–34	282	5	163–194
263	0	9	273	3	35–42	283	5	195–226
264	0	10	274	3	43–50	284	5	227–257
265	1	11,12	275	3	51–58	285	0	258
266	1	13,14	276	3	59–66			

TABLE 5.10 Representations of match length in deflate.

Ind	Bits	Offset	Ind	Bits	Offset	Ind	Bits	Offset
0	0	1	10	4	33–48	20	9	1025–1536
1	0	2	11	4	49–64	21	9	1537–2048
2	0	3	12	5	65–96	22	10	2049–3072
3	0	4	13	5	97–128	23	10	3073–4096
4	1	5,6	14	6	129–192	24	11	4097–6144
5	1	7,8	15	6	193–256	25	11	6145–8192
6	2	9–12	16	7	257–384	26	12	8193–12288
7	2	13–16	17	7	385–512	27	12	12289–16384
8	3	17–24	18	8	513–768	28	13	16385–24576
9	3	25–32	19	8	769–1024	29	13	24577–32768

value pointing out the match length. In the table, the column "bits" represents the length of this binary number. As an example, if the match is 36, it points out the index value 273, which is encoded by the Huffman code. This code should be followed by 3 bits pointing out one value in 35–42. In this case, it should be 1, which is 001 in 3 bits.

The position of the match, the offset, is a number between 1 and 32,768. This number is encoded in a similar way with the help of Table 5.10. Here the offset points out an index, which is encoded using a Huffman code. The number of bits in the table is the number of bits added to the codeword to point out the offset. Hence, there are two Huffman codes used to encode the outcome from the LZ77 encoder, one for the byte/match length and one for the backtrack length, or offset.

X	Y	ℓ_X
A	010	3
B	011	3
C	100	3
D	101	3
E	110	3
F	00	2
G	1110	4
H	1111	4

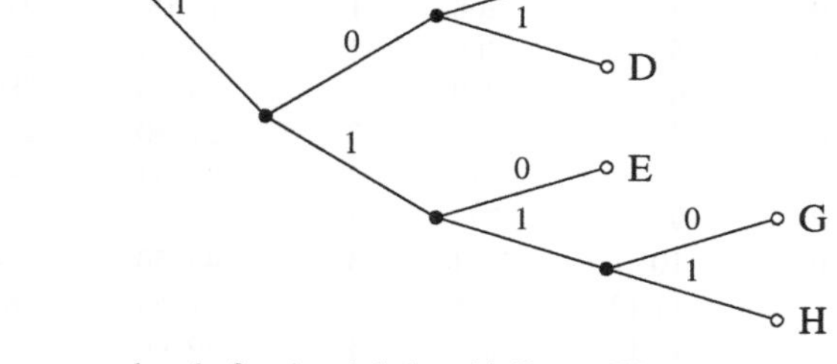

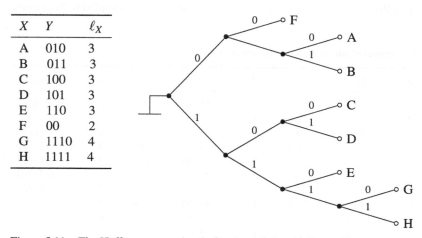

Figure 5.11 The Huffman tree and code for the alphabet (A, B, … , H).

It is seen from the construction of both the tables, that it is expected to have many cases of no matches and short matches with repetitions close to the match. The first case, that it is expected to have many occasions of no matches, is seen since the uncoded bytes are given as input to the Huffman code. In the table for the matches, short matches give codewords followed by few or nonextra bits. These extra bits are added to the codeword lengths and, hence, should be kept low. Finally, for the offset, the same structure is given with short distances giving few extra bits.

As described earlier, the Huffman code can either use a predefined tree or specify one in the file. If it is specified, it should be listed in a tree with shortest paths first and equal lengths paths given in the lexicographical order of the symbols. In that case, the tree can be specified by the sequence of lengths for the codewords for the alphabet. This sequence is also compressed by a Huffman code. As an example given in [29], the alphabet $(A, B, ..., H)$ is given a Huffman tree with the lengths $\ell_x = (3, 3, 3, 3, 3, 2, 4, 4)$. That means starting with a tree with these path lengths, ordered with the shortest first (in the top part of the tree) (see Figure 5.11). Then, the leaves are labeled by the nodes according to lengths. In this example, there is only one leaf with length 2, which should be labeled F. There are five leaves at depth three, where the elements A, B, C, D, and E are positioned in the lexicographical order. Finally, the entries G and H are labels to the depth four leaves, in the lexicographical order. In the figure, the complete code table is also given.

PROBLEMS

5.1 Given the same source sequence as in Problem 4.12, i.e. $\boldsymbol{x}$ = d a c b d a a a d a.

(a) Use a two-pass Huffman coding of the sequence. That is, estimate the distribution from the sequence, construct a Huffman code, and encode the sequence.

(b) Compare the result in (a) with the case when the true distribution is given (see Problem 4.12).

5.2 Use the one-sweep Huffman algorithm to encode the sequence in Problem 5.1.

5.3 Encode the text

```
IF IF = THEN THEN THEN = ELSE ELSE ELSE = IF;
```

using the LZ77 algorithm with $S = 7$ and $B = 7$. How many code symbols were generated? If each letter in the text is translated to a binary form with 8 bits, what is the compression ratio?

5.4 Compress the following text string using LZ77 with $S = 16$ and $B = 8$,

```
I scream, you scream, we all scream for ice cream.
```

Initialise the encoder with the beginning of the string. Give the codewords and the compression ratio.

5.5 A text has been encoded with the LZ78 algorithm, and the following sequence of codewords was obtained,

Index	Codeword
1	$(0,t)$
2	$(0,i)$
3	$(0,m)$
4	$(0,\sqcup)$
5	$(1,h)$
6	$(0,e)$
7	$(4,t)$
8	$(0,h)$
9	$(2,n)$
10	$(7,w)$
11	$(9,\sqcup)$
12	$(1,i)$
13	$(0,n)$
14	$(0,s)$
15	$(3,i)$
16	$(5,.)$

Decode to get the text back.

5.6 Encode the text

```
IF IF = THEN THEN THEN = ELSE ELSE ELSE = IF;
```

using the LZ78 algorithm. How many code symbols were generated? If each letter in the text is translated to a binary form with 8 bits, what is the compression ratio?

5.7 Compress the text

```
Nat the bat swat at Matt the gnat
```

and calculate the compression rate using for the following settings.

(a) LZ77 with $S = 10$ and $B = 3$.

(b) LZSS with $S = 10$ and $B = 3$.

(c) LZ78.

(d) LZW with predefined alphabet of size 256.

5.8 Consider the sequence

```
six sick hicks nick six slick bricks with picks
and sticks
```

(a) What source alphabet should be used?

(b) Use the LZ77 with a window size $N = 8$ to encode and decode the sequence with a binary code alphabet.

(c) Use the LZ78 to encode and decode the sequence with a binary code alphabet.

(d) How many code symbols were generated?

5.9 Use the LZ78 algorithm to encode and decode the string

`the friend in need is the friend indeed`

with a binary code alphabet. What is the minimal source alphabet? How many code symbols were generated?

5.10 Decode the following codeword sequence using the LZW algorithm:

$$y = 102, 114, 101, 115, 104, 108, 121, 32, 256, 105, 101, 100, 263, 257, 259, 263, 108, 258, 104$$

The algorithm is initialized with the ASCII table from index 1 to 255, where the following is interesting for the problem:

Ind:	32	100	101	102	104	105	108	114	115	121
Letter:	space	d	e	f	h	i	l	r	s	y

CHAPTER 6

ASYMPTOTIC EQUIPARTITION PROPERTY AND CHANNEL CAPACITY

TWO OF THE MAIN THEOREMS in information theory are source coding theorem and the channel coding theorem. They present bounds on the overall performance for source coding and channel coding. The source coding theorem is considered in Chapter 4 and states that the symbols from a source with no memory can be compressed down to the entropy. The channel coding theorem states that the possible transmission rate over a channel is determined by the maximum mutual information between the transmitter side and the receiver side, the channel capacity. In this chapter, the asymptotic equipartition property (AEP), a mathematical tool based on the law of large number, is introduced. Based on this property, the source coding theorem will be presented. Then, by applying it to the relation between the transmitted and received symbols in a communication system, the channel coding theorem is presented.

6.1 ASYMPTOTIC EQUIPARTITION PROPERTY

From the law of large numbers, it is possible to get a very efficient tool for classifying very long sequences, called the *asymptotic equipartition property*. The basic idea is that a series of identical and independently distributed (i.i.d.) events are viewed as a vector of events. Then, the probability for each individual vector becomes very small as the length grows, meaning it is pointless individual vectors. Instead, the probability for the type vector becomes interesting. It turns out that it is only a fraction of all possible outcomes that does not have a negligible small probability. To understand what this means, consider 100 consecutive flips with a fair coin. Since, assuming a fair coin, all resulting outcome vectors have the same probability,

$$P(x_1, \dots, x_{100}) = 2^{-100} \approx 8 \cdot 10^{-31} \tag{6.1}$$

The probability for each of the vectors in the outcome is very small, and it is, for example, not very likely that the vector with 100 Heads will occur. However, since

Information and Communication Theory, First Edition. Stefan Höst.

there are $\binom{50}{100} \approx 10^{29}$ vectors with 50 Heads and 50 Tails, the probability for getting one such vector is

$$P(50\text{ Head}, 50\text{ Tail}) = 2^{-100}\binom{50}{100} \approx 0.08 \tag{6.2}$$

which is relatively high. To conclude, it is most likely that the outcome of 100 flips with a fair coin will result in approximately the same numbers of Heads and Tails. This is a direct consequence of the weak law of large numbers, which is recapitalized here.

Theorem 6.1 (The weak law of large numbers) Let $X_1, X_2, \dots, X_n$ be a set of i.i.d. random variables with mean $E[X]$. Then,

$$\frac{1}{n}\sum_i X_i \xrightarrow{p} E[X] \tag{6.3}$$

where $\xrightarrow{p}$ denotes convergence in probability. □

The notation of convergence in probability can also be expressed as

$$\lim_{n\to\infty} P\Big(\Big|\frac{1}{n}\sum_i X_i - E[X]\Big| < \varepsilon\Big) = 1 \tag{6.4}$$

for any arbitrary real number $\varepsilon > 0$. Hence, the arithmetic mean of n i.i.d. random variables approaches the expected value of the distribution as n grows.

Consider instead the logarithmic probability for a vector $\boldsymbol{x} = (x_1, x_2, \dots, x_n)$ of length n, consisting of i.i.d. random variables. From the weak law of large numbers

$$\begin{aligned} -\frac{1}{n}\log p(\boldsymbol{x}) &= -\frac{1}{n}\log\prod_i p(x_i) \\ &= \frac{1}{n}\sum_i -\log p(x_i) \\ &\xrightarrow{p} E\big[-\log p(X)\big] = H(X) \end{aligned} \tag{6.5}$$

or, equivalently,

$$\lim_{n\to\infty} P\Big(\Big|-\frac{1}{n}\log p(\boldsymbol{x}) - E[X]\Big| < \varepsilon\Big) = 1 \tag{6.6}$$

for an arbitrary $\varepsilon > 0$. That is, for all sequences, the mean of the logarithm for the probability approaches the entropy as the length of the sequence grows. For finite sequences, not all will fulfill the criteria set up by the probabilistic convergence. But those that fulfill are the ones that are the most likely to happen, and are called *typical sequences*. This behavior is named the *asymptotic equipartition property* [19]. It is defined as follows.

Definition 6.1 (AEP) The set of *ε-typical sequences* $A_\varepsilon(X)$ is the set of all n-dimensional vectors $\boldsymbol{x} = (x_1, x_2, \dots, x_n)$ of i.i.d. variables with entropy $H(X)$, such that

$$\left| -\frac{1}{n} \log p(\boldsymbol{x}) - H(X) \right| \leq \varepsilon \tag{6.7}$$

for a real number $\varepsilon > 0$. □

The requirement in (6.7) can also be written as

$$-\varepsilon \leq -\frac{1}{n} \log p(\boldsymbol{x}) - H(X) \leq \varepsilon \tag{6.8}$$

which is equivalent to

$$2^{-n(H(X)+\varepsilon)} \leq p(\boldsymbol{x}) \leq 2^{-n(H(X)-\varepsilon)} \tag{6.9}$$

In this way, the AEP can alternatively be defined as follows.

Definition 6.2 (AEP, Alternative definition) The set of *ε-typical sequences* $A_\varepsilon(X)$ is the set of all n-dimensional vectors $\boldsymbol{x} = (x_1, x_2, \dots, x_n)$ of i.i.d. variables with entropy $H(X)$, such that

$$2^{-n(H(X)+\varepsilon)} \leq p(\boldsymbol{x}) \leq 2^{-n(H(X)-\varepsilon)} \tag{6.10}$$

for a real number $\varepsilon > 0$. □

The two definitions above are naturally equivalent, but will be used in different occasions. In the next example, an interpretation of the ε-typical sequences is given.

Example 6.1 Consider a binary vector of length $n = 5$ with i.i.d. elements where $p_X(0) = \frac{1}{3}$ and $p_X(1) = \frac{2}{3}$. The entropy for each symbol is $H(X) = h(1/3) = 0.918$. In Table 6.1, all possible vectors and their probabilities are listed.

As expected, the all zero vector is the least possible vector, while the all one vector is the most likely. However, even this most likely vector is not very likely to happen, it only has a probability of 0.1317. Picking one vector as a guess for what the outcome will be, this should be the one, but even so, it is not likely to make a correct guess. In the case when the order of the symbols is not important, it appears to be better to guess on the *type* of sequence, here meaning the number of ones and zeros. The probability for a vector containing k ones and $5 - k$ zeros is given by the binomial distribution as

$$P(k \text{ ones}) = \binom{n}{k} \left(\frac{2}{3}\right)^k \left(\frac{1}{3}\right)^{n-k} = \binom{n}{k} \frac{2^k}{3^n} \tag{6.11}$$

TABLE 6.1 Probabilities for binary vectors of length 5.

x	$p(x)$		x	$p(x)$		x	$p(x)$	
00000	0.0041		01011	0.0329	$\star$	10110	0.0329	$\star$
00001	0.0082		01100	0.0165		10111	0.0658	$\star$
00010	0.0082		01101	0.0329	$\star$	11000	0.0165	
00011	0.0165		01110	0.0329	$\star$	11001	0.0329	$\star$
00100	0.0082		01111	0.0658	$\star$	11010	0.0329	$\star$
00101	0.0165		10000	0.0082		11011	0.0658	$\star$
00110	0.0165		10001	0.0165		11100	0.0329	$\star$
00111	0.0329	$\star$	10010	0.0165		11101	0.0658	$\star$
01000	0.0082		10011	0.0329	$\star$	11110	0.0658	$\star$
01001	0.0165		10100	0.0165		11111	0.1317	
01010	0.0165		10101	0.0329	$\star$			

which is already discussed in Chapter 2 (see Figure 2.5). Viewing these numbers in a table gives

k	$P(k) = \binom{5}{k}\frac{2^k}{3^5}$
0	0.0041
1	0.0412
2	0.1646
3	0.3292
4	0.3292
5	0.1317

Here it is clear that the most likely vector, the all one vector, does not belong to the most likely type of vector. When guessing of the number of ones, it is more likely to get 3 or 4 ones. This is of course due to the fact that there are more vectors that fulfill this criteria than the single all one vector. So, this concludes that vectors with 3 or 4 ones are the most likely to happen. The question then is how this relates to the definitions of typical sequences. To see this, first a value on ε must be chosen. Here 15% of the entropy is used, which gives $\varepsilon = 0.138$. The interval in Definition 6.2 is given by

$$2^{-n(H(X)+\varepsilon)} = 2^{-5(h(\frac{1}{3})+0.138)} \approx 0.0257 \tag{6.12}$$

$$2^{-n(H(X)-\varepsilon)} = 2^{-5(h(\frac{1}{3})-0.138)} \approx 0.0669 \tag{6.13}$$

Thus, the ε-typical vectors are the ones with probabilities between these numbers. In Table 6.1, these vectors are marked with a $\star$. Notably, it is the same vectors as earlier intuitively concluded to be the most likely to happen, i.e. vectors with 3 or 4 ones.

In the previous example, it was seen that the typical vectors constitute the type of vectors that are most likely to appear. In the example, very short vectors were used to be able to list all of them, but for longer sequences it can be seen that the ε-typical vectors are just a fraction of all vectors. On the other hand, it can also be seen that the probability for a random vector to belong to the typical set is close to one. More formally, the following theorem can be stated.

Theorem 6.2 Consider length n sequences of i.i.d. random variables with entropy $H(X)$. For each $\varepsilon > 0$, there exists an integer n_0 such that, for each $n > n_0$, the set of ε-typical sequences, $A_\varepsilon(X)$, fulfills

$$P\big(\boldsymbol{x} \in A_\varepsilon(X)\big) \geq 1 - \varepsilon \tag{6.14}$$

$$(1-\varepsilon)2^{n(H(X)-\varepsilon)} \leq \big|A_\varepsilon(X)\big| \leq 2^{n(H(X)+\varepsilon)} \tag{6.15}$$

where $|A_\varepsilon(X)|$ denotes the cardinality of the set $A_\varepsilon(X)$. □

The first part of the theorem, (6.14), is a direct consequence of the law of large numbers stating that $-\frac{1}{n}\log p(\boldsymbol{x})$ approaches $H(X)$ as n grows. That means there exists an n_0, such that for all $n \geq n_0$

$$P\Big(\Big|-\frac{1}{n}\log p(\boldsymbol{x}) - H(X)\Big| < \varepsilon\Big) \geq 1 - \delta \tag{6.16}$$

for any δ between zero and one. Replacing δ with ε gives

$$P\Big(\Big|-\frac{1}{n}\log p(\boldsymbol{x}) - H(X)\Big| < \varepsilon\Big) \geq 1 - \varepsilon \tag{6.17}$$

which is equivalent to (6.14). It shows that the probability for a randomly picked sequence being a typical sequence can be made arbitrarily close to 1 by choosing large enough n.

To show the second part, that the number of ε-typical sequence is bounded by (6.15), start with the left-hand side inequality. According to (6.14), for large enough n_0

$$\begin{aligned}1-\varepsilon \leq P\big(\boldsymbol{x} \in A_\varepsilon(X)\big) &= \sum_{\boldsymbol{x}\in A_\varepsilon(X)} p(\boldsymbol{x}) \\ &\leq \sum_{\boldsymbol{x}\in A_\varepsilon(X)} 2^{-n(H(X)-\varepsilon)} = \big|A_\varepsilon(X)\big|2^{-n(H(X)-\varepsilon)}\end{aligned} \tag{6.18}$$

where the second inequality follows directly from the left-hand side inequality in (6.10). Similarly, the right-hand side of (6.15) can be shown by

$$\begin{aligned}1 = \sum_{\boldsymbol{x}} p(\boldsymbol{x}) &\geq \sum_{\boldsymbol{x}\in A_\varepsilon(X)} p(\boldsymbol{x}) \\ &\geq \sum_{\boldsymbol{x}\in A_\varepsilon(X)} 2^{-n(H(X)+\varepsilon)} = \big|A_\varepsilon(X)\big|2^{-n(H(X)+\varepsilon)}\end{aligned} \tag{6.19}$$

The next example, inspired by [32], shows the consequences of the theorem for longer sequences.

Example 6.2 Let $\mathcal{X}^n$ be the set of all binary random sequences of length n with i.i.d. variables, where $p(0) = \frac{1}{3}$ and $p(1) = \frac{2}{3}$. Let $\varepsilon = 0.046$, i.e. 5% of the entropy $h(\frac{1}{3})$. The true number of ε-typical sequences and their bounding functions are given in the next table for lengths of $n = 100$, $n = 500$, and $n = 1000$. As a comparison, the fraction of ε-typical sequences compared to the total number of sequences is also shown. From the table, it is seen that for large n the ε-typical sequences only constitute a small fraction of the total number of sequences.

n	$(1-\varepsilon)2^{n(H(X)-\varepsilon)}$	$\lvert A_\varepsilon(X)\rvert$	$2^{n(H(X)+\varepsilon)}$	$\lvert A_\varepsilon(X)\rvert / \lvert\mathcal{X}^n\rvert$
100	1.17×10^{26}	7.51×10^{27}	1.05×10^{29}	5.9×10^{-3}
500	1.90×10^{131}	9.10×10^{142}	1.34×10^{145}	2.78×10^{-8}
1000	4.16×10^{262}	1.00×10^{287}	1.79×10^{290}	9.38×10^{-15}

Next, the probability for the ε-typical sequences is given together with the probability for the most probable sequence, the all-one sequence. Here it is clearly seen that the most likely sequence has a very low probability and is in fact very unlikely to happen. Instead, the most likely event is that a random sequence is taken from the typical sequences, for which the probability approaches one.

n	$P(A_\varepsilon(X))$	$P(\boldsymbol{x} = 11\dots1)$
100	0.660	2.4597×10^{-18}
500	0.971	9.0027×10^{-89}
1000	0.998	8.1048×10^{-177}

6.2 SOURCE CODING THEOREM

In Chapter 4, it was shown that a symbol from a source can, on average, be represented by a binary vector, or sequences, of the same length as the entropy, which is stated as the source coding theorem. It was also shown that the entropy is a hard limit and that the average length cannot be shorter if the code should be uniquely decodable. In Chapter 4, the result was shown using the Kraft inequality and Shannon–Fano coding. In this section, the source coding theorem will be shown using AEP.

To construct a source code, consider source symbols as n-dimensional vectors of i.i.d. variables. In the previous section, it was seen that the typical sequences constitute a small fraction of all sequences, but they are also the most likely to happen. The list of all sequences can be partitioned into two parts, one with the typical

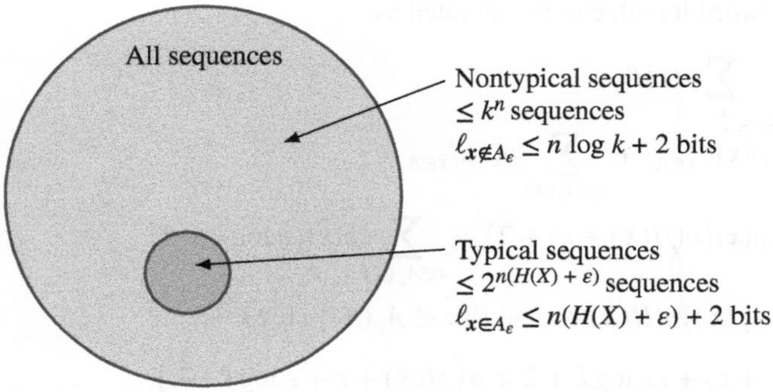

Figure 6.1 Principle of Shannon's source-coding algorithm.

sequences and other with the nontypical. Each of the sets is listed and indexed by binary vectors. To construct the codewords, use a binary prefix stating which set the codeword belongs to, e.g., use 0 for the typical sequences and 1 for the nontypical. The following two tables show the idea of the lookup tables:

Typical

x	Index vec
x_0	0_0 ... 00
x_1	0_0 ... 01
$\vdots$	$\vdots$

Nontypical

x	Index vec
x_a	1_0 00
x_b	1_0 01
$\vdots$	$\vdots$
$\vdots$	$\vdots$

The number of the typical sequences is bounded by $|A_\varepsilon(X)| \leq 2^{n(H(X)+\varepsilon)}$, and, hence, the length of the codewords is

$$\ell_{x \in A_\varepsilon} = \left\lceil \log \left| A_\varepsilon(X) \right| \right\rceil + 1 \leq \log \left| A_\varepsilon(X) \right| + 2 \leq n\big(H(X) + \varepsilon\big) + 2 \quad (6.20)$$

Similarly, the length for codewords corresponding to the nontypical sequences can be bounded by

$$\ell_{x \notin A_\varepsilon} = \lceil \log k^n \rceil + 1 \leq \log k^n + 2 = n \log k + 2 \quad (6.21)$$

where k is the number of outcomes for the variables in the source vectors. In Figure 6.1, the procedure is shown graphically. The idea is that most of the time the source vector will be a typical sequence, which leads to a short codeword. In some rare occasions, it can happen that a nontypical sequence occurs and then there has to be a codeword for this as well. But since the nontypical source vectors are very rare, this will have a negligible effect on the average value of the codeword length.

The average codeword length can be bounded as

$$
\begin{aligned}
L = E\big[\ell_x\big] &= \sum_x p(x)\ell_x \\
&= \sum_{x\in A_\varepsilon(X)} p(x)\ell_{x\in A_\varepsilon} + \sum_{x\notin A_\varepsilon(X)} p(x)\ell_{x\notin A_\varepsilon} \\
&\le \sum_{x\in A_\varepsilon(X)} p(x)\big(n(H(X)+\varepsilon)+2\big) + \sum_{x\notin A_\varepsilon(X)} p(x)\big(n\log k + 2\big) \\
&= P\big(x\in A_\varepsilon(X)\big)n(H(X)+\varepsilon) + P\big(x\notin A_\varepsilon(X)\big)n\log k + 2 \\
&\le n(H(X)+\varepsilon) + \varepsilon n\log k + 2 = n\Big(\underbrace{H(X) + \varepsilon + \varepsilon\log k + \frac{2}{n}}_{\delta}\Big) \\
&= n\big(H(X)+\delta\big)
\end{aligned}
\tag{6.22}
$$

where δ can be made arbitrary small for sufficiently large n. Hence, the average codeword length per symbol in the source vector can be chosen arbitrarily close to the entropy. This can be stated as the source coding theorem.

Theorem 6.3 Let X be a length n vector of i.i.d. random variables and probability function $p(x)$. Then, there exists a code that maps the outcome x into binary sequences such that the mapping is invertible and the average codeword length per symbol is

$$
\frac{1}{n}E\Big[\ell_x\Big] \le H(X) + \delta \tag{6.23}
$$

where δ can be made arbitrarily small for sufficiently large n, □

In the above reasoning, it is assumed that the symbols in the vectors are i.i.d. If, however, there is a dependency among the symbols random processes have to be considered. In [19], it was shown that every ergodic source has the AEP. Hence, if $x = x_1, x_2, \dots, x_n$ is a sequence from an ergodic source, then

$$
-\frac{1}{n}\log p(x) \xrightarrow{p} H_\infty(X), \quad n\to\infty \tag{6.24}
$$

The set of typical sequences should then be defined as sequences such that

$$
\Big|-\frac{1}{n}\log p(x) - H_\infty(X)\Big| \le \varepsilon \tag{6.25}
$$

This leads to the source coding theorem for ergodic sources [19], which is stated here without further proof.

Theorem 6.4 Let X be a stationary ergodic process of length n and x a vector from it. Then there exists a code that maps the outcome x into binary sequences such that the mapping is invertible and the average codeword length per symbol

$$\frac{1}{n}E\left[\ell(x)\right] \leq H_\infty(X) + \delta \tag{6.26}$$

where δ can be made arbitrarily small for sufficiently large n. □

Generally, the class of ergodic processes is the largest class of random processes where the law of large numbers is satisfied.

6.3 CHANNEL CODING

In Figure 1.2, the block diagram of a communication system is shown. It was stated that the source coding and the channel coding can be separated, which is one important result of information theory. In Chapters 4 and 5, source coding was treated, and in the previous section AEP was used to show the source coding theorem. In this section, the analysis will be concentrated on the channel coding of the communication system. In Figure 6.2, a block diagram with the channel encoder, the channel, and the channel decoder is shown. The channel is a mathematical model representing everything that can occur in the actual transmission including, e.g., background noise, scratches on the surface of a CD, or erasures due to overflow in router buffers. The aim of the channel encoder is to introduce redundancy in such way that the decoder can detect, or even correct, errors that occurred on the channel.

The encoding scheme can be described as follows:

- The information symbols U are assumed to be taken from a set $\mathcal{U} = \{u_1, u_2, \dots, u_M\}$ of M symbols.
- The encoding function $x : \mathcal{U} \to \mathcal{X}$ is a mapping from the set $\mathcal{U}$ to the set of codewords $\mathcal{X}$. Denote the codewords as $x(u_i)$, $i = 1, 2, \dots, M$. In the most cases in this text, the codewords are binary vectors of length n and $M = 2^k$ for an integer k.
- As the codeword is transmitted over the channel, errors occur and the received vector is $y \in \mathcal{Y}$. In many situations, the received symbols are taken from a larger alphabet than the code symbols. It can be that $\mathcal{Y}$ are real values detected by the receiver, whereas the code symbols $\mathcal{X}$ are discrete, e.g., binary.
- The decoding function is then a (typically nonlinear) mapping from the received word to an estimate of the transmitted codeword or the initial set, $g : \mathcal{Y} \to \mathcal{U}$.

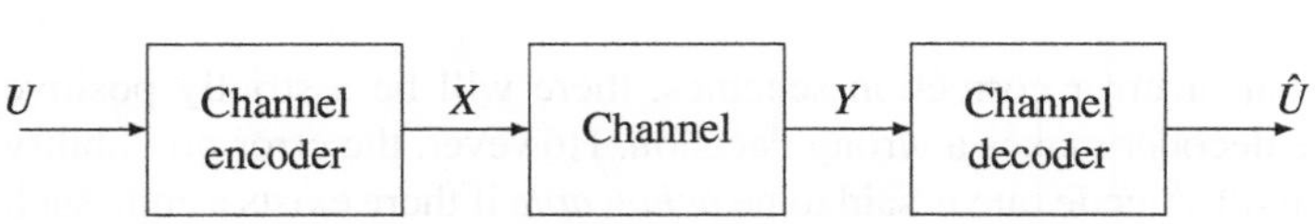

Figure 6.2 A model of the channel-coding system.

If there are M possible information symbols, $|\mathcal{U}| = M = 2^k$, and the codewords are n-dimensional vectors, it is said to be an (n, k) code. In this environment, the code rate is defined as $R = \frac{k}{n}$. The code rate is a positive real number less than one, $0 < R \leq 1$. If $R < 1$ not all possible vectors of length n are codewords and there is a built in redundancy in the system. More formally, the set of codewords spans a k-dimensional subspace of the n-dimensional vector space. The idea with the decoding function is to use this redundancy when estimating the transmitted codeword. The simplest such code is the repetition code, which will be described in the next example.

Example 6.3 Consider a system where the information is 1 bit, $U \in \{0, 1\}$. The encoder maps this 1 bit to a length 3 vector of identical bits, i.e. $0 \rightarrow 000$ and $1 \rightarrow 111$. The code rate is $R = \frac{1}{3}$, meaning that one information bit is represented by three code bits on the channel, or for each codebit transmitted there is $\frac{1}{3}$ information bit.

As the code bits are transmitted over the channel, errors might occur. Assume that each bit has an error probability of ε, and that it is a small number (i.e., $\varepsilon \ll 0.5$). The most likely event is that the codeword will be received unchanged, in which case the decoding back to information bit is straightforward.

If there is an error event during the transmission of a codeword, the most likely error is to alter one of the 3 bits. If, for example, the information bit $u = 1$ should be transmitted, the codeword is $\boldsymbol{x} = 111$. Assuming that the channel corrupts the second bit, the received word is $\boldsymbol{y} = 101$. The receiver can directly see that this is not a codeword and conclude that there has been an error event during the transmission, i.e. it has detected an error. Furthermore, since the error probability ε is assumed to be small, the receiver can conclude that the most likely transmitted codeword was $\hat{\boldsymbol{x}} = 111$, which maps to the estimated information bit $\hat{u} = g(101) = 1$, and the decoder has corrected the error.

On the other hand, in the case when two code bits are erroneously received, the decoder will make a wrong decision and a decoding error occurs. Again assume that the codeword $\boldsymbol{x} = 111$ is transmitted, and that both the second and the third bits are altered, i.e. the received word is $\boldsymbol{y} = 100$. Then the decoder should assume the most likely event that one error in the first bit has occurred and correct to $\hat{\boldsymbol{x}} = 000$, which maps back to the estimated information bit $\hat{u} = g(100) = 0$.

The error probability, i.e. the probability that the decoder makes an erroneous decision, is an important measure of the efficiency of a code

$$P_e = P\big(g(Y) \neq u | U = u\big) \tag{6.27}$$

In all error detection or error correction schemes, there will be a strictly positive probability that the decoder makes a wrong decision. However, the error probability can be arbitrarily small. A code rate is said to be *achievable* if there exists a code such that the error probability tends to zero as n grows, $P_e \rightarrow 0,\ n \rightarrow \infty$.

An important result for bounding the error probability is given in Fano's lemma. It upper bounds the uncertainty about the information symbol when the estimate from the decoder is given.

Lemma 6.1 (Fano's Lemma) Let U and $\widehat{U}$ be two random variables with the same alphabet of size M, and $P_e = P(\widehat{U} \neq U)$ the error probability. Then

$$H(U|\widehat{U}) \leq h(P_e) + P_e \log(M-1) \tag{6.28}$$

□

To show this result, first introduce a binary random variable Z that describes the error,

$$Z = \begin{cases} 0, & U = \widehat{U} \\ 1, & U \neq \widehat{U} \end{cases} \tag{6.29}$$

Then the conditioned entropy of U is given by

$$\begin{aligned} H(U|\widehat{U}) &= H(UZ|\widehat{U}) = H(Z|\widehat{U}) + H(U|\widehat{U}Z) \\ &\leq H(Z) + H(U|\widehat{U}Z) \\ &= \underbrace{H(Z)}_{h(P_e)} + \underbrace{H(U|\widehat{U}, Z=0)}_{=0} P(Z=0) + \underbrace{H(U|\widehat{U}, Z=1)}_{\leq \log(M-1)} \underbrace{P(Z=1)}_{P_e} \\ &\leq h(P_e) + P_e \log(M-1) \end{aligned} \tag{6.30}$$

where in the first inequality it is used that

$$H(UZ|\widehat{U}) = H(U|\widehat{U}) + H(Z|U\widehat{U}) = H(U|\widehat{U}) \tag{6.31}$$

since given both U and $\widehat{U}$, there is no uncertainty about Z. Furthermore, it is used that conditioned on $\widehat{U}$ and $Z = 1$, U can take only $M - 1$ values and the entropy is upper bounded by $\log(M-1)$.

To get an understanding of the interpretation of the lemma, first plot the function

$$F(p) = h(p) + p \log(M-1) \tag{6.32}$$

and set the derivative equal to zero

$$\frac{\partial}{\partial p} F(p) = \log \frac{1-p}{p} + \log(M-1) = 0 \tag{6.33}$$

which gives an optima for $p = \frac{M-1}{M}$, where $F(\frac{M-1}{M}) = \log M$. At the end points, $F(0) = 0$ and $F(1) = \log(M-1)$. To see that this is really a maximum, take the derivative ones again to get

$$\frac{\partial^2}{\partial p^2} F(p) = \frac{1}{p(p-1)\ln 2} \leq 0, \quad 0 \leq p \leq 1 \tag{6.34}$$

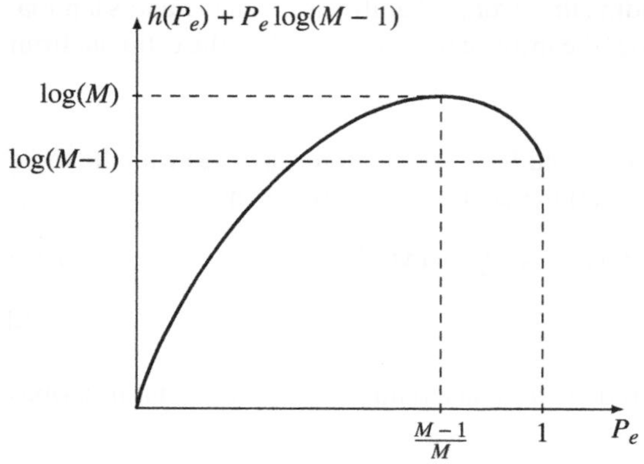

Figure 6.3 A plot of the inequality of Fano's lemma.

Since the second derivative is negative in the whole interval, it is concluded that the function is concave; hence it must be a global maximum. In Figure 6.3, the function is plotted. Notice that the function is nonnegative and that it spans the same interval as $H(U|\hat{U})$,

$$0 \leq H(U|\hat{U}) \leq \log M \tag{6.35}$$

which is interpreted as the uncertainty about the transmitted symbol U when receiving $\hat{U}$. Therefore, whatever value this uncertainty takes, it is possible to map it to a positive value for the error probability. One consequence is that the only time it is possible to get a zero error probability is when the uncertainty is zero, i.e. when there are either no disturbances on the channel or completely known disturbances. And this is never the case in a real communication situation.

In the next section, the channel coding theorem will be introduced, which gives the necessary and sufficient condition for when it is possible to get arbitrarily small P_e. It will also be seen that it is possible to give a bound on how much information is possible to transmit over a given channel.

6.4 CHANNEL CODING THEOREM

Shannon defined communication as transmitting information from one place and time to another place and time. This describes a lot of scenarios, for example, a telephone call, saving, and extracting data from a USB stick, but also a normal face-to-face conversation or even a text like this document. In all of those scenarios, there is a probability of errors along the actual transmission. In the case of telephone call, there can be disturbances along the line, for a normal conversation there are typically background noise and in most texts there are typos. The *channel* of the transmission is a statistical model representing all these disturbances. There are of course numerous types of channels, and they can be made arbitrarily complicated depending on the

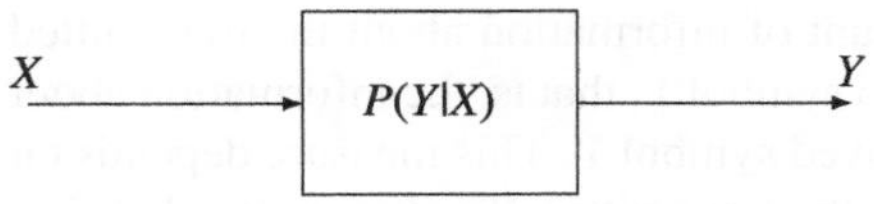

Figure 6.4 A discrete memoryless channel.

level of the modeling. This chapter will concentrate on discrete memoryless channels (DMC).

Definition 6.3 A *discrete channel* is a mathematical system $(\mathcal{X}, P(Y|X), \mathcal{Y})$, where $\mathcal{X}$ is the input alphabet and $\mathcal{Y}$ the output alphabet. The actual transmission is described by a transition probability distribution $P(Y|X)$. The channel is *memoryless* if the probability distribution is independent of previous input symbols. □

In Figure 6.4, a block diagram of a discrete memoryless channel is shown.

One of the most well-known channels, the *binary symmetric channel* (BSC), is described in the next example.

Example 6.4 [BSC] Consider a channel where both the input and the output alphabets are binary, $X \in \{0, 1\}$ and $Y \in \{0, 1\}$, and that the error probability equals p, i.e. the transition probabilities are described by the following table:

	$P(Y\|X)$	
X	$Y = 0$	$Y = 1$
0	$1 - p$	p
1	p	$1 - p$

The probability for having an error, i.e. transmitting 0 and receiving 1 or transmitting 1 and receiving 0, is p and the probability for no error, i.e. receiving the same as transmitting, is $1 - p$. This channel model is often denoted by the *binary symmetric channel*, where the symmetry reflects that the probability of error is independent of transmitted symbol. The channel can be viewed graphically as in Figure 6.5, where X is the transmitter side and Y the receiver side.

A measure of the amount of information that is possible to transmit over a channel can be obtained by the mutual information between the receiver and transmitter

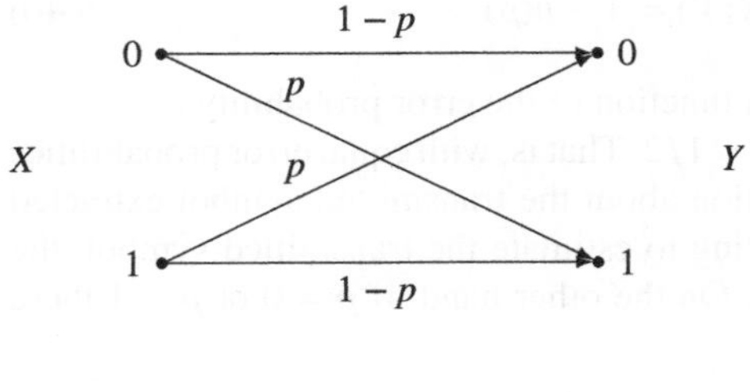

Figure 6.5 A graphical interpretation of the BSC.

sides, $I(X;Y)$. It measures the average amount of information about the transmitted symbol X obtained by observing the received symbol Y, that is, the information about X that the decoder can extract from the received symbol Y. This measure depends on the distribution of X, which is controlled by the transmitter. By altering the distribution, the mutual information might be changed, and the maximum over all distributions is defined as the *information channel capacity*.

Definition 6.4 The *information channel capacity* of a discrete memoryless channel is

$$C = \max_{p(x)} I(X;Y) \tag{6.36}$$

where the maximum is taken over all input distributions. □

The information channel capacity for the binary symmetric channel is derived in the next example.

Example 6.5 For the binary symmetric channel, the mutual information between transmitted and received variable is

$$\begin{aligned} I(X;Y) &= H(Y) - H(Y|X) \\ &= H(Y) - \sum_{i=0}^{1} P(X=i)H(Y|X=i) \\ &= H(Y) - \sum_{i=0}^{1} P(X=i)h(p) \\ &= H(Y) - h(p) \le 1 - h(p) \end{aligned} \tag{6.37}$$

where there is equality if and only if $P(Y=0) = P(Y=1) = \frac{1}{2}$. Since the capacity is obtained by maximizing over all input distributions, the probability of Y should be viewed in terms of the probability of X,

$$P(Y=0) = (1-p)P(X=0) + pP(X=1) \tag{6.38}$$

$$P(Y=1) = pP(X=0) + (1-p)P(X=1) \tag{6.39}$$

From symmetry, it is seen that $P(Y=0) = P(Y=1) = \frac{1}{2}$ is equivalent to $P(X=0) = P(X=1) = \frac{1}{2}$. Since there is a distribution of X giving the maximizing distribution for Y, the capacity for this channel becomes

$$C = \max_{p(x)} I(X;Y) = 1 - h(p) \tag{6.40}$$

In Figure 6.6, the capacity is plotted as a function of the error probability p.

It is seen that the capacity is 0 for $p = 1/2$. That is, with equal error probabilities on the channel there can be no information about the transmitted symbol extracted from the received symbol, and when trying to estimate the transmitted symbol, the decoder is not helped by the received Y. On the other hand, if $p = 0$ or $p = 1$ there

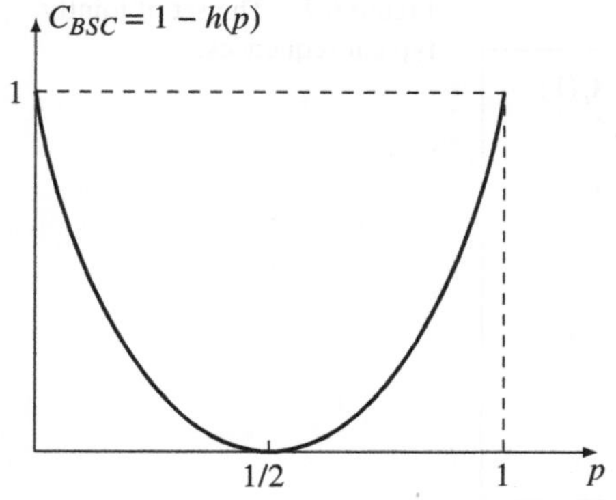

Figure 6.6 Capacity of a BSC as a function of the error probability p.

are no uncertainty about the transmitted symbol when given the received Y, and the capacity is 1 bit per transmission, or channel use.

To upper bound the capacity, it can be used that the mutual information is upper bounded by

$$I(X;Y) = H(Y) - H(Y|X) \leq H(Y) \leq \log |k_Y| \tag{6.41}$$

where k_Y is the number of possible outcomes for Y. Similarly, $I(X;Y) \leq \log |k_X|$, where k_X is the number of possible outcomes of X. To summarize, the capacity can be bounded as in the next theorem.

Theorem 6.5 For a DMC $(\mathcal{X}, P(Y|X), \mathcal{Y})$, the channel capacity is bounded by

$$0 \leq C \leq \min\{\log |k_X|, \log |k_Y|\} \tag{6.42}$$

□

Often the transmitted symbols are binary digits, and then the capacity is limited by $C \leq 1$.

In the following, Shannon's channel coding theorem will be introduced, which relates the information channel capacity to the coding rate. For this purpose, the AEP and typical sequences need to be extended as *jointly typical sequences*. The idea is to consider a sequence of pairs of $\boldsymbol{X}$ and $\boldsymbol{Y}$ and say that each of the sequences should be typical and that the sequence of pairs, viewed as a random variable, should also be typical. Then the set of sequences of pairs $(\boldsymbol{X}, \boldsymbol{Y})$ that are the most likely to actually happen can be used for decoding and the achievable code rates can be derived.

Definition 6.5 The set of all *jointly typical* sequences $A_\varepsilon(X, Y)$ is the set of all pairs of n-dimensional vectors of i.i.d. variables

$$\boldsymbol{x} = (x_1, x_2, \dots, x_n) \quad \text{and} \quad \boldsymbol{y} = (y_1, y_2, \dots, y_n)$$

All sequences $(\boldsymbol{x}, \boldsymbol{y})$

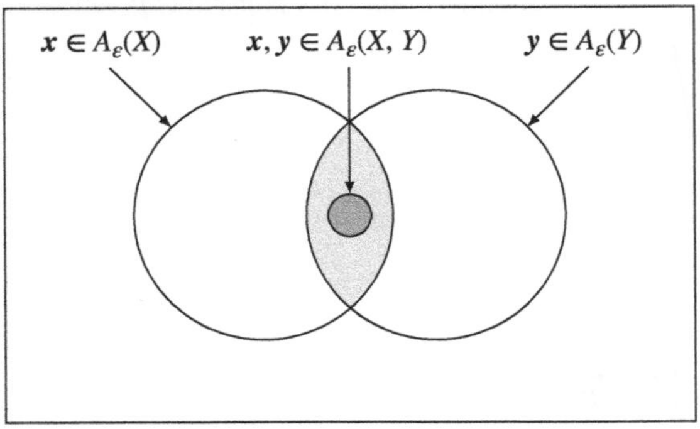

Figure 6.7 The set of jointly typical sequences.

such that each of $\boldsymbol{x}, \boldsymbol{y}$ and $(\boldsymbol{x}, \boldsymbol{y})$ are ε-typical, i.e.

$$\left| -\frac{1}{n} \log p(\boldsymbol{x}) - H(X) \right| \leq \varepsilon, \tag{6.43}$$

$$\left| -\frac{1}{n} \log p(\boldsymbol{y}) - H(Y) \right| \leq \varepsilon, \tag{6.44}$$

$$\left| -\frac{1}{n} \log p(\boldsymbol{x}, \boldsymbol{y}) - H(X, Y) \right| \leq \varepsilon \tag{6.45}$$

where $p(\boldsymbol{x}, \boldsymbol{y}) = \prod_i p(x_i, y_i)$, $p(\boldsymbol{x}) = \sum_{\boldsymbol{y}} p(\boldsymbol{x}, \boldsymbol{y})$ and $p(\boldsymbol{y}) = \sum_{\boldsymbol{x}} p(\boldsymbol{x}, \boldsymbol{y})$. □

In Figure 6.7, the big rectangle represents the set of all sequences $(\boldsymbol{x}, \boldsymbol{y})$. Then all sequences where $\boldsymbol{x}$ is typical are gathered in the left subset, $\boldsymbol{x} \in A_\varepsilon(X)$, and all sequences where $\boldsymbol{y}$ is typical in the right subset, $\boldsymbol{y} \in A_\varepsilon(Y)$. The intersection, marked as light gray, containing all sequences where both $\boldsymbol{x}$ or $\boldsymbol{y}$ are typical. Among those there are sequences where also the sequence of the pairs is typical, which is represented by the dark gray area, $\boldsymbol{x}, \boldsymbol{y} \in A_\varepsilon(X, Y)$.

As before, the definition of typical sequences can be given in an alternative form.

Definition 6.6 (Equivalent definition) The set of all *jointly typical* sequences $A_\varepsilon(X, Y)$ is the set of all pairs of n-dimensional vectors of i.i.d. variables

$$\boldsymbol{x} = (x_1, x_2, \dots, x_n) \quad \text{and} \quad \boldsymbol{y} = (y_1, y_2, \dots, y_n)$$

such that each of $\boldsymbol{x}$, $\boldsymbol{y}$, and $(\boldsymbol{x}, \boldsymbol{y})$ are ε-typical, i.e.

$$2^{-n(H(X)+\varepsilon)} \leq p(\boldsymbol{x}) \leq 2^{-n(H(X)-\varepsilon)} \tag{6.46}$$

$$2^{-n(H(Y)+\varepsilon)} \leq p(\boldsymbol{y}) \leq 2^{-n(H(Y)-\varepsilon)} \tag{6.47}$$

$$2^{-n(H(X,Y)+\varepsilon)} \leq p(\boldsymbol{x}, \boldsymbol{y}) \leq 2^{-n(H(X,Y)-\varepsilon)} \tag{6.48}$$

□

As for the one-dimensional case, the set of typical sequences is a small fraction of all sequences but their probability is close to one. In the next theorem, this is described.

Theorem 6.6 Let $(\boldsymbol{X}, \boldsymbol{Y})$ be sequences of length n drawn i.i.d. vectors according to $p(\boldsymbol{x},\boldsymbol{y}) = \prod_i p(x_i, y_i)$. Then, for sufficiently large n,

1. $P\Big((\boldsymbol{x},\boldsymbol{y}) \in A_\varepsilon(X, Y)\Big) \geq 1 - \varepsilon$.
2. $(1-\varepsilon)2^{n(H(X,Y)-\varepsilon)} \leq |A_\varepsilon(X, Y))| \leq 2^{n(H(X,Y)+\varepsilon)}$.
3. Let $(\tilde{\boldsymbol{X}}, \tilde{\boldsymbol{Y}})$ be distributed according to $p(\boldsymbol{x})p(\boldsymbol{y})$, i.e. $\tilde{\boldsymbol{X}}$ and $\tilde{\boldsymbol{Y}}$ are independent with the marginals derived from $p(\boldsymbol{x},\boldsymbol{y})$. Then

$$(1-\varepsilon)2^{-n(I(X;Y)+3\varepsilon)} \leq P\Big((\tilde{\boldsymbol{x}}, \tilde{\boldsymbol{y}}) \in A_\varepsilon(X, Y)\Big) \leq 2^{-n(I(X;Y)-3\varepsilon)}$$

□

To show the first part of the theorem, use the weak law of large numbers. Then, there exists an n_1 such that for all $n \geq n_1$

$$P_1 = P\Big(\Big|-\tfrac{1}{n}\log p(\boldsymbol{x}) - H(X)\Big| > \varepsilon\Big) < \frac{\varepsilon}{3} \tag{6.49}$$

Similarly, there exists an n_2 such that for all $n \geq n_2$

$$P_2 = P\Big(\Big|-\tfrac{1}{n}\log p(\boldsymbol{y}) - H(Y)\Big| > \varepsilon\Big) < \frac{\varepsilon}{3} \tag{6.50}$$

and there exists an n_3 such that for all $n \geq n_3$

$$P_3 = P\Big(\Big|-\tfrac{1}{n}\log p(\boldsymbol{x},\boldsymbol{y}) - H(X, Y)\Big| > \varepsilon\Big) < \frac{\varepsilon}{3} \tag{6.51}$$

Then, for $n \geq \max\{n_1, n_2, n_3\}$

$$\begin{aligned}
P\big((\boldsymbol{x},\boldsymbol{y}) \notin A_\varepsilon(X, Y)\big) &= P\Big(\Big|-\tfrac{1}{n}\log p(\boldsymbol{x}) - H(X)\Big| > \varepsilon, \\
&\qquad \cup\Big|-\tfrac{1}{n}\log p(\boldsymbol{y}) - H(Y)\Big| > \varepsilon, \\
&\qquad \cup\Big|-\tfrac{1}{n}\log p(\boldsymbol{x},\boldsymbol{y}) - H(X, Y)\Big| > \varepsilon\Big) \\
&\leq P_1 + P_2 + P_3 < \varepsilon
\end{aligned} \tag{6.52}$$

where in the second last inequality follows from the union bound[1] was used.

[1] The union bound states that for the events $\mathcal{A}_1, \dots, \mathcal{A}_n$ the probability that at least one is true is

$$P\Big(\bigcup_{i=1}^{n} \mathcal{A}_i\Big) \leq \sum_{i=1}^{n} P(\mathcal{A}_i)$$

To show the second part of the theorem, a similar argument as for the single-variable case can be used. The right-hand side of the inequality can be shown by

$$\begin{aligned} 1 &\geq \sum_{\boldsymbol{x},\boldsymbol{y} \in A_\varepsilon} p(\boldsymbol{x},\boldsymbol{y}) \\ &\geq \sum_{\boldsymbol{x},\boldsymbol{y} \in A_\varepsilon} 2^{-n(H(X,Y)+\varepsilon)} = |A_\varepsilon(X,Y)| 2^{-n(H(X,Y)+\varepsilon)} \end{aligned} \tag{6.53}$$

The left-hand side can be shown, for sufficiently large n, by

$$\begin{aligned} 1-\varepsilon &\leq \sum_{\boldsymbol{x},\boldsymbol{y} \in A_\varepsilon} p(\boldsymbol{x},\boldsymbol{y}) \\ &\leq \sum_{\boldsymbol{x},\boldsymbol{y} \in A_\varepsilon} 2^{-n(H(X,Y)-\varepsilon)} = |A_\varepsilon(X,Y)| 2^{-n(H(X,Y)-\varepsilon)} \end{aligned} \tag{6.54}$$

For $(\tilde{\boldsymbol{x}},\tilde{\boldsymbol{y}})$ distributed according to $p(\boldsymbol{x})p(\boldsymbol{y})$, where $p(\boldsymbol{x}) = \sum_{\boldsymbol{y}} p(\boldsymbol{x},\boldsymbol{y})$ and $p(\boldsymbol{y}) = \sum_{\boldsymbol{x}} p(\boldsymbol{x},\boldsymbol{y})$, respectively, the probability for $(\tilde{\boldsymbol{x}},\tilde{\boldsymbol{y}})$ to be a jointly typical sequence

$$\begin{aligned} P\big((\tilde{\boldsymbol{x}},\tilde{\boldsymbol{y}}) \in A_\varepsilon(X,Y)\big) &= \sum_{\boldsymbol{x},\boldsymbol{y} \in A_\varepsilon} p(\boldsymbol{x})p(\boldsymbol{y}) \\ &\leq \sum_{\boldsymbol{x},\boldsymbol{y} \in A_\varepsilon} 2^{-n(H(X)-\varepsilon)} 2^{-n(H(Y)-\varepsilon)} \\ &\leq 2^{-n(H(X)-\varepsilon)} 2^{-n(H(Y)-\varepsilon)} 2^{n(H(X,y)+\varepsilon)} \\ &= 2^{-n(H(X)+H(Y)-H(X,Y)-3\varepsilon)} \\ &= 2^{-n(I(X;Y)-3\varepsilon)} \end{aligned} \tag{6.55}$$

which shows the right-hand side of the third property. The left-hand side can be obtained by

$$\begin{aligned} P\big((\tilde{\boldsymbol{x}},\tilde{\boldsymbol{y}}) \in A_\varepsilon(X,Y)\big) &= \sum_{\boldsymbol{x},\boldsymbol{y} \in A_\varepsilon} p(\boldsymbol{x})p(\boldsymbol{y}) \\ &\geq \sum_{\boldsymbol{x},\boldsymbol{y} \in A_\varepsilon} 2^{-n(H(X)+\varepsilon)} 2^{-n(H(Y)+\varepsilon)} \\ &\geq 2^{-n(H(X)+\varepsilon)} 2^{-n(H(Y)+\varepsilon)} (1-\varepsilon) 2^{n(H(X,Y)-\varepsilon)} \\ &= (1-\varepsilon) 2^{-n(H(X)+H(Y)-H(X,Y)+3\varepsilon)} \\ &= (1-\varepsilon) 2^{-n(I(X;Y)+3\varepsilon)} \end{aligned} \tag{6.56}$$

which completes the proof of Theorem 6.6.

The jointly typical sequences play an important role when showing the channel coding theorem. It relates the necessary coding rate used in the communication system used with the channel information capacity.

Theorem 6.7 (Channel coding theorem) For a given code rate R, there exists a code with probability of error approaching zero if and only if

$$R < C = \max_{p(x)} I(X;Y) \tag{6.57}$$

□

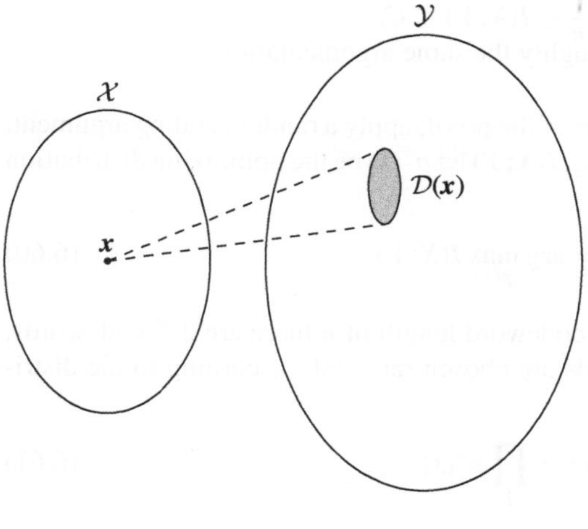

Figure 6.8 The decoding region of $\boldsymbol{y}$ mapping to the codeword $\boldsymbol{x}$ via the decoding regions $\mathcal{D}(\boldsymbol{x})$.

What this theorem means is that for a discrete memoryless channel with capacity C, and a (n, nR) code is used, where R is the code rate (information symbols per code symbols), then as n grows,

- it is possible to find a code such that the error probability is arbitrarily low, if $R < C$.
- it is not possible to find a code such that the error probability is arbitrarily low, if $R > C$.

The proof of the theorem is quite extensive. Before going into details of this an intuitive explanation is given. In Figure 6.8, the sets of n-dimensional transmit vectors $\boldsymbol{x}$ and received vectors $\boldsymbol{y}$ are shown. The transmitted codeword is taken from the left set and the received vector from the right set. The decoding rule is decided by decoding regions. If the received vector $\boldsymbol{y}$ is in the decoding region for $\boldsymbol{x}$, $\mathcal{D}(\boldsymbol{x})$, which is a subset of the received vectors, the gray area in the figure, the estimated transmitted codeword is $\boldsymbol{x}$. One way to define the decoding regions for a received vector $\boldsymbol{y}$ is to find an $\boldsymbol{x}$ such that the pair $(\boldsymbol{x}, \boldsymbol{y}) \in A_\varepsilon(X, Y)$ is jointly typical. If such $\boldsymbol{x}$ exists and is unique, decode to $\boldsymbol{x}$.

Since typical sequences are used in the decoding, when neglecting ε, the number of such possible mappings is $|(\boldsymbol{x}, \boldsymbol{y}) \in A_\varepsilon(X, Y)| = 2^{nH(X,Y)}$. On the other hand, transmitting random sequences, the number of typical sequences is $|\boldsymbol{x} \in A_\varepsilon(X)| = 2^{nH(X)}$. That means, in each decoding region $\mathcal{D}(\boldsymbol{x})$ the number of possible received decodable sequences is

$$N = \frac{2^{nH(X,Y)}}{2^{nH(X)}} = 2^{n(H(X,Y)-H(X))} \tag{6.58}$$

But, the number of typical received sequences is $|\boldsymbol{y} \in A_\varepsilon(Y)| = 2^{nH(Y)}$. Optimally, with disjoint decoding regions the number of codewords is

$$2^k = \frac{|\boldsymbol{y} \in A_\varepsilon(Y)|}{N} \leq 2^{n(H(X)+H(Y)-H(X,Y))} = 2^{nI(X;Y)} \tag{6.59}$$

That is, the optimal code rate is $R = \frac{k}{n} \leq I(X;Y) \leq C$.

The more formal proof uses roughly the same argumentation.

Proof. Starting with the existence part of the proof, apply a random coding argument. With the capacity given by $C = \max_{p(x)} I(X;Y)$ let $p^*(x)$ be the optimizing distribution of X,

$$p^*(x) = \arg\max_{p(x)} I(X;Y) \tag{6.60}$$

Consider a code with rate R. With a codeword length of n there are 2^{nR} codewords, hence an (n, nR) code. The codewords are chosen randomly according to the distribution

$$p(\boldsymbol{x}) = \prod_i p^*(x) \tag{6.61}$$

The decoding of a received vector $\boldsymbol{y}$ is done by finding a codeword $\boldsymbol{x}$ such that $(\boldsymbol{x}, \boldsymbol{y}) \in A_\varepsilon(X, Y)$. If such codeword does not exist or is not unique, an error has occurred. Introduce the event that the pair $(\boldsymbol{x}, \boldsymbol{y})$ is a jointly typical sequence

$$E_{\boldsymbol{x},\boldsymbol{y}} = \Big\{ (\boldsymbol{x}, \boldsymbol{y}) \in A_\varepsilon(X, Y) \Big\} \tag{6.62}$$

For a pair $(\boldsymbol{x}_0, \boldsymbol{y})$ of a codeword and a received vector, an error occurs if they are not jointly typical, $(\boldsymbol{x}_0, \boldsymbol{y}) \in E^c_{\boldsymbol{x}_0,\boldsymbol{y}}$, or if the decoding is not unique, $(\boldsymbol{x}_0, \boldsymbol{y}) \in E_{\boldsymbol{x}_0,\boldsymbol{y}}$ and $(\boldsymbol{x}, \boldsymbol{y}) \in E_{\boldsymbol{x},\boldsymbol{y}}$ for $\boldsymbol{x} \neq \boldsymbol{x}_0$ (i.e., two equally likely codewords). In the previous case, $(\cdot)^c$ denotes the complementary set. An error event for codeword $\boldsymbol{x}_0$ can be denoted as

$$E_e = E^c_{\boldsymbol{x}_0,\boldsymbol{y}} \cup \Big(E_{\boldsymbol{x}_0,\boldsymbol{y}} \cap \Big(\bigcup_{\boldsymbol{x} \neq \boldsymbol{x}_0} E_{\boldsymbol{x},\boldsymbol{y}} \Big) \Big) \tag{6.63}$$

The mathematical structure $(\mathcal{P}(\mathcal{U}), \mathcal{U}, \emptyset, \cap, \cup, ^c)$, where $\mathcal{P}(\mathcal{U})$ is the power set, i.e. the set of all subsets of $\mathcal{U}$, is a Boolean algebra. Hence, the error event can be rewritten as[2]

$$E_e = E^c_{\boldsymbol{x}_0,\boldsymbol{y}} \cup \Big(\bigcup_{\boldsymbol{x} \neq \boldsymbol{x}_0} E_{\boldsymbol{x},\boldsymbol{y}} \Big) \tag{6.64}$$

The probability of error is then the probability of the event E_e and from the union bound

$$P_e = P(E_e) = P\Big(E^c_{\boldsymbol{x}_0,\boldsymbol{y}} \cup \Big(\bigcup_{\boldsymbol{x} \neq \boldsymbol{x}_0} E_{\boldsymbol{x},\boldsymbol{y}} \Big) \Big) \leq P\big(E^c_{\boldsymbol{x}_0,\boldsymbol{y}}\big) + \sum_{\boldsymbol{x} \neq \boldsymbol{x}_0} P\big(E_{\boldsymbol{x},\boldsymbol{y}}\big) \tag{6.65}$$

[2] In a Boolean algebra $(\mathcal{B}, 1, 0, \wedge, \vee, ')$, where $\wedge$ is AND, $\vee$ is OR, and $'$ complement, the two rules consensus and absorption give

$$a' \vee (a \wedge b) = a' \vee (a \wedge b) \vee b = a' \vee b$$

The result is obtained by letting $a = E_{\boldsymbol{x}_0,\boldsymbol{y}}$ and $b = \bigcup_{\boldsymbol{x} \neq \boldsymbol{x}_0} E_{\boldsymbol{x},\boldsymbol{y}}$.

The two probabilities included can be bounded as

$$P(E^c_{x_0,y}) = P\big((x_0, y) \notin A_\varepsilon(X, Y)\big) \to 0, \quad n \to \infty \tag{6.66}$$

$$P(E_{x,y}) = P\big((x, y) \in A_\varepsilon(X, Y)\big) \le 2^{-n(I(X;Y)-3\varepsilon)} = 2^{-n(C-3\varepsilon)} \tag{6.67}$$

where the last equality is obtained since X is distributed according to the maximizing distribution $p^*(x)$. In the limit as $n \to \infty$, (6.65) becomes

$$\begin{aligned} P_e &= \sum_{x \neq x_0} P\big(E_{x,y}\big) \le \sum_{x \neq x_0} 2^{-n(C-3\varepsilon)} \\ &= (2^{nR} - 1)2^{-n(C-3\varepsilon)} < 2^{n(R-C+3\varepsilon)} \end{aligned} \tag{6.68}$$

To achieve reliable communication, it is required that $P_e \to 0$ as $n \to \infty$. That is,

$$\left(2^{R-C+3\varepsilon}\right)^n \to 0 \tag{6.69}$$

which is equivalent to $2^{R-C+3\varepsilon} < 1$, or $R - C + 3\varepsilon < 0$. This gives

$$R < C - 3\varepsilon \tag{6.70}$$

Since a random code was used, there exists at least one code that fulfills this requirement.

To show the converse, i.e. that it is not possible to achieve reliable communication if the coding rate exceeds the capacity, first assume that $R > C$ and that the 2^{nR} codewords are equally likely. The latter assumption implies the information symbols are equally likely, and that the channel coding scheme is preceded by perfect source coding. Also assume that the codewords and received words are n-dimensional vectors. Let $\hat{X}$ be a random variable describing the estimated transmitted codeword. If this uniquely maps to an estimated information symbol, the error probability is $P_e = P(X \neq \hat{X})$, where X is the transmitted codeword. According to Fano's lemma

$$H(X|\hat{X}) \le h(P_e) + P_e \log\big(2^{nR} - 1\big) \tag{6.71}$$

On the other hand,

$$I(X;\hat{X}) = H(X) - H(X|\hat{X}) \tag{6.72}$$

which leads to an expression for the left-hand side of Fano's inequality

$$H(X|\hat{X}) = H(X) - I(X;\hat{X}) \ge H(X) - I(X;Y) \tag{6.73}$$

where the data-processing lemma is used in the last equality. The mutual information over the channel can be written as

$$
\begin{aligned}
I(X;Y) &= H(Y) - H(Y|X) \\
&= H(Y) - \sum_{1=1}^{n} H(Y_i|Y_1 \dots Y_{i-1}X) \\
&= H(Y) - \sum_{1=1}^{n} H(Y_i|X_i) \\
&= \sum_{1=1}^{n} H(Y_i) - H(Y_i|X_i) = \sum_{1=1}^{n} I(X_i;Y_i) \leq \sum_{1=1}^{n} C = nC \qquad (6.74)
\end{aligned}
$$

where the third equality follows since the channel is memoryless. From the assumption of equally likely codewords, the entropy of X is $H(X) = \log 2^{nR} = nR$. Hence, (6.73) can be bounded as

$$H(X|\hat{X}) \geq nR - nC = n(R - C) \qquad (6.75)$$

With Fano's inequality in (6.71)

$$h(P_e) + P_e \log\left(2^{nR} - 1\right) \geq n(R - C) > 0 \qquad (6.76)$$

where the strict inequality follows from the assumption that $R > C$. This means that the left-hand side is strictly positive and therefor $P_e > 0$. Hence, if $R > C$ the error probability will not go to zero as n goes to infinity, whatever code is chosen. This concludes the proof of the channel coding theorem. ■

The first part of theorem only shows the existence of a code that meets the bound. It does not say anything about how it should be found. Since 1948 when the results were published, there has been a lot of research on error-correcting codes, and today there are codes that can come very close to the capacity limit. The focus then becomes to reduce the computational complexity in the system. In Chapter 7, a short introduction to channel coding is given.

An interesting extension of the channel coding theorem is to have a system with a dedicated feedback channel from the receiver to the transmitter, as shown in Figure 6.9. Then the transmitter can see the previous received symbol Y_{i-1}, so the transmitted symbol is $X_i = X(U, Y_{i-1})$.

In reality, the feedback can have a significant meaning, making the decoding easier, but the transmission rate is not improved compared to the case without feedback. To derive the capacity for this case, the above proof has to be adjusted a bit. The

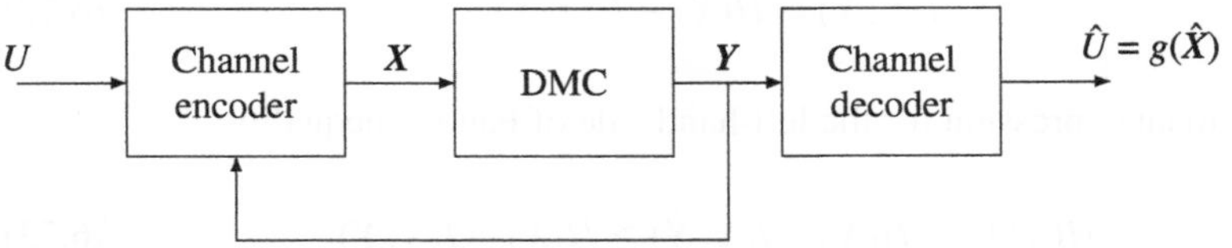

Figure 6.9 A channel with feedback.

first part, the existence, does not depend on the feedback and can be reused entirely. But for the converse part, it cannot be assumed $I(X;Y) \leq nC$. Instead, consider the error probability $P_e = P(U \neq \hat{U})$. Then Fano's lemma can be written as

$$H(U|\hat{U}) \leq h(P_e) + P_e \log\left(2^{nR} - 1\right) \tag{6.77}$$

but it can also be seen

$$H(U|\hat{U}) = H(U) - I(U;\hat{U}) \geq H(U) - I(U;Y) \tag{6.78}$$

where the inequality follows from the data-processing lemma. The mutual information between the information symbol and the received vector can be written as

$$\begin{aligned} I(U;Y) = H(Y) - H(Y|U) &\leq \sum_i H(Y_i) - \sum_i H(Y_i|Y_1 \dots Y_{i-1}U) \\ &= \sum_i H(Y_i) - \sum_i H(Y_i|Y_1 \dots Y_{i-1}UX_i) \\ &= \sum_i H(Y_i) - \sum_i H(Y_i|X_i) = \sum_i I(X_i;Y_i) \leq nC \end{aligned} \tag{6.79}$$

Then, similar to the case with no feedback,

$$H(U|\hat{U}) \geq H(U) - I(U;Y) \geq nR - nC = n(R - C) \tag{6.80}$$

which leads back to the same argument as before, and the code rate is not achievable if $R > C$. Summarizing, the following theorem has been shown.

Theorem 6.8 The capacity for a feedback channel is equal to the nonfeedback channel,

$$C_{FB} = C = \max_{p(x)} I(X;Y) \tag{6.81}$$

□

6.5 DERIVATION OF CHANNEL CAPACITY FOR DMC

In Example 6.5, the information capacity for the BSC was derived as $C = 1 - h(p)$, where p is the bit error probability on the channel. From the channel coding theorem, it is seen that this is a hard limit for what coding rates are possible to achieve reliable communication. For example, if the bit error probability on the channel is $p = 0.1$ the capacity is $C = 1 - h(0.1) = 0.9192$. Then reliable communication is possible if and only if $R < 0.9192$. In this section, the capacity will be derived for some of the most common channels.

Example 6.6 [Binary erasure channel] In the binary erasure channel (BEC), there is one more output symbol, Δ, than for the BSC, interpreted as an erasure. The idea is that the decoder will get the information that the received signal was not trustworthy. The probability of an erasure is α, and the probability for correct received symbol is

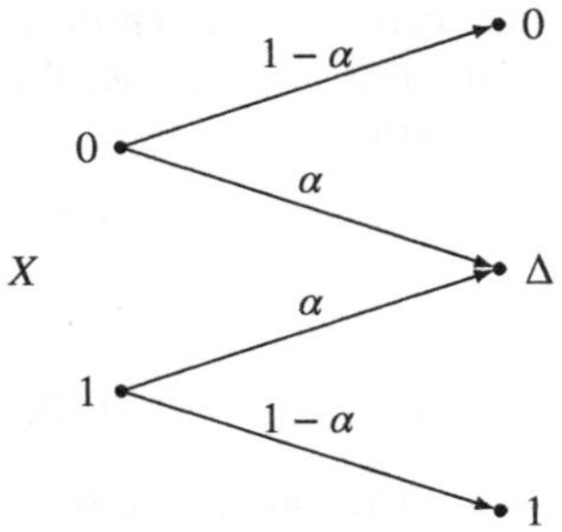

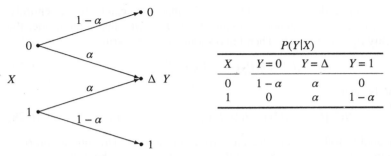

$P(Y\|X)$			
X	$Y = 0$	$Y = \Delta$	$Y = 1$
0	$1 - \alpha$	α	0
1	0	α	$1 - \alpha$

Figure 6.10 The binary erasure channel.

$1 - \alpha$. It is assumed that the probability for incorrect received symbol, i.e. transmitting a 0 and receiving a 1, and vice versa, is negligible and therefore set to zero. The graphical representation of the channel and its probability distribution is shown in Figure 6.10.

The mutual information between X and Y can be written as

$$\begin{aligned} I(X;Y) &= H(Y) - H(Y|X) \\ &= H(Y) - \sum_x H(Y|X=x)P(X=x) \\ &= H(Y) - h(\alpha) \end{aligned} \tag{6.82}$$

since $H(Y|X = x) = H(1 - \alpha, \alpha, 0) = h(\alpha)$, both for $x = 0$ and $x = 1$. That is, maximizing $I(X; Y)$ is equivalent to maximizing $H(Y)$ for varying $p = P(X = 1)$. The maximum entropy is $H(Y) \leq \log 3$, but this requires equiprobable Y, and contrary to the case for the BSC, this cannot be guaranteed.

To derive $H(Y)$, first get the distribution of Y, expressed in terms of the distribution of X. Assume that $P(X = 1) = p$, then with $p(x, y) = p(x)p(y|x)$ and $p(y) = \sum_x p(x, y)$ the distributions are obtained as

$P(X,Y)$			
X	$Y = 0$	$Y = \Delta$	$Y = 1$
0	$(1-p)(1-\alpha)$	$(1-p)\alpha$	0
1	0	$p\alpha$	$p(1-\alpha)$

and

$Y:$	0	Δ	1
$P(Y):$	$(1-p)(1-\alpha)$	α	$p(1-\alpha)$

Hence, the entropy of Y is $H(Y) = H((1 - p)(1 - \alpha), \alpha, p(1 - \alpha))$. Naturally, this function can be optimized by letting the derivative to be equal to zero. But, it can

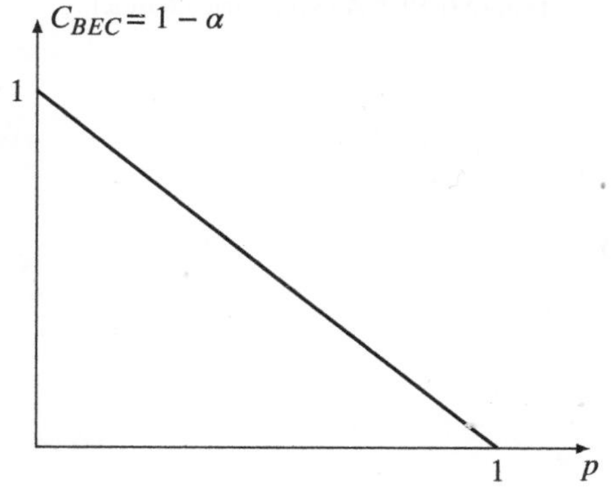

Figure 6.11 The capacity of the BEC as a function of the erasure probability.

also be noted that

$$\begin{aligned} H(Y) &= -(1-p)(1-\alpha)\log(1-p)(1-\alpha) \\ &\quad -\alpha\log\alpha - p(1-\alpha)\log p(1-\alpha) \\ &= (1-\alpha)\Big(-p\log p - (1-p)\log(1-p)\Big) \\ &\quad -\alpha\log\alpha - (1-\alpha)\log(1-\alpha) \\ &= (1-\alpha)h(p) + h(\alpha) \end{aligned} \tag{6.83}$$

which is maximized by $p = \frac{1}{2}$. The capacity is

$$C = \max_p H(Y) - h(\alpha) = \max_p (1-\alpha)h(p) = 1 - \alpha \tag{6.84}$$

In Figure 6.11, the capacity for the BEC is plotted as a function of the erasure probability α. If $\alpha = 0$, the capacity equals one, since then there are not any errors. On the other hand, if $\alpha = 1$ all of the received symbols will be erasures and there is no information about the transmitted sequence and hence the capacity is zero.

In the general case, to find the capacity, standard optimization techniques must be used, i.e. taking the derivative of $I(X, Y)$ equal to zero. However, there are also many cases where there are built-in symmetries in the channel, which can be used to ease the calculations. In the previous examples, the capacity was derived by considering the entropy of the received variable as a function of the distribution of the transmitted variable. To formalize these derivations, a definition of a *symmetric channel*[3] is needed.

Definition 6.7 (Symmetric channels) A discrete memoryless channel, with N input alternatives $X \in \{x_1, \dots, x_N\}$ and M output alternatives $Y \in \{y_1, \dots, y_M\}$, is *symmetric* if, in the graphical representation

[3] In the literature, there is no unified naming for different types of channel symmetries. In this text, the notation given in [11] will be followed

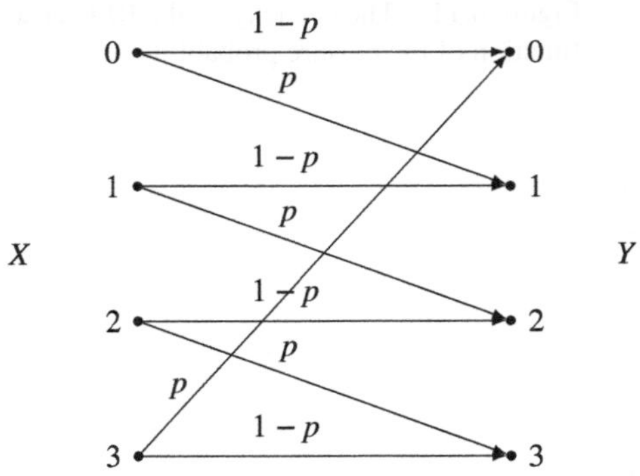

Figure 6.12 A symmetric channel.

- The set of branches leaving a symbol x_i have the same set of probabilities $\{q_1, q_2, \dots, q_M\}$ for all $i = 1, \dots, N$.
- The set of branches entering a symbol y_j has the same set of probabilities $\{p_1, p_2, \dots, p_N\}$ for all $j = 1, \dots, M$.

Seen from the transition matrix perspective, all rows are permutations of each other and all columns are permutations of each other. □

The BSC is an example of a symmetric channel. In the next example, another popular channel, also symmetric, is introduced.

Example 6.7 Consider a channel with four different inputs $X \in \{0, 1, 2, 3\}$, and the same set of outputs $Y \in \{0, 1, 2, 3\}$. The transmitted value is received correctly with probability $1 - p$ and as the next following symbol (modulo 4) with probability p. In Figure 6.12, the channel is shown.

The transition probability table is

$P(Y|X)$

X	$Y = 0$	$Y = 1$	$Y = 2$	$Y = 3$
0	$1 - p$	p	0	0
1	0	$1 - p$	p	0
2	0	0	$1 - p$	p
3	p	0	0	$1 - p$

The mutual information between X and Y is given by

$$I(X; Y) = H(Y) - H(Y|X) = H(Y) - \sum_x \underbrace{H(Y|X = x)}_{h(p)} p(x)$$

$$= H(Y) - h(p) \leq \log 4 - h(p) = 2 - h(p) \qquad (6.85)$$

with equality if and only if $p(y) = \frac{1}{4}$. Assume that $p(x) = \frac{1}{4}$ for all x, then

$$p(y) = \sum_x p(y|x)p(x) = \sum_x \frac{1}{4}p(y|x) = \frac{1}{4}p + \frac{1}{4}(1-p) = \frac{1}{4} \tag{6.86}$$

hence the distribution of X maximizing $I(X;Y)$ is the uniform distribution, $p(x) = \frac{1}{4}$, and the capacity for the channel is

$$C = 2 - h(p) \tag{6.87}$$

The derivation of the capacity in the previous example is typical for a symmetric channel. Assume a symmetric channel with N inputs and M outputs. Then, following the previous example, the mutual information is

$$I(X;Y) = H(Y) - H(Y|X) = H(Y) - \sum_x H(Y|X=x)p(x) \tag{6.88}$$

Since the channel is symmetric, the outgoing transitions from a given input symbol x is the same independent of the x. The entropy function does not take the order of the probabilities, i.e. the semantics of the message, into consideration. Therefore, the entropy of Y conditioned on x is the same for all x,

$$H(Y|X=x) = H(p_1, p_2, \dots, p_N) = H(\boldsymbol{r}) \tag{6.89}$$

where $\boldsymbol{r} = (p_1, p_2, \dots, p_N)$ is one row in the table of $P(Y|X)$. The mutual information can then be written as

$$I(X;Y) = H(Y) - H(\boldsymbol{r}) \leq \log M - H(\boldsymbol{r}) \tag{6.90}$$

with equality if and only if $p(y) = \frac{1}{M}$. As in Example 6.7, assume that the distribution of X is uniform, i.e. $p(x) = \frac{1}{N}$. Then the probability of Y becomes

$$p(y) = \sum_x p(x)p(y|x) = \frac{1}{N}\sum_x p(y|x) = \frac{1}{N}A \tag{6.91}$$

where the constant value $A = \sum_x p(y|x)$ follows from the symmetry. Summing this over Y gives

$$\sum_y p(y) = \sum_y A\frac{1}{N} = A\frac{M}{N} = 1 \tag{6.92}$$

which gives that $p(y) = \frac{1}{M}$. Since it is possible to find a distribution on X such that Y will have a uniform distribution, the capacity is

$$C_{\text{Sym}} = \max_{p(x)} I(X;Y) = \log M - H(\boldsymbol{r}) \tag{6.93}$$

With this at hand, the symmetry in Example 6.7 can be used to get the capacity as

$$C = \log 4 - H(1-p, p, 0, 0) = 2 - h(p) \tag{6.94}$$

Also the BSC is symmetric, with $M = 2$ and $\boldsymbol{r} = (1-p, p)$. Hence, the capacity is

$$C_{BSC} = \log 2 - H(1-p, p) = 1 - h(p) \tag{6.95}$$

Actually, in the previous derivation of C_{Sym} it is not used that the incoming transitions for all the receiving symbols y have the same distribution, it is only used that the sum is constant for all y. Therefore, a weaker definition of symmetry is given below, still giving the same result on the capacity.

Definition 6.8 (Weakly symmetric channels) A discrete memoryless channel is *weakly symmetric* if

- The set of branches leaving a symbol x_i has the same set of probabilities $\{q_1, q_2, \dots, q_M\}$ for all $i = 1, \dots, N$.
- The set of branches entering a symbol y_j has the same sum of probabilities $\sum_x p_x$ for all $j = 1, \dots, M$.

Seen from the transition probability table perspective, all rows are permutations of each other and all columns have the same sum $\sum_x p(y|x)$. Naturally, all symmetric channels are also weakly symmetric. □

The result on the capacity is then stated as a theorem.

Theorem 6.9 If a discrete memoryless channel is symmetric, or weakly symmetric, the channel capacity is

$$C = \log M - H(\boldsymbol{r}) \tag{6.96}$$

where $\boldsymbol{r}$ is the set of probabilities labeling branches leaving an input symbol X, or, equivalently, one row in the transition probability table. The capacity is reached for the uniform distribution $p(x) = \frac{1}{N}$. □

In the next example, a weakly symmetric channel is considered.

Example 6.8 In Figure 6.13, a channel with two erasure symbols, one closer to symbol 0 and one closer to symbol 1, is shown.

The corresponding transition probability table is

$P(Y|X)$

X	$Y=0$	$Y=\Delta_0$	$Y=\Delta_1$	$Y=3$
0	1/3	1/4	1/4	1/6
1	1/6	1/4	1/4	1/3

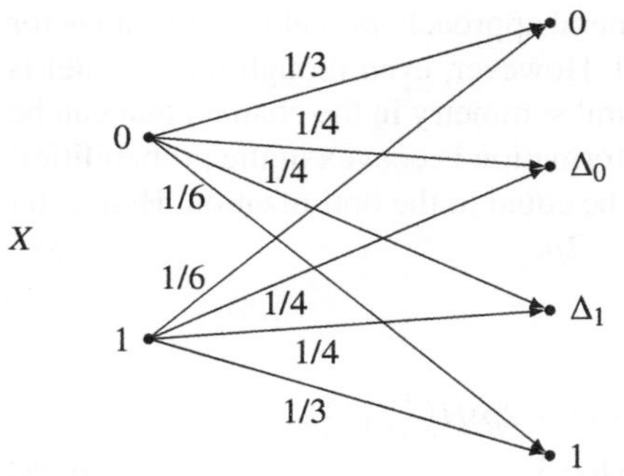

Figure 6.13 A binary double erasure channel that is weakly symmetric.

The two rows of the matrix have the same set of probabilities. Summing each column gives the constant value $1/2$, concluding that the channel is weakly symmetric. There are $M = 4$ output alternatives, and the capacity for this channel is calculated as

$$C = \log 4 - H\left(\tfrac{1}{3}, \tfrac{1}{4}, \tfrac{1}{4}, \tfrac{1}{6}\right) \approx 2 - 1.9591 = 0.0409 \tag{6.97}$$

This is a very low value on the capacity, which is not surprisingly since all the crossover probabilities are in the same order. To reach reliable communication, it will require a large overhead in terms of a low code rate.

In all of the channels considered above, $H(Y|X = x)$ has been constant. This is true when the distribution for the outgoing branches in the channel model has the same set of probabilities.

This section will be concluded by considering two channels that do not have the symmetry property. In this case, the derivations fall back to standard optimization methods. If the channel is even more complicated, it might be better to solve the calculations with numerical methods. In that case, it should be noted that the mutual information is a convex function, which simplify calculations.

Example 6.9 Consider the three input and three output DMC described in Figure 6.14. To derive the capacity, a distribution for X has to be assigned over which the

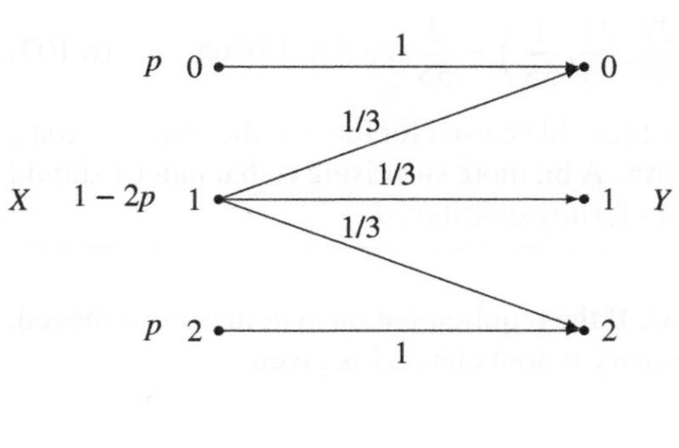

Figure 6.14 A ternary channel.

mutual information can be optimized. In a general approach, probability variables for two of the values for X would be introduced. However, even though the channel is not symmetric by definition, there is a structural symmetry in the channel that can be used together with the fact that the mutual information is convex in the probabilities. Therefore, the probability for 0 and 2 should be equal to the optimization. Hence, let $P(X = 0) = P(X = 2) = p$ and $P(X = 1) = 1 - 2p$.

The mutual information equals

$$\begin{aligned} I(X;Y) &= H(Y) - H(Y|X) \\ &= H(Y) - 2ph(0) - (1-2p)H\left(\tfrac{1}{3}, \tfrac{1}{3}, \tfrac{1}{3}\right) \\ &= H(Y) - (1-2p)\log 3 \end{aligned} \tag{6.98}$$

The distribution of Y can be found by $P(Y = 0) = P(Y = 2) = p + \frac{1}{3}(1 - 2p) = \frac{1+p}{3}$ and $P(Y = 1) = \frac{1-2p}{3}$, leading to an expression for the mutual information as

$$\begin{aligned} I(X;Y) &= H\left(\frac{1+p}{3}, \frac{1+p}{3}, \frac{1-2p}{3}\right) - (1-2p)\log 3 \\ &= -2\frac{1+p}{3}\log\frac{1+p}{3} - \frac{1-2p}{3}\log\frac{1-2p}{3} - (1-2p)\log 3 \end{aligned} \tag{6.99}$$

To maximize, set the derivative equal to zero,

$$\begin{aligned} \frac{\partial}{\partial p} I(X;Y) &= -\frac{2}{3\ln 2} - \frac{2}{3}\log\frac{1+p}{3} + \frac{2}{3\ln 2} + \frac{2}{3}\log\frac{1-2p}{3} + 2\log 3 \\ &= \frac{2}{3}\log\frac{1-2p}{1+p} + 2\log 3 = 0 \end{aligned} \tag{6.100}$$

where it is used that $\frac{\partial}{\partial p}\big(f(p)\log f(p)\big) = \frac{\partial f(p)}{\partial p}\big(\frac{1}{\ln 2} + \log f(p)\big)$. The above equation is equivalent to

$$\frac{1-2p}{1+p} = \frac{1}{27} \tag{6.101}$$

which leads to

$$p = \frac{26}{55} \tag{6.102}$$

So the optimizing distribution of X is $P(X = 0) = P(X = 2) = \frac{26}{55}$ and $P(X = 1) = \frac{3}{55}$. To get the capacity insert in $I(X; Y)$, to get

$$C = I(X;Y)\Big|_{p=\frac{26}{55}} = H\left(\frac{27}{55}, \frac{27}{55}, \frac{1}{55}\right) - \frac{3}{55}\log 3 \approx 1.0265 \tag{6.103}$$

The derivation shows that inputs 0 and 2 should be used for most of the transmissions, which is intuitive since they are error-free. A bit more surprising is that input 1 should be used, even though the outputs are uniformly distributed.

The final example is related to the BSC. If the requirement on symmetry is removed, a general form of a binary input and binary output channel is given.

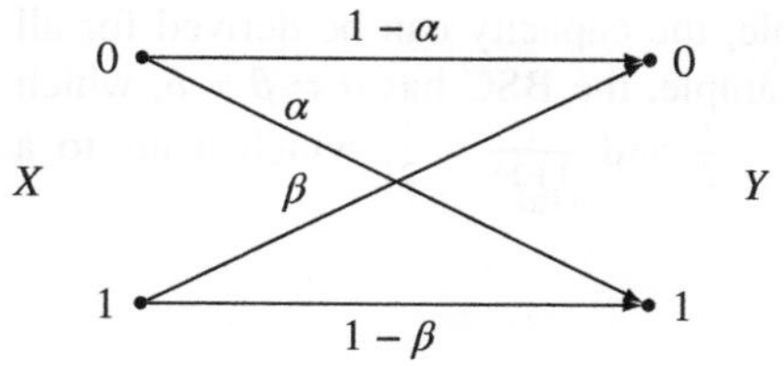

Figure 6.15 A graphical interpretation of a BSC.

Example 6.10 [Binary asymmetric channel] In a binary transmission scheme, the received symbol Y is estimated based on detecting the transmitted signal in the presence of noise from the transmission. If everything is tuned and the system is working properly, the thresholds in the receiver should be such that the error probability is symmetric, which is modeled by a BSC. If, for some reason, everything is not as it should, there might be an asymmetry in the error probability and the crossover probability for 0 and 1 are not equal, a binary asymmetric channel can be modeled. Assume the error probabilities are $P(Y = 1|X = 0) = \alpha$ and $P(Y = 0|X = 1) = \beta$. In Figure 6.15, the graphical view of such channel is shown.

To derive the capacity for this channel start by assuming a probability for X, so lets assign $P(X = 0) = p$. Then the probability for Y is given by

$$P(Y = 0) = p(1 - \alpha) + (1 - p)\beta = p(1 - \alpha - \beta) + \beta \tag{6.104}$$

$$P(Y = 1) = p\alpha + (1 - p)(1 - \beta) = (1 - p)(1 - \alpha - \beta) + \alpha \tag{6.105}$$

The mutual information can be written as

$$\begin{aligned} I(X;Y) &= H(Y) - H(Y|X) \\ &= h\Big(p(1 - \alpha - \beta) + \beta\Big) - ph(\alpha) - (1 - p)h(\beta) \end{aligned} \tag{6.106}$$

Using that $\frac{\partial}{\partial x}h(x) = \log\frac{1-x}{x}$, the optimizing p is given by the following equation:

$$\frac{\partial}{\partial p}I(X;Y) = (1 - \alpha - \beta)\log\frac{(1 - p)(1 - \alpha - \beta) + \alpha}{p(1 - \alpha - \beta) + \beta} - h(\alpha) + h(\beta) = 0 \tag{6.107}$$

To simplify notations, the variable $A = \frac{h(\alpha)-h(\beta)}{1-\alpha-\beta}$ is introduced and the optimizing p is given by

$$p^* = \frac{1 - \beta\big(1 + 2^A\big)}{(1 - \alpha - \beta)\big(1 + 2^A\big)} \tag{6.108}$$

The argument of the binary entropy function in the expression of $I(X;Y)$ is then $p^*(1 - \alpha - \beta) + \beta = \frac{1}{1+2^A}$ and the capacity becomes

$$C = I(X;Y)\Big|_{p=p^*} = h\Big(\frac{1}{1 + 2^A}\Big) + p^*h(\alpha) + (1 - p^*)h(\beta) \tag{6.109}$$

With the results in the previous example, the capacity can be derived for all DMC with binary inputs and outputs. For example, the BSC has $\alpha = \beta = \delta$, which gives $A = 0$, and consequently $p^* = \frac{1-2\delta}{(1-2\delta)2} = \frac{1}{2}$ and $\frac{1}{1+2^A} = \frac{1}{2}$, which leads to a capacity of $C = h(\frac{1}{2}) - h(\delta) = 1 - h(\delta)$.

PROBLEMS

6.1 Consider a binary memoryless source where $P(0) = p$ and $P(1) = q = 1 - p$. For large n, the number of 1s in a sequence of length n tends to nq.

(a) How many sequences of length n has the number of ones equal to nq?

(b) How many bits per source symbol is required to represent the sequences in (a).

(c) Show that as $n \to \infty$ the number bits per source symbol required to represent the sequences in (a) equals the entropy, $h(q) = h(p)$.

Hint: Use Stirling's formula to approximate $n! \approx \sqrt{2\pi n}(\frac{n}{e})^n$.

6.2 Show that for all jointly ε-typical sequences, $(\boldsymbol{x}, \boldsymbol{y}) \in A_\varepsilon(X, Y)$,

$$2^{-n(H(X|Y)+2\varepsilon)} \leq p(\boldsymbol{x}|\boldsymbol{y}) \leq 2^{-n(H(X|Y)-2\varepsilon)}$$

6.3 A binary memoryless source with $P(X = 0) = \frac{49}{50}$ and $P(X = 1) = \frac{1}{50}$ generates vectors of length $n = 100$. Let $\varepsilon = \frac{1}{50}\log 7$.

(a) What is the probability for the most probable vector?

(b) Is the most probable vector ε-typical?

(c) How many ε-typical vectors are there?

6.4 A string is 1 m long. It is split in two pieces where one is twice as long as the other. With probability 3/4 the longest part is saved and with probability 1/4 the short part is saved. Then, the same split is done with the saved part, and this continues the same way with a large number of splits. How large share of the string is in average saved at each split during a long sequence of splits?

Hint: Consider the distribution of saved parts for the most common type of sequence.

6.5 In Shannon's original paper from 1948, the following discrete memoryless channels are given. Calculate their channel capacities.

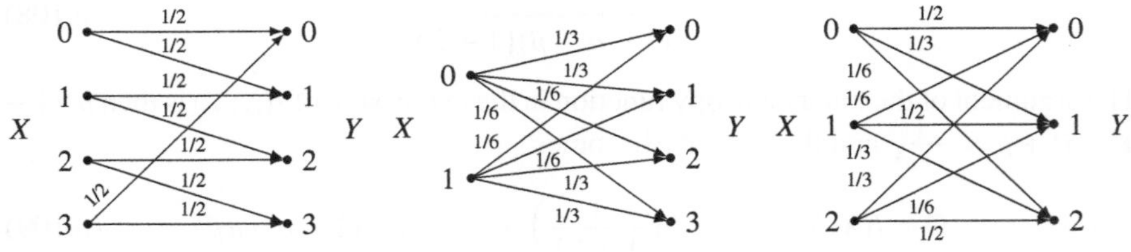

(a) Noisy typewriter (b) Soft decoding (c) 3-ary channel

6.6 Calculate the channel capacity for the extended binary erasure channel shown below.

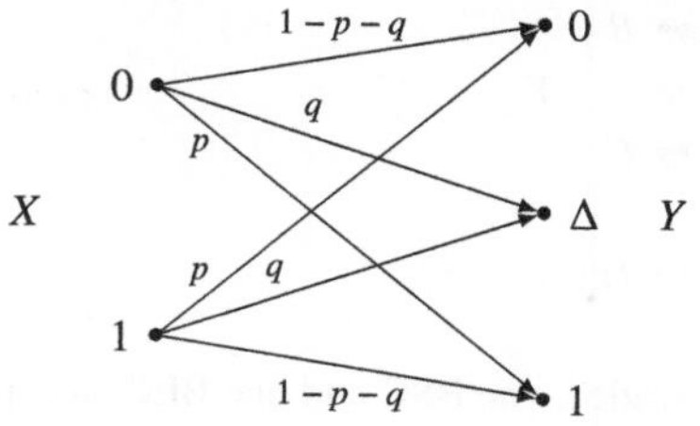

6.7 Determine the channel capacity for the following channel.

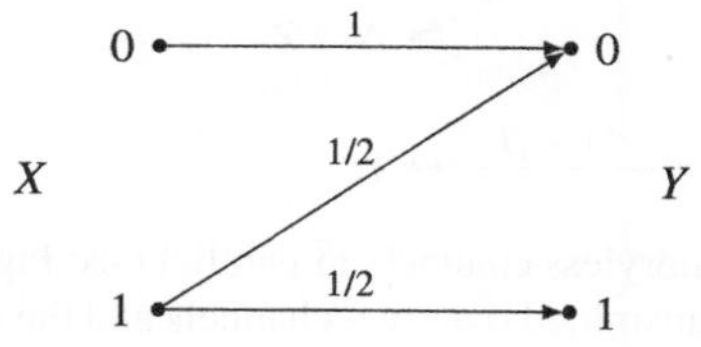

Hint: $D(h(p)) = D(p)\log(\frac{1-p}{p})$.

6.8 The random variable $X \in \{0, 1, \dots, 14\}$ is transmitted over an additive channel,

$$Y = X + Z, \qquad \text{mod } 15$$

where $p(Z = 1) = p(Z = 2) = p(Z = 3) = \frac{1}{3}$. What is the capacity for the channel and for what distribution $p(x)$ is it reached?

6.9 Cascade two binary symmetric channels as in the following picture. Determine the channel capacity.

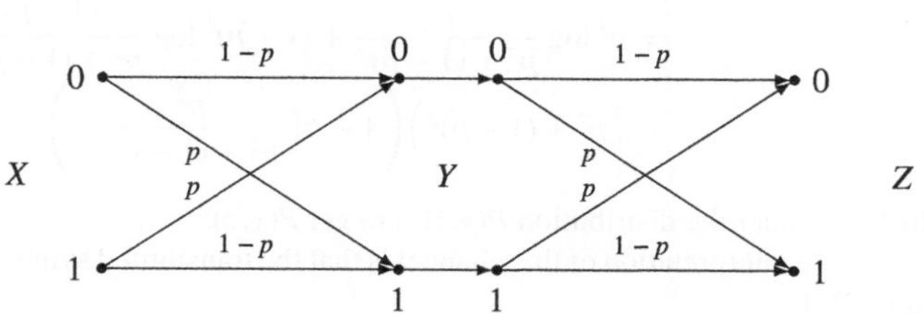

6.10 A discrete memoryless channel is shown in Figure 6.16.

(a) What is the channel capacity and for what distribution on X is it reached?

(b) Assume that the probability for X is given by $P(X = 0) = 1/6$ and $P(X = 1) = 5/6$, and that the source is memoryless. Find an optimal code to compress the sequence Y. What is the average codeword length?

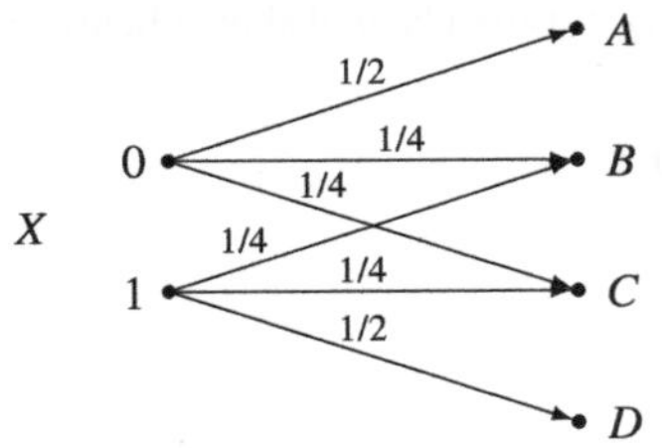

Figure 6.16 A discrete memoryless channel.

6.11 Two channels are cascaded, one BSC and one BEC, according to Figure 6.17. Derive the channel capacity.

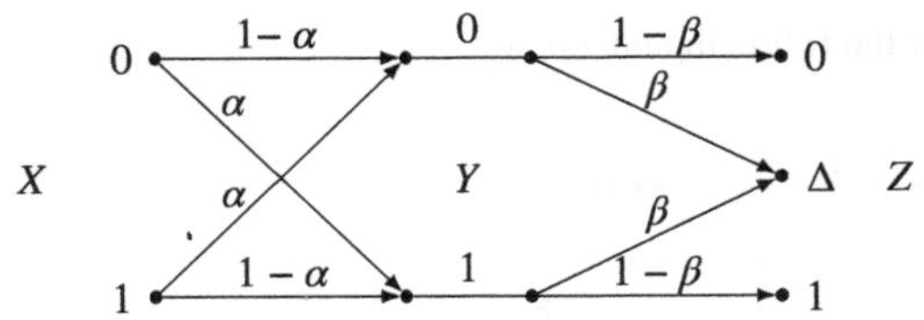

Figure 6.17 One BSC and one BEC in cascade.

6.12 Use two discrete memoryless channels in parallel (see Figure 6.18). That is, the transmitted symbol X is transmitted over two channels and the receiver gets two symbols, Y and Z. The two channels work independently in parallel, i.e. $P(y|x)$ and $P(z|x)$ are independent distributions, and hence, $P(y|x,z) = P(y|x)$ and $P(z|x,y) = P(z|x)$). However, it does not mean Y and Z are independent, so in general $p(y|z) \neq p(y)$.

(a) Consider the information about X the receiver gets by observing Y and Z, and show that

$$I(X;Y,Z) = I(X;Y) + I(X;Z) - I(Y;Z)$$

(b) Consider the case when both the channels are BSC with error probability p, and let $P(X=0) = P(X=1) = \frac{1}{2}$. Use the result in a and show that

$$\begin{aligned} I(X;Y,Z) &= H(Y,Z) - 2h(p) \\ &= p^2 \log \frac{2p^2}{p^2+(1-p)^2} + (1-p)^2 \log \frac{2(1-p)^2}{p^2+(1-p)^2} \\ &= \left(p^2+(1-p)^2\right)\left(1 - h\left(\frac{p^2}{p^2+(1-p)^2}\right)\right) \end{aligned}$$

Hint: Consider the distribution $P(y,z|x)$ to get $P(y,z)$.

One interpretation of this channel is that the transmitted symbol is sent twice over a BSC.

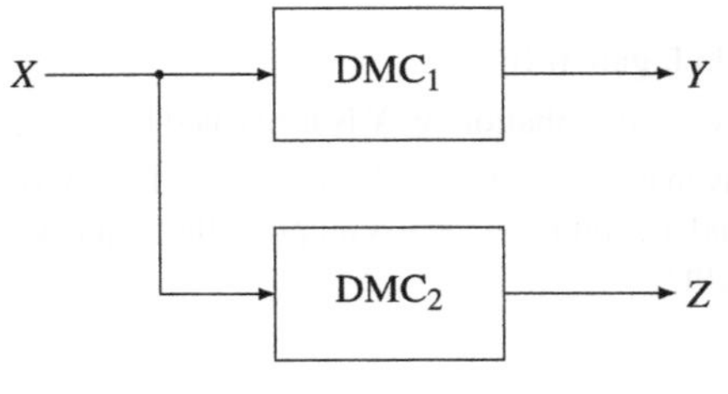

Figure 6.18 Two DMC used in parallel.

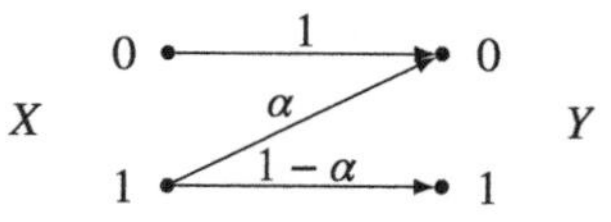

Figure 6.19 One BSC and one BEC in cascade.

6.13 In Figure 6.19, a general Z-channel is shown. Plot the capacity as a function of the error probability α.

6.14 In Figure 6.20, a discrete memoryless channel is given.

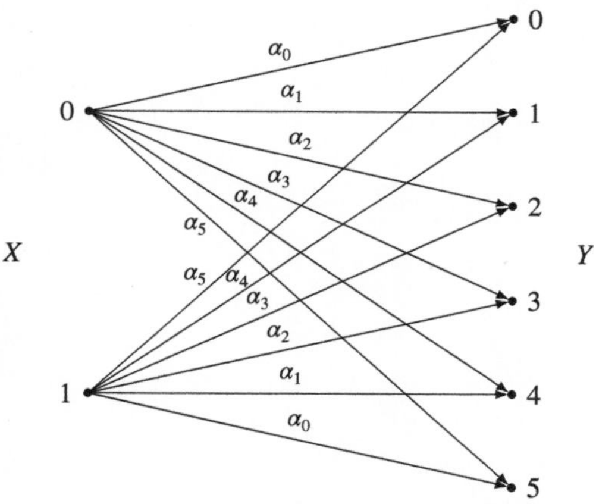

Figure 6.20 A channel and its probability function.

(a) Show that the maximizing distribution giving the capacity

$$C_6 = \max_{p(x)} I(X;Y)$$

is given by $P(X=0) = \frac{1}{2}$ and $P(X=1) = \frac{1}{2}$.
Verify that the capacity is given by

$$C_6 = 1 + H(\alpha_0 + \alpha_5, \alpha_1 + \alpha_4, \alpha_2 + \alpha_3) - H(\alpha_0, \alpha_1, \alpha_2, \alpha_3, \alpha_4, \alpha_5)$$

(b) Split the outputs in two sets, $\{0, 1, 2\}$ and $\{3, 4, 5\}$ and construct a binary symmetric channel (BSC) with error probability $p = \alpha_3 + \alpha_4 + \alpha_5$. Denote the capacity of the corresponding BSC as C_{BSC} and show that

$$C_{\text{BSC}} \le C_6 \le 1$$

where C_6 is the capacity of the channel in Figure 6.20.

CHAPTER 7

CHANNEL CODING

IN THE PREVIOUS chapter, the capacity for a discrete memoryless channels was introduced as the maximum information that can be transmitted over it. Through the channel coding theorem, this also gives the requirement on the code rate to be able to achieve reliable communication. That is, for each code rate lower than the capacity there exists a code such that the decoding error probability approaches zero. The question now is how these codes should be designed. Even though the channel coding theorem is based on random coding, which suggests most codes are good as long as they are long enough, this question has defied the research community ever since Shannon published his results. Especially finding practically usable codes where the rate is approaching the capacity is a difficult task.

The problem is that to reach the capacity the codeword length will tend to be infinity and in most cases the computational complexity for decoding grows close to exponentially with it. The most promising codes today are the so-called low-density parity check (LDPC) codes, and there are examples where the rate is very close to the capacity limit for a low decoding error probability. One main advantage with these codes is that the decoding complexity grows only linearly with the length.

In this chapter, the principal idea of channel coding will be described. This will be done by considering some well-known codes representing the two main code families: block codes and convolutional codes. Then codes can be combined, or concatenated, in different ways to build larger and more efficient codes by using small constituent codes. The main part of the chapter will be devoted to error-correcting codes, but there is also a section on error-detection codes, represented by cyclic redundancy check (CRC) codes.

In a communication system, after the source encoding the redundancy of the data should be, if not removed, so at least significantly reduced. Then the information symbols are close to uniformly distributed. On the other hand, during the transmission over the channel there will be disturbances that will either alter or erase some of the transmitted symbols. The idea with channel coding is to add redundancy in a controlled way such that it is possible to correct some of the errors at the receiver side.

Information and Communication Theory, First Edition. Stefan Höst.

7.1 ERROR-CORRECTING BLOCK CODES

Assuming the information symbols are taken from an infinite sequence, in a block code the encoding is performed by blocking the sequence into blocks, or vectors, of k symbols each. Then each block is individually encoded into a length n vectors. The first example of such code was already considered in Example 6.3, the repetition code. It is a very simple and intuitive code that describes the main idea of coding, recapitulated below.

Example 7.1 [Repetition code] Instead of transmitting each bit uncoded, transmit three copies of each bit. That is, the encoding rule is described in the following table:

u	x
0	000
1	111

The code has $2^k = 2$ codeword vectors of length $n = 3$, so the repetition code is a $(n, k) = (3, 1)$ code. The coding rate is $R = \frac{1}{3}$, meaning there are on average $1/3$ bit information in each code bit. It was seen in Example 6.3 that the code can correct 1 bit error occurred on the channel.

The repetition code is the simplest example that shows it is possible to correct channel errors. However, it is not a very efficient way to do this, and the remaining error probability after decoding of the sequence is not really improved. There are much more efficient ways to so this, and in this chapter some of the ideas will be introduced. Still, to get codes that can be realized in real communication systems a good book in coding theory is recommended, e.g., [33, 34, 35, 36, 70, 90].

A code is defined as the set of codewords, and not dependent on the mapping from the information words. In this view, there is a clear distinction in the way source codes and channel codes are defined. The next example introduces a code with four codewords.

Example 7.2 The block code $\mathcal{B}_1 = \{0000, 1011, 0110, 1101\}$ has four codewords, meaning $k = \log 4 = 2$. There are four bits in each codewords, so $n = 4$. Hence, it is an $(n, k) = (4, 2)$ code and the code rate is $R = \frac{2}{4}$.

It is desirable to have a linear encoding rule, which put constraints on the codewords. Before continuing to the mapping used by the encoder, a linear code is defined.

Definition 7.1 A code $\mathcal{B}$ is *linear* if for every pair of codewords, $\boldsymbol{x}_i$ and $\boldsymbol{x}_j$, their sum is also a codeword,

$$\boldsymbol{x}_i, \boldsymbol{x}_j \in \mathcal{B} \Rightarrow \boldsymbol{x}_i + \boldsymbol{x}_j \in \mathcal{B} \tag{7.1}$$

where the addition denotes positionswise addition modulo 2. □

Here, and in the sequel of this chapter, it is assumed that the code is binary. There are, however, very powerful and widely used codes derived over higher order fields, e.g., the class of Reed–Solomon codes that are used in many applications, e.g., CD and DVD, as well as the communication standards asymmetric digital subscriber line (ADSL) [37] and very high bit rate digital subscriber line (VDSL) [38].

It can directly be seen that the repetition code is linear. Also the code $\mathcal{B}_1$ in the previous example can easily be verified, by viewing the addition between all pairs of codewords, to be linear.

A codeword added to itself is the all-zero vector, $\boldsymbol{x} + \boldsymbol{x} = \boldsymbol{0}$. Thus, the all-zero vector is a codeword in all linear codes, $\boldsymbol{0} \in \mathcal{B}$.

From algebra, it is known that a binary linear (n, k) code $\mathcal{B}$ spans a k-dimensional subspace of the binary n-dimensional space $\mathbb{F}_2^n$. Then each codeword is a linear combination of k linearly independent codewords, $\boldsymbol{g}_1, \dots, \boldsymbol{g}_k$, where $\boldsymbol{g}_i \in \mathcal{B}$. Since the k codewords are linearly independent, all different linear combinations of them will give different codewords as results. Using the binary vector $\boldsymbol{u} = (u_1, \dots, u_k)$ as coefficients for the linear combinations, 2^k codewords can be derived as

$$\begin{aligned}
\boldsymbol{x} &= u_1\boldsymbol{g}_1 \oplus \cdots \oplus u_k\boldsymbol{g}_k \\
&= (u_1 \dots u_k)\begin{pmatrix} \boldsymbol{g}_1 \\ \vdots \\ \boldsymbol{g}_k \end{pmatrix} \\
&= \underbrace{(u_1 \dots u_k)}_{\boldsymbol{u}} \underbrace{\begin{pmatrix} g_{11} & \cdots & g_{1n} \\ \vdots & & \vdots \\ g_{k1} & \cdots & g_{kn} \end{pmatrix}}_{G}
\end{aligned} \tag{7.2}$$

The above equation can be used as an encoding rule for the (n, k) linear code, where $\boldsymbol{u}$ is the information word and $\boldsymbol{x}$ the codeword. The matrix G is named the *generator matrix* for the code and determines the mapping. Of course, by choosing another order or another set of the codewords in the generator matrix, the mapping will be altered, but the code, i.e. the set of codewords, will be the same.

Example 7.3 In the code $\mathcal{B}_1$, the two codewords $\boldsymbol{x}_1 = 0110$ and $\boldsymbol{x}_2 = 1101$ are linearly independent. Therefore, the generator matrix can be formed as

$$G = \begin{pmatrix} 0 & 1 & 1 & 0 \\ 1 & 1 & 0 & 1 \end{pmatrix} \tag{7.3}$$

The mapping between the information words and the codewords then becomes

$\boldsymbol{u}$	$\boldsymbol{x} = \boldsymbol{u}G$
00	0000
01	1101
10	0110
11	1011

Assuming the codewords are transmitted over a binary symmetric channel (BSC) with error probability $p \ll 0.5$, one error in a codeword is the most probable error event. After that comes two errors, and so on. One direct first decoding rule is to choose the codeword that is the most likely to be sent, conditioned on the received word. This decoding rule is called the *maximum a posteriori* (MAP) decoder and is stated as

$$\hat{\boldsymbol{x}} = \arg\max_{\boldsymbol{x}\in\mathcal{B}} P(\boldsymbol{x}|\boldsymbol{y}) \tag{7.4}$$

Earlier, the case was considered when the source coding, e.g., Huffman coding, preceded the channel coding. Then it is reasonable to assume that all codewords, or information words, are equally likely, $P(\boldsymbol{x}) = 2^{-k}$. The MAP rule can then be expanded according to

$$\hat{\boldsymbol{x}} = \arg\max_{\boldsymbol{x}\in\mathcal{B}} P(\boldsymbol{x}|\boldsymbol{y}) = \arg\max_{\boldsymbol{x}\in\mathcal{B}} P(\boldsymbol{y}|\boldsymbol{x})\frac{P(\boldsymbol{x})}{P(\boldsymbol{y})} = \arg\max_{\boldsymbol{x}\in\mathcal{B}} P(\boldsymbol{y}|\boldsymbol{x}) \tag{7.5}$$

since $\frac{P(\boldsymbol{x})}{P(\boldsymbol{y})}$ can be considered as a constant, i.e., $p(\boldsymbol{x}) = 2^{-k}$ and $\boldsymbol{y}$ is the received vector and thus not varying. This decoding rule is called *maximum likelihood* (ML) decoding.

For a BSC, both the transmitted and received vectors are binary. Then the number of errors is the same as the number of positions in which they differ. Intuitively, the decoding then is the same as finding the codeword that differs from $\boldsymbol{y}$ in least positions. It will be useful to first define the Hamming distance between two vectors as the number of positions in which they differ. A closely related function is the Hamming weight, as stated in the next definition [79].

Definition 7.2 The *Hamming distance*, $d_H(\boldsymbol{x},\boldsymbol{y})$, between two vectors $\boldsymbol{x}$ and $\boldsymbol{y}$, is the number of positions in which they differ. The *Hamming weight*, $w_H(\boldsymbol{x})$, of a vector $\boldsymbol{x}$, is the number of nonzero positions of $\boldsymbol{x}$. □

For binary vectors, the Hamming distance can be derived from the Hamming weight as

$$d_H(\boldsymbol{x},\boldsymbol{y}) = w_H(\boldsymbol{x}+\boldsymbol{y}) \tag{7.6}$$

where the addition is taken positionwise in the vectors. The equality follows since addition and subtraction are identical over the binary field, $a + a = 0$, mod 2.

Example 7.4 The Hamming distance between the vectors 0011010 and 0111001 is

$$d_H(0\underline{0}11\underline{01}0, 0\underline{1}11\underline{00}1) = 3 \tag{7.7}$$

It can also be derived as the weight of the difference,

$$w_H(0011010 + 0111001) = w_H(0100011) = 3 \tag{7.8}$$

It should be noted that the Hamming distance is a metric. That means it is a function such that

- $d_H(\boldsymbol{x},\boldsymbol{y}) \geq 0$ with equality if and only if $\boldsymbol{x} = \boldsymbol{y}$ (nonnegative).
- $d_H(\boldsymbol{x},\boldsymbol{y}) = d_H(\boldsymbol{y},\boldsymbol{x})$ (symmetry).
- $d_H(\boldsymbol{x},\boldsymbol{y}) + d_H(\boldsymbol{y},\boldsymbol{z}) \geq d_H(\boldsymbol{x},\boldsymbol{z})$ (triangular inequality).

It follows directly from the definition of the Hamming distance that the first two conditions hold. To show the third condition, consider three vectors $\boldsymbol{x}$, $\boldsymbol{y}$, and $\boldsymbol{z}$. Then, to go from vector $\boldsymbol{x}$ to vector $\boldsymbol{y}$ there are $d_H(\boldsymbol{x},\boldsymbol{y})$ positions needed to be changed, and to go from vector $\boldsymbol{y}$ to vector $\boldsymbol{z}$ there are $d_H(\boldsymbol{x},\boldsymbol{y})$ positions need to be changed. Then there might be some positions that are equal in $\boldsymbol{x}$ and $\boldsymbol{z}$, but not in $\boldsymbol{y}$. In that case, first it must be changed when going from $\boldsymbol{x}$ to $\boldsymbol{y}$ and then changed back when going from $\boldsymbol{y}$ to $\boldsymbol{z}$, which gives the result. This means that the Hamming distance can be viewed as a distance between two points in an n-dimensional space.

Going back to the ML decoding criterion, instead of maximizing the probability, the logarithm of the probability is maximized. Since the logarithm is a strictly increasing function, this does not change the result.

$$\begin{aligned}\hat{\boldsymbol{x}} &= \arg\max_{\boldsymbol{x}\in\mathcal{B}} \log P(\boldsymbol{y}|\boldsymbol{x}) \\ &= \arg\max_{\boldsymbol{x}\in\mathcal{B}} \log \prod_i P(y_i|x_i)\end{aligned} \tag{7.9}$$

where it is assumed that the errors on the channel occur independent of each other, i.e., the BSC. Then each transmitted bit is inverted with probability p. Hence, if the codeword $\boldsymbol{x}$ is transmitted and the vector $\boldsymbol{y}$ received there are $d_H(\boldsymbol{x},\boldsymbol{y})$ errors in the transmission and $n - d_H(\boldsymbol{x},\boldsymbol{y})$ positions with no errors. The decoding criteria can therefor be written as

$$\begin{aligned}\hat{x} &= \arg\max_{x\in B} \log\Big(p^{d_H(x,y)}(1-p)^{n-d_H(x,y)}\Big)\\ &= \arg\max_{x\in B} d_H(x,y)\log p + (n - d_H(x,y))\log(1-p)\\ &= \arg\max_{x\in B} d_H(x,y)\log\frac{p}{1-p} + n\log 1 - p = \arg\min_{x\in B} d_H(x,y)\end{aligned} \quad (7.10)$$

where in the last equality it is assumed that $p < \frac{1}{2}$, or equivalently, that $\log\frac{p}{1-p} < 0$ which gives a minimum instead of maximum. This decoding rule is called *minimum distance* (MD) decoding. For a BSC, this is equivalent to an ML decoder and an attractive alternative. Below the three different decoding methods are summarized in one definition.

Theorem 7.1 A decoder that receives the vector y estimates the transmitted codeword according to

- the MAP decoder

$$\hat{x} = \arg\max_{x\in B} P(x|y) \quad (7.11)$$

- the ML decoder

$$\hat{x} = \arg\max_{x\in B} P(y|x) \quad (7.12)$$

  when the codewords are equally likely, this is equivalent to MAP decoding.

- the MD decoder

$$\hat{x} = \arg\min_{x\in B} d_H(y,x) \quad (7.13)$$

  when the codewords are transmitted aver a BSC, this is equivalent to ML decoding.

□

The MAP decoding rule is the most demanding seen from computational complexity in the receiver, since it has to take the a priori probabilities into account.

One important consequence of the MD decision rule is that the separation in Hamming distance between the codewords gives a measure of the error correcting capability of a code. That is, if the distance from one codeword to the nearest other codeword is large, there will be room for more errors in between. Therefore, the MD between two codewords is an important measure of how good a code is.

Definition 7.3 The *MD* for a code B is the minimum Hamming distance between two different codewords,

$$d_{\min} = \min_{\substack{x_1,x_2\in B\\ x_1\neq x_2}} d_H(x_1,x_2) \quad (7.14)$$

□

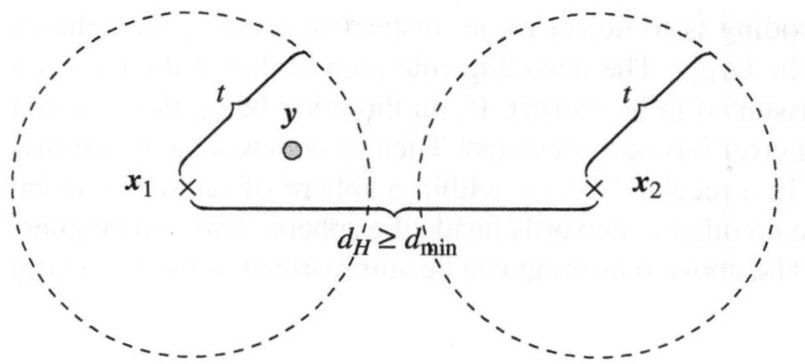

Figure 7.1 Two codewords, x_1 and x_2, in the n-dimensional binary space, projected into two dimensions.

Since, for a linear code the sum of two codewords is again a codeword, and that the Hamming distance can be derived as the Hamming weight of the sum, the MD for a linear code can be derived as

$$d_{\min} = \min_{\substack{x \in \mathcal{B} \\ x \neq 0}} w_H(x) \tag{7.15}$$

Example 7.5 The MD for the repetition code is

$$d_{\min}^{(\text{Rep})} = \min\{w_H(111)\} = 3 \tag{7.16}$$

and for $\mathcal{B}_1$

$$d_{\min}^{(1)} = \min\{w_H(1101), w_H(0110), w_H(1011)\} = \min\{3, 2, 3\} = 2 \tag{7.17}$$

The codewords are n-dimensional vectors, and, hence, they can be viewed as 2^k points in an n-dimensional binary space. In Figure 7.1, two codewords, x_1 and x_2, in the n-dimensional space are shown (projected in two dimensions). After transmission of a codeword over a BSC, the possible received vectors are represented by all points in the space. The Hamming distance between two codewords is at least $d_{\min}$, and the decoding rule is assumed to be the MD decoding. Then, surrounding each codeword with a sphere of radius t bits, such that there are no vectors in the sphere that is in more than one sphere. All received sequences within such sphere should be decoded to the codeword in its center. For example, the received vector y in the figure is closest to the codeword x_1, which should be the estimate of the transmitted codeword, $\hat{x} = x_1$. Since it is a discrete space and there must be no overlapping of the spheres, t must satisfy

$$2t + 1 \leq d_{\min} \tag{7.18}$$

or, in other words

$$t \leq \frac{d_{\min} - 1}{2} \tag{7.19}$$

In this view, t is the number of errors that can be corrected, which should be as large as possible.

If the aim of the decoding is to detect errors instead of correct, the spheres around the codewords can be larger. The decoding rule then is that if the received vector is a codeword it is assumed to be correct. If, on the other hand, the received vector is not a codeword an error has been detected. Then, if codeword $\boldsymbol{x}$ was transmitted, all errors resulting in a received vector within a sphere of radius t will be detected as long as there are no other codewords inside the sphere. This can be guaranteed as long as $t < d_{\min}$. The above reasoning can be summarized in the following theorem.

Theorem 7.2 When using a code with minimum distance $d_{\min}$, it is always possible to either *detect* an error $\boldsymbol{e}$ if

$$w_H(\boldsymbol{e}) < d_{\min} \tag{7.20}$$

or *correct* an error $\boldsymbol{e}$ if

$$w_H(\boldsymbol{e}) \leq \frac{d_{\min} - 1}{2} \tag{7.21}$$

□

Example 7.6 The repetition code has $d_{\min} = 3$ and can therefore either detect two errors or correct one error. The code $\mathcal{B}_1$ has $d_{\min} = 2$ and can detect all errors of weight one, but there is no guarantee of correcting errors. For example if the received vector is $\boldsymbol{y} = 1001$, this differs 1 bit from both the codewords 1101 and 1011. Since there are two codewords that are closest, the probability is $\frac{1}{2}$ to choose the correct one.

7.1.1 Hamming Codes

The codes considered so far, the repetition code and $\mathcal{B}_1$, are not very efficient in the transmission. Next, a slightly larger code will be considered, that will give some more insight in how the decoding can be performed. As seen in the previous section for a linear code, it is possible to find a generator matrix G such that the mapping between information word $\boldsymbol{u}$ and the codeword $\boldsymbol{x}$ is $\boldsymbol{x} = \boldsymbol{u}G$. The generator matrix with dimensions $k \times n$ spans a k-dimensional subspace of the binary n-dimensional space $\mathbb{F}_2^n$. Then the null space of it is spanned by the *parity check matrix H* of size $n - k \times n$, defined by

$$GH^T = 0 \tag{7.22}$$

where ^{T} denotes the matrix transpose. Since a length n vector $\boldsymbol{x}$ is a codeword if and only if $\boldsymbol{u}G$ for some vector $\boldsymbol{u}$, then

$$\boldsymbol{x}H^T = \boldsymbol{u}GH^T = \boldsymbol{u}0 = 0 \tag{7.23}$$

This result gives the following theorem.

Theorem 7.3 Let H be a parity check matrix for an (n, k) linear block code. Then an n-dimensional vector $\boldsymbol{x}$ is a codeword if and only if

$$\boldsymbol{x}H^T = 0 \tag{7.24}$$

□

Since the MD of a linear code is the minimum weight of a nonzero codeword, the parity check matrix can be used to find it. The requirements in (7.24) means that using the coefficients in $\boldsymbol{x} = (x_1 \dots x_n)$ the linear combination of the columns in H equals zero. For a codeword of least weight, this corresponds to a linear combination of $d_{\min}$ columns of H giving a zero. It also means there is no linear combination of less number of columns summing to zero. Since that would represent a codeword with weight less than $d_{\min}$.

Theorem 7.4 The minimum distance $d_{\min}$ of a linear code equals the minimum number of linearly dependent columns in the parity check matrix H. □

Theorems 7.3 and 7.4 give a way to design a code for a given $d_{\min}$ through the parity check matrix. The *Hamming code* of order m is defined from a parity check matrix containing all nonzero vectors of length m as columns. The length of the codewords is the number of columns in H, i.e. $n = 2^m - 1$, and since $m = n - k$ the length of the information vector is $k = 2^m - 1 - m$. That is, the Hamming code of order m is a $(2^m - 1, 2^m - 1 - m)$ binary linear code with the rate

$$R = \frac{2^m - 1 - m}{2^m - 1} \tag{7.25}$$

Since all columns of H are different, there are not two linearly dependent columns. On the other hand, it is easy to find three columns that are linearly dependent, e.g., $0 \dots 001 + 0 \dots 010 + 0 \dots 011 = 0 \dots 000$. That is, the minimum number of linearly dependent columns is 3, and thus the minimum distance of a Hamming code is $d_{\min} = 3$.

Example 7.7 A Hamming code of order $m = 3$ is a $(7, 4)$ code. The parity check matrix is formed by all nonzero (column-)vectors of length $m = 3$ in some order, e.g.,

$$H = \begin{pmatrix} 1 & 0 & 0 & 0 & 1 & 1 & 1 \\ 0 & 1 & 0 & 1 & 0 & 1 & 1 \\ 0 & 0 & 1 & 1 & 1 & 0 & 1 \end{pmatrix} \tag{7.26}$$

The codewords can be found from extensive search by using $xH^T = 0$,

x	w_H	x	w_H	x	w_H	x	w_H
0000000	0	0100101	3	1000111	4	1100010	3
0001110	3	0101011	4	1001001	3	1101100	4
0010011	3	0110110	4	1010100	3	1110001	4
0011101	4	0111000	3	1011010	4	1111111	7

From the table of codewords, it is easily verified that $d_{\min} = \min_{x \neq 0} w_H(x) = 3$. Hence, this code can correct all single errors that occur on the channel.

From the parity check matrix, it is possible to check if a vector is a codeword and thereby list all codewords. The next step is to determine a mapping between an information vector and a code vector. It is often desirable to use a linear mapping, and therefore a suitable generator matrix should be found. Naturally, this can be formed by choosing k linearly independent codewords. Since the parity check matrix for a Hamming code consists of all nonzero vectors, it is possible to order them such that the unit matrix can be found in H. That is, collect all weight one vectors such that they form the unit matrix I. In the example given above for the Hamming code of order $m = 3$, the unit matrix is the three leftmost columns. The remaining of the matrix is denoted by P^T. Hence, the parity check matrix can be chosen as

$$H = (I \quad P^T) \tag{7.27}$$

where I is a unit matrix of size m and P^T an $m \times 2^m - 1 - m$ binary matrix. A generator matrix can be found from

$$G = (P \quad I) \tag{7.28}$$

where P is the transpose of P^T and I a unit matrix of size $2^m - 1 - m$. Since over the binary field addition and subtraction are equivalent,

$$GH^T = (P \quad I)\begin{pmatrix} I \\ P \end{pmatrix} = P + P = 0 \tag{7.29}$$

which concludes that G is a valid generator matrix.

Example 7.8 In the parity check matrix for the (7, 4) Hamming code identify

$$P^T = \begin{pmatrix} 0 & 1 & 1 & 1 \\ 1 & 0 & 1 & 1 \\ 1 & 1 & 0 & 1 \end{pmatrix} \tag{7.30}$$

which gives the generator matrix

$$G = \begin{pmatrix} 0 & 1 & 1 & 1 & 0 & 0 & 0 \\ 1 & 0 & 1 & 0 & 1 & 0 & 0 \\ 1 & 1 & 0 & 0 & 0 & 1 & 0 \\ 1 & 1 & 1 & 0 & 0 & 0 & 1 \end{pmatrix} \tag{7.31}$$

Then, by using $\boldsymbol{x} = \boldsymbol{u}G$, the mapping between information and code vectors is determined according to the following table:

$\boldsymbol{u}$	$\boldsymbol{x}$	$\boldsymbol{u}$	$\boldsymbol{x}$	$\boldsymbol{u}$	$\boldsymbol{x}$	$\boldsymbol{u}$	$\boldsymbol{x}$
0000	0000000	0100	1010100	1000	0111000	1100	1101100
0001	1110001	0101	0100101	1001	1001001	1101	0011101
0010	1100010	0110	0110110	1010	1011010	1110	0001110
0011	0010011	0111	1000111	1011	0101011	1111	1111111

Notice that since the generator matrix has the identity matrix as the last part, the four last digits of the codeword constitute the information word.

Assuming the codewords are sent over a BSC, an optimal decoder can be constructed as an MD decoder. The errors on a BSC can be viewed as an additive error vector, where an error is represented by a 1 and no error by a 0. Then, the received vector is

$$\boldsymbol{y} = \boldsymbol{x} + \boldsymbol{e} \tag{7.32}$$

For example, if the codeword $\boldsymbol{x} = (0110110)$ is transmitted and there are errors at the fourth and sixth positions, the error vector is $\boldsymbol{e} = (0001010)$. The received vector is $\boldsymbol{y} = (0110110) + (0001010) = (0111100)$.

By using the parity check matrix, it should be noticed that

$$\boldsymbol{y}H^T = (\boldsymbol{x} + \boldsymbol{e})H^T = \boldsymbol{x}H^T + \boldsymbol{e}H^T = \boldsymbol{e}H^T \tag{7.33}$$

since $\boldsymbol{x}$ is a codeword. That means there is a way to obtain a fingerprint function from the error pattern occurred on the channel, that is independent of the codeword. This is defined as the syndrome of the received vector.

Definition 7.4 (Syndrome) Let $\boldsymbol{x}$ be a codeword and H the corresponding parity check matrix. Then if the received vector is $\boldsymbol{y} = \boldsymbol{x} + \boldsymbol{e}$, the *syndrome* is formed as

$$\boldsymbol{s} = \boldsymbol{y}H^T = \boldsymbol{e}H^T. \tag{7.34}$$

□

It is now possible to make a table that maps the syndrome to the least weight error vectors and subtract it from the received vector. By considering the syndrome of the received vector and mapping to the least weight error vector, an estimate of the error vector $\hat{e}$ can be achieved. The estimated codeword is

$$\hat{x} = y + \hat{e} = x + e + \hat{e} \tag{7.35}$$

Naturally, if a correct decoding is done, $\hat{e} = e$, and $\hat{x} = x$. Since the most probable error vector has the least weight, this is the most probable event and the estimated codeword is identical to the transmitted. A *syndrome table* consisting of all possible syndromes mapped to each least weight error vector gives an easy to use mapping from the syndrome to the estimated error vector.

Example 7.9 For the (16, 7) Hamming code the syndrome table is

e	$s = eH^T$	e	$s = eH^T$
0000000	000	0001000	011
1000000	100	0000100	101
0100000	010	0000010	110
0010000	001	0000001	111

Then, by forming the syndrome for a received vector an estimate of the corresponding least weight error vector can be found by a table lookup function. With the estimated error vector, $\hat{e}$, the corresponding most probable transmitted codeword is $\hat{x} = y + \hat{e}$.

Assume that the codeword $x = (0110110)$ is transmitted and that there is an error in the third bit, i.e. $e = (0010000)$. Then the received vector is $y = (0100110)$. The syndrome for this vector is $s = yH^T = (001)$, which translates to the estimated error vector $\hat{e} = (001000)$ and the estimated codeword $\hat{x} = (0110110)$. In the codeword, the last 4 bits equals the estimated information word, and hence $\hat{u} = (0110)$. Since there is a unique syndrome for each single error, this procedure will be able to correct all single errors.

If instead an error pattern with two errors is introduced, say $e = (0001010)$, the received vector is $y = (0111100)$. The syndrome becomes $s = (101)$, which according to the table corresponds to the error $\hat{e} = (0000100)$ and the estimated codeword is $\hat{x} = (0111000)$, which gives the information word $\hat{u} = (1000)$. Since this was not the transmitted information vector, a decoding error has occurred.

To conclude, the Hamming code can correct all single errors but it cannot correct double errors.

To summarize *syndrome decoding,* the steps are given in the following algorithm.

Algorithm 7.1 (Syndrome decoding)

Initialization

Form a syndrome table with the most probable (least Hamming weight) error patterns for each possible syndrome.

Decoding

1. For a received vector, $\boldsymbol{y}$, form the syndrome according to $\boldsymbol{s} = \boldsymbol{y}H^T$.
2. Use the syndrome table to look up the corresponding least weight error pattern $\hat{\boldsymbol{e}}$.
3. Get the most likely transmitted codeword as $\hat{\boldsymbol{x}} = \boldsymbol{y} + \hat{\boldsymbol{e}}$.
4. Derive the estimated information word $\hat{\boldsymbol{x}} \rightarrow \hat{\boldsymbol{u}}$.

7.1.2 Bounds on Block Codes

For the Hamming code, it is relatively easy to find the code parameters like the number of codewords, length, and minimum distance. However, it is not that easy to do this in the general case. Therefore, several bounds on the maximum number of codewords for a code of length n and minimum distance d has been developed. There are both upper and lower bounds, and in this description first three upper bound are described and then two lower bounds. The bounds will also be considered in the limit as the codeword length tends to infinity.

The Hamming code of order m has the codeword length $n = 2^m - 1$ and the minimum distance $d_{\min} = 3$. In the binary n-dimensional space, there are in total 2^n possible vectors. Group all vectors with the Hamming distance one from a codeword in a sphere. Then, since the minimum distance is three, all the spheres will be disjoint. In each sphere, there are $n + 1$ vectors, including the codeword in the center. An upper bound on the total number of codewords, M, is then the same as the total number of spheres, which is the total number of vectors divided by the number of vectors in a sphere,

$$M \leq \frac{2^n}{n+1} = \frac{2^{2^m-1}}{2^m} = 2^{2^m-1-m} \tag{7.36}$$

which is, in fact, equal to the number of codewords in the Hamming code.

In the general case, define a sphere of radius t around a codeword, containing all vectors with Hamming distance to the codeword not exceeding t. For a code with codeword length n and minimum distance d, the largest sphere around each codeword such that they are disjoint has radius $t = \lfloor \frac{d-1}{2} \rfloor$ and the total number of vectors in it is $\sum_{i=0}^{t} \binom{n}{i}$. The number of codewords can then be upper bounded by the total number of vectors divided by the number of vectors in the spheres as in the following theorem. The bound is often called the *Hamming bound* or the *sphere packing bound*.

Theorem 7.5 (Hamming bound) A binary code with codeword length n and minimum distance $d_{\min} = d$ can have at most M codewords, where

$$M \leq \frac{2^n}{\sum_{i=0}^{t} \binom{n}{i}}, \quad \text{where } t = \left\lfloor \frac{d-1}{2} \right\rfloor \tag{7.37}$$

□

In general, not all vectors are included in the spheres and the bound is not fulfilled with equality. In the case when there is equality, the code is called a *perfect code*. Apparently, from the above the Hamming code is such perfect code. Another example of a perfect code is the repetition code with an odd number of repetitions. The code $\mathcal{B}_1$ in the above, however, is an example of a code where there is not equivalence. Here the number vectors is 16 and the minimum distance 2. That means the spheres contains only the codewords, and the bound is $M_1 = 16$.

An important parameter for determining the efficiency of a code is the covering radius. This is the minimum radius of spheres centered around each codeword such that the complete n-dimensional space is covered. If there is equality in the Hamming bound, the covering radius is t, but in general it is larger. For the code $\mathcal{B}_1$, the covering radius is 1, while $t = 0$.

In the next bound, the maximum number of codewords will be derived. Denote by $M(n, d)$ the maximum number of codewords in a code with codeword length n and minimum distance d. Assume a code $\mathcal{B}$ with $M(n, d)$ codewords of length n and minimum distance d. Then by setting position n equal to 0 or 1, two new codes are created, $\mathcal{B}(0)$ and $\mathcal{B}(1)$, both with minimum distance d. Together they have $M(n, d)$ codewords, meaning one of them has at least $M(n, d)/2$ codewords, or

$$M(n, d) \leq 2M(n-1, d) \tag{7.38}$$

The observation $M(d, d) = 2$ together with the above gives the following induction formula:

$$M(n, d) = 2M(n-1, d) = \cdots = 2^{n-d}M(d, d) = 2^{n-d+1} \tag{7.39}$$

which is named the Singleton bound.

Theorem 7.6 (Singleton bound) A binary code with codeword length n and minimum distance $d_{\min} = d$ can have at most M codewords, where

$$M \leq 2^{n-d+1} \tag{7.40}$$

□

A code that fulfills the Singleton bound with equality is called *maximum distance separable* (MDS). This means no other code with that length and that MD has more codewords. Equivalently, it can be interpreted as no other $(n, \log M)$ code has higher MD.

The third upper bound to consider in this text is due to Plotkin. First define a number S as the sum of all distances between pairs of codewords,

$$S = \sum_{x \in \mathcal{B}} \sum_{y \in \mathcal{B}} d_H(\boldsymbol{x}, \boldsymbol{y}) \geq M(M-1)d_{\min} \tag{7.41}$$

where the inequality follows since $d(\boldsymbol{x}, \boldsymbol{x}) = 0$ and that all other pairs have at least distance $d_{\min}$. The idea is to list all codewords in a table with M rows and n columns, where each row is a codeword. Let n_i be the number of ones in column i. Then the total contribution to S from this column is first the distance from all ones to all zeros, and then the distance from all zeros to the ones, which gives $2n_i(M - n_i)$. This function has a maximum for $n_i = M/2$. Summarizing over all columns gives

$$S = \sum_{i=1}^{n} 2n_i(M - n_i) \leq \sum_{i=1}^{n} 2\frac{M}{2}\left(M - \frac{M}{2}\right) = n\frac{M^2}{2} \tag{7.42}$$

Putting things together gives

$$n\frac{M}{2} \geq (M-1)d_{\min} \tag{7.43}$$

which is expressed in the next theorem

Theorem 7.7 (Plotkin bound) For an $(n, \log M)$ block code $\mathcal{B}$ the MD is bounded by

$$d_{\min} \leq \frac{nM}{2(M-1)} \tag{7.44}$$

□

For the case when $d_{\min} > \frac{n}{2}$, the number of codewords in a block code with length n and minimum distance $d_{\min}$ is bounded by

$$M \leq \frac{2d_{\min}}{2d_{\min} - n} \tag{7.45}$$

The bounds due to Hamming, Singleton, and Plotkin are all very well known and common in the literature. They are upper bounding the number of codewords, which also means they are upper bounding the minimum distance. Next consider two important lower bounds on the maximum number of codewords in a code with length n and minimum distance d. When deriving the Hamming bound, spheres were centered around each codeword such that they were disjoint. Their maximum radius is $t = \lfloor \frac{d-1}{2} \rfloor$. Then, in the case of nonperfect codes, there can be vectors in the space that are not included in any of the spheres.

If the covering radius is considered in the same way, all vectors are covered. It might be that some vectors are included in more than one sphere meaning that it gives a lower bound on the maximum number of codewords. If the minimum distance is $d_{\min} = d$, the maximum covering radius is $d - 1$. To see this, assume that it is more than $d - 1$. Then there are at least one vector at a distance of at least d from every codeword and, hence, this vector can also be regarded as a codeword, and the number of codewords from the beginning was not maximized. This gives the Gilbert lower bound on the maximum number of codewords stated below.

Theorem 7.8 (Gilbert bound) There exists a binary code with M codewords of length n and minimum distance $d_{\min} = d$, where

$$M \geq \frac{2^n}{\sum_{i=0}^{d-1} \binom{n}{i}} \tag{7.46}$$

□

The second lower bound to consider in this section is due to Varshamov. The result is similar to the Gilbert bound, even though the path leading to it is different. Consider again a code construction based on choosing the columns of the parity check matrix such that the least linearly dependent columns is d. Then the minimum distance is $d_{\min} = d$.

The parity check matrix for an (n, k) code, where $M = 2^k$, is a binary $(n-k) \times n$ matrix. To construct it, choose the first column as any nonzero column of length n. Then the next columns should be chosen, one by one, such that they are not linearly dependent on $d-2$ other columns. To see that this is always possible, assume that j columns are chosen and that $d \leq j \leq n-1$. Then there are $\sum_{i=0}^{d-2} \binom{j}{i}$ linear combinations of the columns. If it is possible to find a different vector, then this can be used as the next column. Hence, to be able to choose the nth column

$$2^{n-k} > \sum_{i=0}^{d-2} \binom{n-1}{i} \tag{7.47}$$

Equivalently, this means there exists an (n, k) code with minimum distance d if

$$k \leq n - \log\left(1 + \sum_{i=0}^{d-2} \binom{n-1}{i}\right) \tag{7.48}$$

The largest k satisfying this is $n - \left\lceil \log\left(1 + \sum_{i=0}^{d-2} \binom{n-1}{i}\right)\right\rceil$.

Theorem 7.9 (Varshamov bound) There exists a binary code with M codewords of length n and minimum distance $d_{\min} = d$, where

$$M \geq 2^{n - \left\lceil \log\left(1 + \sum_{i=0}^{d-2} \binom{n-1}{i}\right)\right\rceil} \tag{7.49}$$

□

By rewriting the Gilbert bound as

$$M \geq 2^{n - \log \sum_{i=0}^{d-1} \binom{n}{i}} \tag{7.50}$$

their relation is more visible. Similarly, the Hamming bound can be written as

$$M \leq 2^{n - \log \sum_{i=0}^{\lfloor \frac{d-1}{2} \rfloor} \binom{n}{i}} \tag{7.51}$$

Naturally, the above bounds on the maximum number of codewords in a code will grow to infinity as n grows. Instead the code rate $R = \frac{k}{n}$ can be considered as a function of the relative minimum distance $\delta = \frac{d}{n}$. For these derivations, it is easiest to start with is the Singleton bound which gives the asymptotic bound

$$R \leq \frac{\log 2^{n-d-1}}{n} = 1 - \frac{d}{n} - \frac{1}{n} \rightarrow 1 - \delta, \quad n \rightarrow \infty \tag{7.52}$$

Hence, the following theorem can be established.

Theorem 7.10 (Asymptotic Singleton bound) Consider a code with length n and relative minimum distance $\delta = \frac{d_{\min}}{n}$. Then, as $n \to \infty$, the code rate is upper bounded by $R \leq 1 - \delta$. □

It turns out that the other bounds are all functions of the volume, or the number of codewords, of a sphere with radius αn in the binary n-dimensional space,

$$V(n, \alpha n) = \sum_{i=0}^{\alpha n} \binom{n}{i} \tag{7.53}$$

The problem to get asymptotic versions then goes back to getting an asymptotic value of the sphere. Consider the function $\frac{1}{n} \log V(n, \alpha n)$ and derive its limit value as $n \to \infty$. This can be done by deriving upper and lower bounds, which sandwich the function as n grows. To get an upper bound, consider the following derivations. Consider a number α, in the interval $0 \leq \alpha \leq \frac{1}{2}$, then

$$\begin{aligned}
1 = \big(\alpha + (1-\alpha)\big)^n &= \sum_{i=0}^{n} \binom{n}{i} \alpha^i (1-\alpha)^{n-i} \\
&\geq \sum_{i=0}^{\alpha n} \binom{n}{i} \alpha^i (1-\alpha)^{n-i} \\
&= \sum_{i=0}^{\alpha n} \binom{n}{i} \Big(\frac{\alpha}{1-\alpha}\Big)^i (1-\alpha)^n \\
&\geq \sum_{i=0}^{\alpha n} \binom{n}{i} \Big(\frac{\alpha}{1-\alpha}\Big)^{\alpha n} (1-\alpha)^n \\
&= \sum_{i=0}^{\alpha n} \binom{n}{i} \alpha^{\alpha n} (1-\alpha)^{(1-\alpha)n} = \sum_{i=0}^{\alpha n} \binom{n}{i} 2^{-nh(\alpha)}
\end{aligned} \tag{7.54}$$

where the first inequality comes from limiting the summation and the second follows since $\frac{\alpha}{1-\alpha} \leq 1$ and $\alpha n \geq 1$ for large n. The last equality follows from the fact that

$$2^{-nh(\alpha)} = \alpha^{\alpha n} (1-\alpha)^{(1-\alpha)n} \tag{7.55}$$

Rearranging in the above calculation gives

$$\frac{1}{n} \log V(n, \alpha n) \leq h(\alpha) \tag{7.56}$$

To get a lower bound on the volume, refer to Stirling's approximation. There are different versions of this, and a commonly used[1] is $n! \approx \sqrt{2\pi n}\big(\frac{n}{e}\big)^n$. By lower bounding

[1] More accurately, it is shown in [39] that the factorial function is bounded by

$$\sqrt{2\pi n}\Big(\frac{n}{e}\Big)^n e^{\frac{1}{12n+1}} \leq n! \leq \sqrt{2\pi n}\Big(\frac{n}{e}\Big)^n e^{\frac{1}{12n}}$$

the volume with the last term of the sum

$$V(n,\alpha n)=\sum_{i=0}^{\alpha n}\binom{n}{i}\geq\binom{n}{\alpha n}=\frac{n!}{\alpha n!(1-\alpha)n!}$$

$$\geq\frac{\sqrt{2\pi n}\left(\frac{n}{e}\right)^{n}e^{\frac{1}{12n+1}}}{\sqrt{2\pi\alpha n}\left(\frac{\alpha n}{e}\right)^{\alpha n}e^{\frac{1}{12\alpha n}}\sqrt{2\pi(1-\alpha)n}\left(\frac{(1-\alpha)n}{e}\right)^{(1-\alpha)n}e^{\frac{1}{12(1-\alpha)n}}}$$

$$=\frac{2^{-\frac{1}{2}\log 2\pi\alpha(1-\alpha)n+\left(\frac{1}{12n+1}-\frac{1}{12\alpha n}-\frac{1}{12(1-\alpha)n}\right)\log e}}{\alpha^{\alpha n}(1-\alpha)^{(1-\alpha)n}}$$

$$=2^{nh(\alpha)-O(\log n)} \tag{7.57}$$

where $O(\log n)$ denotes a function growing in the order of $\log n$. Again, consider the logarithm of the volume per dimension to get

$$\frac{1}{n}\log V(n,\alpha n)\geq h(\alpha)-\frac{1}{n}O(\log n)\rightarrow h(\alpha),\quad n\rightarrow\infty \tag{7.58}$$

Hence, as n grows toward infinity the normalized volume is sandwiched between the upper and lower bounds, both approaching the binary entropy function. That gives

$$\lim_{n\rightarrow\infty}V(n,\alpha n)=h(\alpha) \tag{7.59}$$

Going back to the Hamming bound, the spheres considered have radius $t=\left\lceil\frac{d-1}{2}\right\rceil$. For large n, and consequently also large d, it can be written as

$$t=n\frac{\left\lceil\frac{d-1}{2}\right\rceil}{n}\approx\frac{\delta}{2}n \tag{7.60}$$

Therefore, the Hamming bound gives the following asymptotic bound on the coding rate

$$R=\frac{1}{n}\log M\leq\frac{1}{n}\left(n-\log V\left(n,\tfrac{\delta}{2}n\right)\right)\rightarrow 1-h\left(\tfrac{\delta}{2}\right),\quad n\rightarrow\infty \tag{7.61}$$

For code constructions with rates not approaching zero,[2] the following theorem is stated.

Theorem 7.11 (Asymptotic Hamming bound) Consider a code with length n and relative minimum distance $\delta=\frac{d_{\min}}{n}$. Then, as $n\rightarrow\infty$ the code rate is upper bounded by $R\leq 1-h\left(\frac{\delta}{2}\right)$. □

[2] An obvious example on such code construction is the repetition code, where the rate $R=1/n\rightarrow 0$, as $n\rightarrow\infty$.

As the codeword length n grows, the number of codewords $M = 2^{nR}$ will also grow. Then according the Plotkin bound on the MD in Theorem 7.7,

$$\delta = \lim_{n\to\infty} \frac{d_{\min}}{n} \leq \lim_{n\to\infty} \frac{1}{2}\frac{M}{M-1} = \frac{1}{2} \tag{7.62}$$

The result is stated in the following theorem.

Theorem 7.12 For a code construction where the rate does not tend to zero the relative minimum distance $\delta = \frac{d_{\min}}{n}$ is bounded by $\delta \leq \frac{1}{2}$, as $n \to \infty$. □

Next, the result in (7.45) is reviewed as the codeword length grows. Even though the derivations require slightly deeper treatment of coding theory than given here, it is included for completeness. Consider a block code $\mathcal{B}$ with minimum distance $d_{\min}$ and codeword length n. Let $w = n - 2d_{\min} + 1$ and consider a binary vector $\boldsymbol{\alpha}$ of this length. Form a subcode, $\mathcal{B}_{\boldsymbol{\alpha}}$ consisting of the codewords in $\mathcal{B}$ starting with $\boldsymbol{\alpha}$. Since $\mathcal{B}_{\boldsymbol{\alpha}}$ is a subcode of $\mathcal{B}$ the minimum distance is bounded by $d_{\min}^{(\boldsymbol{\alpha})} \geq d_{\min}$. There are 2^w such subcodes partitioning $\mathcal{B}$, and according to (7.45) each of them can have the number of codewords at most

$$M_{\boldsymbol{\alpha}} \leq \frac{2d_{\min}}{2d_{\min} - n + w} = 2d_{\min} \tag{7.63}$$

Hence, the total number of codewords in $\mathcal{B}$ is

$$M = \sum_{\boldsymbol{\alpha}} M_{\boldsymbol{\alpha}} \leq 2d_{\min} 2^{n-2d_{\min}+1} = d_{\min} 2^{n-2d_{\min}+2} \tag{7.64}$$

and the rate

$$\begin{aligned} R &= \frac{\log M}{n} \leq \frac{\log d_{\min} + n - 2d_{\min} + 2}{n} \\ &= 1 - 2\frac{d_{\min}}{n} + \frac{\log d_{\min} + 2}{n} \to 1 - 2\delta, \quad n \to \infty \end{aligned} \tag{7.65}$$

The result is formulated in the next theorem.

Theorem 7.13 (Plotkin asymptotic bound) Consider a code with length n and relative minimum distance $\delta = \frac{d_{\min}}{n}$, where $0 \leq \delta \leq 1/2$. Then, as $n \to \infty$ the code rate is upper bounded by $R \leq 1 - 2\delta$. □

For large n, the differences between Gilbert's bound and Varshamov's bound become negligible, and asymptotically they are the same. Normally, this bound is called the Gilbert–Varshamov bound. For both cases, the radius of the considered sphere is approximately δn, and the lower bound on the code rate can be derived as

$$R = \frac{1}{n}\log M \geq \frac{1}{n}(n - \log V(n, \delta n)) \to 1 - h(\delta), \quad n \to \infty \tag{7.66}$$

which gives the next theorem.

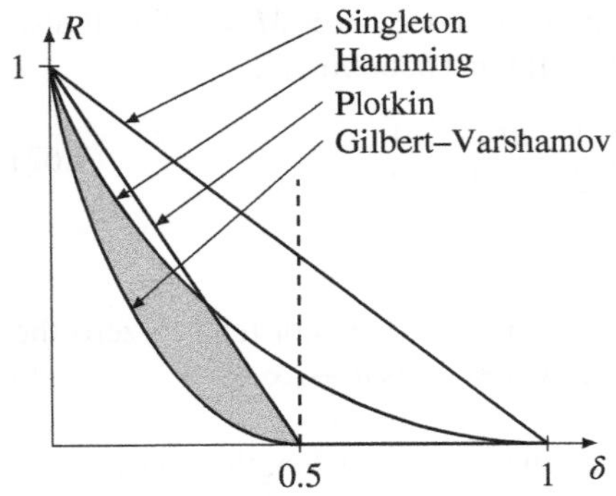

Figure 7.2 A view of Singleton bound, Hamming bound, Plotkin bound, and Gilbert–Varshamov bound for large codeword lengths. The gray-shaded area is the gap between the upper and lower bounds.

Theorem 7.14 (Gilbert–Varshamov bound) There exists a code with rate R and relative minimum distance $\delta = \frac{d_{\min}}{n}$ such that, as $n \to \infty$ the code rate satisfies $R \geq 1 - h(\delta)$. □

The considered asymptotic bounds are shown in Figure 7.2. Here the gray-shaded area represents the gap between the upper and lower bounds. What it means is that rates above the gray area are known to be impossible to reach. It is also known that there exist codes for the rates that are below the gray area, i.e. the gray area marks the unknown region. In the literature, there are several more bounds that can narrow this gap (see, e.g., [33, 36]).

7.2 CONVOLUTIONAL CODE

The second class of codes treated in this text is the class of convolutional codes. The main idea of error-correcting codes is that a sequence of symbols representing the pure information should be represented by data with a higher dimensionality, in such a way that errors can be corrected. In the previous section, the sequence was first blocked in k-tuples and each of these treated independently from each other to form length n codewords. This is the block coding approach. If the input sequence is instead viewed as an infinite sequence and the redundancy is formed along the way, the convolutional coding approach arise. It was first presented by Elias in 1955 [40]. The information sequence is processed by a linear system where the number of outputs exceeds the number of inputs. In this way, the resulting sequence, the code sequence, has a built in redundancy related to the difference in input and output length as well as the memory of the linear system. In this text, mainly binary sequences are considered as well as encoders with one input sequence and two output sequences.

Consider a binary (infinite) sequence

$$\boldsymbol{x} = x_0 x_1 x_2 x_3 \ldots \tag{7.67}$$

Then the sequence is fed to a linear circuit, or an encoder, with one input and two outputs. For each input bit in the sequence, the output consists of 2 bits. That is, the

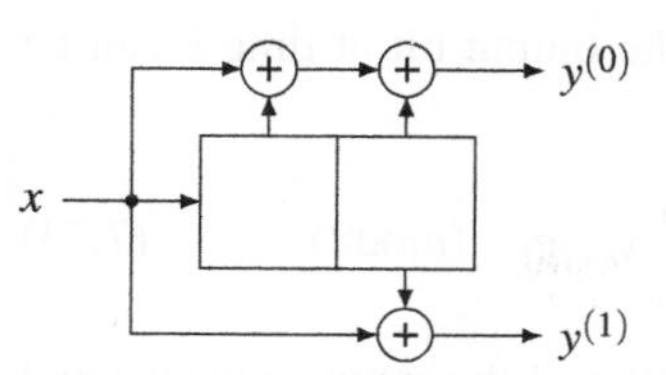

Figure 7.3 Circuit for the (7, 5) encoder.

output can be written as

$$y = y_0^{(0)} y_0^{(1)}\ y_1^{(0)} y_1^{(1)}\ y_2^{(0)} y_2^{(1)}\ y_3^{(0)} y_3^{(1)} \dots \tag{7.68}$$

In the next example, an encoder for such system is shown.

Example 7.10 In Figure 7.3, one of the most common examples of a convolutional encoder is shown. Assuming that the encoder starts in the all-zero state, the input sequence

$$\boldsymbol{x} = 10100000\dots \tag{7.69}$$

will give the output sequence

$$\boldsymbol{y} = 11\ 10\ 00\ 10\ 11\ 00\ 00\ 00\dots \tag{7.70}$$

From system theory, it is well known that the output sequences can be derived as the convolution of the input sequence and the impulse responses,

$$\boldsymbol{y}^{(0)} = \boldsymbol{x} * (111) \tag{7.71}$$

$$\boldsymbol{y}^{(1)} = \boldsymbol{x} * (101) \tag{7.72}$$

hence the name *convolutional codes*. The impulse responses (111) and (101) are often described in an octal form giving 7 and 5. This specific encoder is therefore often mentioned as the (7, 5)-encoder.

By assuming a length m shift register instead of 2 as in the previous example, a more general relation between the input sequence and the output sequence can be obtained (see Figure 7.4). The input sequence is fed to the shift register, and at each level the symbol is multiplied with a constant g_i, $i = 1, 2, \dots, m$. For the binary case, when having one input and one output, the multiplier means either a connection or

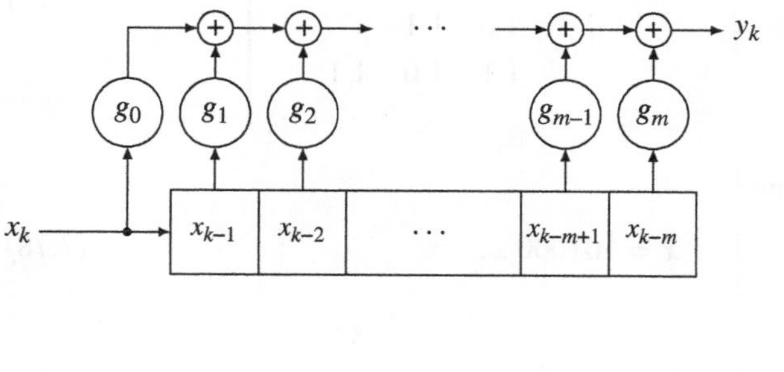

Figure 7.4 A linear circuit as the encoder.

no connection for the values 1 and 0, respectively. The output bit at time k can be derived as

$$y_k = x_k g_0 + x_{k-1} g_1 + \cdots + x_{k-m} g_m = \sum_{i=0}^{m} x_{k-i} g_i \quad (\text{mod}2) \tag{7.73}$$

This relation shows the output sequence as a convolution of the input sequence and the impulse response, i.e.

$$\boldsymbol{y} = \boldsymbol{x} * (g_0 g_1 g_2 \dots g_{m-1} g_m) \tag{7.74}$$

as was used also in the previous example. In general, the circuit of Figure 7.4 can be used to describe a code where at each time instant b information bits are encode into c code bits. At time k, x_k is a binary row vector of length b and y_k a binary row vector of length c. The coefficients in the figure are represented by $b \times c$ binary matrices. Hence, the (7, 5) encoder in Example 7.10 is described by the coefficient matrices

$$g_0 = (1 \quad 1) \quad g_1 = (1 \quad 0) \quad g_2 = (1 \quad 1) \tag{7.75}$$

The convolution identified in (7.74) can be derived by a matrix multiplication. Assuming that the encoder starts in the all-zero state at time $k = 0$ the output from the information sequence $\boldsymbol{x} = (x_0 x_1 x_2 \dots)$

$$\boldsymbol{y} = \boldsymbol{x}G$$

$$= (x_0 x_1 x_2 \dots) \begin{pmatrix} g_0 & g_1 & g_2 & \cdots & g_{m-1} & g_m & & & & \\ & g_0 & g_1 & g_2 & \cdots & g_{m-1} & g_m & & & \\ & & g_0 & g_1 & g_2 & \cdots & g_{m-1} & g_m & & \\ & & & g_0 & g_1 & g_2 & \cdots & g_{m-1} & g_m & \\ & & & & \ddots & \ddots & \ddots & & \ddots & \ddots \end{pmatrix} \tag{7.76}$$

where G is the time domain generator matrix.

Example 7.11 For the (7, 5) encoder in Example 7.10, the generator matrix is

$$G = \begin{pmatrix} 11 & 10 & 11 & & & & & \\ & 11 & 10 & 11 & & & & \\ & & 11 & 10 & 11 & & & \\ & & & 11 & 10 & 11 & & \\ & & & & 11 & 10 & 11 & \\ & & & & & \ddots & \ddots & \ddots \end{pmatrix} \tag{7.77}$$

Hence, encoding the sequence

$$\boldsymbol{x} = 101000\dots \tag{7.78}$$

is equivalent to adding row one and three in the generator matrix to get

$$y = 11 \quad 10 \quad 00 \quad 10 \quad 11 \quad 00 \quad 00 \quad \ldots \tag{7.79}$$

The rate of a convolutional code is the ratio between the number of inputs and the number of outputs,

$$R = \frac{b}{c} \tag{7.80}$$

As an example, the $(7,5)$ encoder gives a rate $R = \frac{1}{2}$ code.

7.2.1 Decoding of Convolutional Codes

So far, it is the encoder circuit that has been treated. The code is, as for block codes, the set of codewords. Since the information sequences are infinite sequences, so are the codewords. This also means that the number of codewords is infinite. This fact might be seen as an obstacle when it comes to decoding, as for an ML decoder the received sequence should compare all possible code sequences. It turns out that there is a very clever structure to compare all code sequences and with that a simple method to perform ML decoding. The decoding algorithm is the Viterbi algorithm [41], which was published in April 1967. However, at that point it was not fully understood that the algorithm was neither optimal nor practically implementable. In [42], Forney introduced the *trellis* structure, which makes the algorithm much more understandable. In the same paper, it was shown that the algorithm indeed performs an ML decoding.

Since the complexity of the algorithm grows exponentially with the memory in the encoder, it was still not seen as a practical alternative. It was not until late 1968 when Heller published the first simulation results for relatively short convolutional codes [43] this view was changed. Today there are many systems containing convolutional codes that rely on the trellis structure and the Viterbi algorithm, in one way or another. Convolutional codes are also used for concatenation of codes, e.g., turbo codes. Then, an iterative decoding procedure, based on a MAP decoding algorithm [44] is used, often called the BCJR algorithm from the inventors. This MAP algorithm also uses a the trellis structure as a base for the probability derivations. In the next, the Trellis structure will be introduced first and then the Viterbi algorithm.

To start describing the trellis structure, again assume the $(7,5)$ encoder in Example 7.10. The two memory elements in this circuit represent the memory of the code, which is called the state. This state represents everything the encoder needs to know about the past symbols in the sequence. The output and the next state at a certain time are both functions of the current state and the current input. These two functions can be viewed in a state transition graph, as depicted in Figure 7.5.

If the current state is the all-zero state and the input is 0, the next state is the all-zero state and the output 00. If, on the other hand, the input is 1, the next state is 10 and the output 11. Continuing with the other three states in the same way completes the graph. In this way, the graph describes the behavior of the encoder circuit. Each

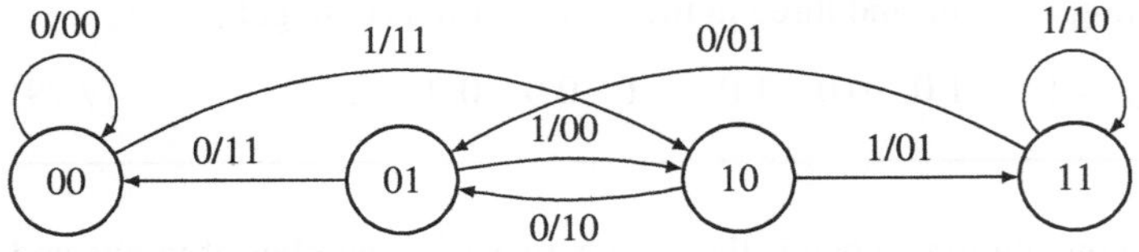

Figure 7.5 State transition graph for the (7, 5) encoder.

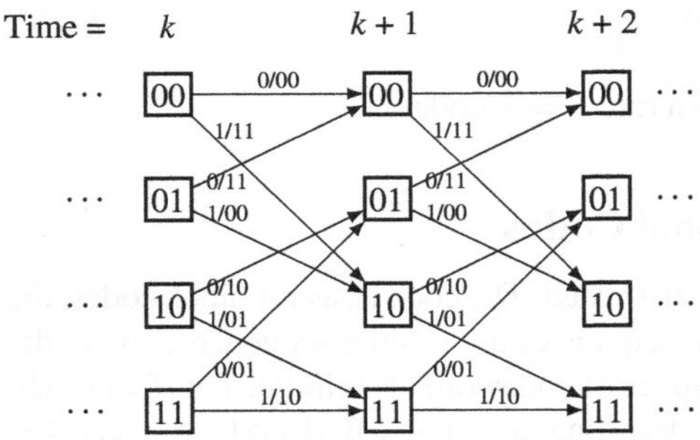

Figure 7.6 Trellis segments for the (7, 5) encoder.

information sequence represents a path in the graph, giving both a state sequence and a code sequence. Even though the graph gives a good overview and a description of how the encoder works, the description of the sequences needs one more dimension, the time. To include this, a two-dimensional structure with the possible states listed vertically and the time horizontal can be considered. That mean each state will exist once in every time instant (see Figure 7.6). This picture resembles the structure of a garden trellis, hence the name trellis [72].

In Figure 7.6, three trellis segments for the (7, 5) encoder are shown. This gives the efficient description of all possible state sequences. Since there is a one-to-one mapping between the input sequences and the state sequences, and between the state sequences and the code sequences, this also gives a graphical view of *all possible code sequences*. Previously, in Examples 7.10 and 7.11, the encoder was assumed to start in the all-zero state. Then at time $k = 0$ the state is known to be 00, and other states do not need to be listed (see Figure 7.7). At time $k = 1$, there are two possible states,

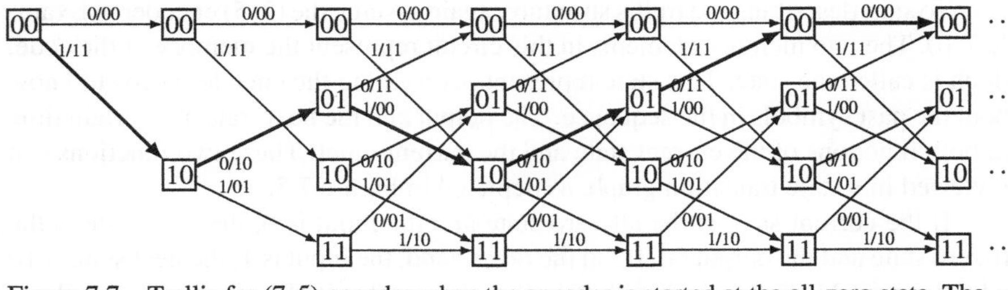

Figure 7.7 Trellis for (7, 5) encoder when the encoder is started at the all-zero state. The path marked with bold edges corresponds to the information sequence $\boldsymbol{x} = 101000\ldots$.

00 and 10. First in time $k = 2$, when the memory of the encoder has been filled, all states are possible. So the trellis in Figure 7.7 describes all codewords generated by a (7, 5) encoder that starts in the zero state. In Example 7.10, the information sequence $\boldsymbol{x} = 101000\ldots$ was used. By following this sequence in the trellis, it is seen that the corresponding state sequence starts in state 00. As the first input is 1, the next state is 10 and the output is 11. The second input is 0, which gives the next state 01 and the output 10. Continuing this through the trellis, the state and code sequences are

$$\boldsymbol{\sigma} = 00\ 10\ 01\ 10\ 01\ 00\ 00\ldots \tag{7.81}$$

$$\boldsymbol{y} = 11\ 10\ 00\ 10\ 11\ 00\ 00\ldots \tag{7.82}$$

The above sequences are marked as a path with bold edges in Figure 7.7.

The error-correcting capability of a convolutional code is determined by the minimum Hamming distance between two code sequences. In the next definition, the free distance of the code is introduced as a direct counterpart to the minimum distance for block codes.

Definition 7.5 The *free distance* for a convolutional code C is the minimum Hamming weight between two different code sequences

$$d_{\text{free}} = \min_{\substack{y_1, y_2 \in C \\ y_1 \neq y_2}} d_H(y_1, y_2) \tag{7.83}$$

□

Since the mapping between the information sequences and code sequences is determined by a convolution, which is a linear mapping, the code is linear. That means, to derive the free distance it is not necessary to compare all code sequences with each other, it is enough to compare one sequence with all the other. If that fixed sequence is chosen as the all-zero sequence, the free distance can be derived as

$$d_{\text{free}} = \min_{\substack{y \in C \\ y \neq 0}} w_H(y) \tag{7.84}$$

With the free distance as the measure of separation between code sequences the same arguing as for block codes gives the following theorem on the error correction and detection capabilities for a convolutional code.

Theorem 7.15 When using a convolutional code C with free distance d_{free}, it is always possible to either *detect* an error $\boldsymbol{e}$ if

$$w_H(\boldsymbol{e}) < d_{\text{free}} \tag{7.85}$$

or *correct* an error $\boldsymbol{e}$ if

$$w_H(\boldsymbol{e}) \leq \frac{d_{\text{free}} - 1}{2} \tag{7.86}$$

□

Example 7.12 By counting the number of ones along the nonzero paths in the trellis of Figure 7.7, the minimal weight is found to be 5. That gives the free distance $d_{\text{free}} = 5$, which is answered by, e.g., the code sequence

$$y = 11\ 10\ 11\ 00\ 00\ 00 \ldots \tag{7.87}$$

Hence, by Theorem 7.15 it is seen that two errors can always be corrected by the code. Alternatively, any four errors can always be detected by the code.

In the description above, it is assumed that the code sequences have infinite duration. Even though this exist also in practical implementations, e.g., some space applications, the information sequence is often split into finite duration vectors (or blocks). Then, each vector is encoded separately by first setting the encoder in the all-zero state and then feeding the information vector. To preserve the error-correcting capability of the code, the encoder is driven back to the all-zero state after encoding the vector. With vectors of length K and an encoder with memory m the code sequences will be of length $K + m$. The trellis will then have one starting state at time $k = 0$ and one ending state at time $k = K + m$. For the $(7, 5)$ encoder such trellis is shown in Figure 7.8. To simplify the figure, the labels of the branches are omitted.

Assuming the code vector is transmitted bitwise over a BSC, the ML decoder can be implemented as a minimum distance decoder. Thus, the received (binary) vector should be compared to the possible transmitted vectors. The code symbols of the branches in the trellis are compared with the received symbols by using the Hamming distance. In this sense, the trellis is a directed graph with the property that all states at a specific time has the same length from the starting state.

If both the starting state and ending state are known, e.g., the all-zero state, Viterbi's idea is as follows. Start in time $k = 0$ and state $\sigma = 00$ and let the metric for this state be $\mu_{00} = 0$. Then, for each time instance $k = \tau$ in the trellis, label all branches to the next time instance $k = \tau + 1$ with the Hamming distance between the corresponding branch output and the received bits. For all the states at time $k = \tau$, there is a cumulative metric μ_σ for the lowest weight path from the starting state to this state. Then, for each state in time $k = \tau + 1$ there are two alternative paths from time $k = \tau$. The total weight for the path from the starting state is the metric for the

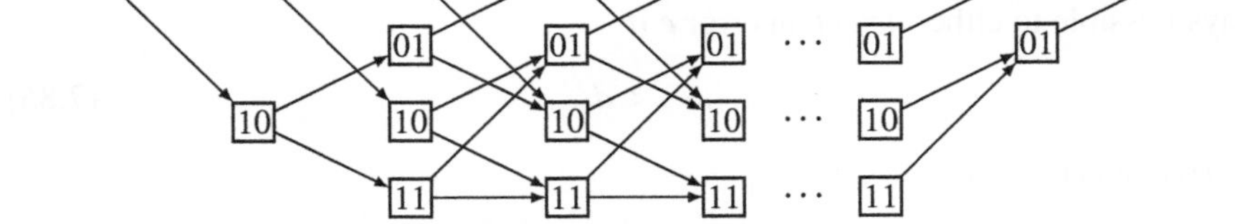

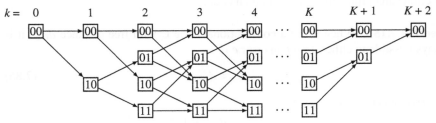

Figure 7.8 Trellis for $(7, 5)$ encoder when the encoder starts and terminates in the all-zero state. The information vector is K bits long and the code sequence $2(K + 2)$ bits.

previous state and branch weight. Keep only the branch corresponding to the lowest weight path for each state. Continuing this will at time $k = K + m$ result in one path with the least weight through the trellis. This represents the codeword with the least Hamming distance to the received vector.

Algorithm 7.2 (Viterbi)

Let $\boldsymbol{y}$ be a code vector of length $K + m$, starting and terminating in the all-zero state, and let S_k be the possible states in time k. The code vector is transmitted over a BSC, and the received vector is denoted by $\boldsymbol{r}$. Let $\mathcal{X}$ be the possible input vectors for the encoder and $S^{+}(\sigma, x)$ and $Y(\sigma, x)$ be the next state function and the output function when current state is σ and input is x. In the trellis, let each node contains two variables: the metric $\mu_k(\sigma)$ and the backtrack state $BT_k(\sigma)$. The $BT_k(\sigma)$ contains a pointer back to the state at time $k - 1$ from where the minimum weight path comes. Then the Viterbi algorithm can be performed according to

1. Initialization:
   Let $\mu_0(0) = 0$
2. Expand:
   FOR $k = 0, 1, \dots, K + m - 1$:
   FOR EACH $\sigma \in S_{k-1}$ and $x \in \mathcal{X}$
   $\mu = \mu_k(\sigma) + d_H(Y(\sigma, x), r_k)$
   IF $\mu < \mu_{k+1}(S^{+}(\sigma, x))$
   $\mu_{k+1}(S^{+}(\sigma, x)) = \mu$
   $BT_{k+1}(S^{+}(\sigma, x)) = \sigma$

Backtrack from end state to starting state using BT-path to get $\hat{\boldsymbol{y}}$.

In the case when there are two equally likely paths entering a state, the surviving path should be chosen randomly.

The procedure of the Viterbi algorithm is best shown through an example. In the next example, it is assumed a length four information vector is encoded by a (7,5) encoder. This is suitable for a textbook example, but in a real implementation the length of the vector should be much longer than the memory of the encoder. Otherwise, the effective code rate will be considerably lowered. In the example, the rate $R = 1/2$ encoder is used to encode four information bits to 12 code bits, giving an effective rate of $R_{\text{Eff}} = 4/12 \approx 0.33$. If instead the length of the information vector is 500, the effective rate becomes $R_{\text{Eff}} = 500/1004 \approx 0.5$.

Example 7.13 Assume the information vector $\boldsymbol{x} = 1011$ should be transmitted. To drive the encoder back to the zero state at the end two dummy zeros are appended to form $\tilde{\boldsymbol{x}} = 1011\ 00$. Continuing with the (7,5) encoder, the code vector is

$$\boldsymbol{y} = \tilde{\boldsymbol{x}} G = 11\ 10\ 00\ 01\ 01\ 11 \tag{7.88}$$

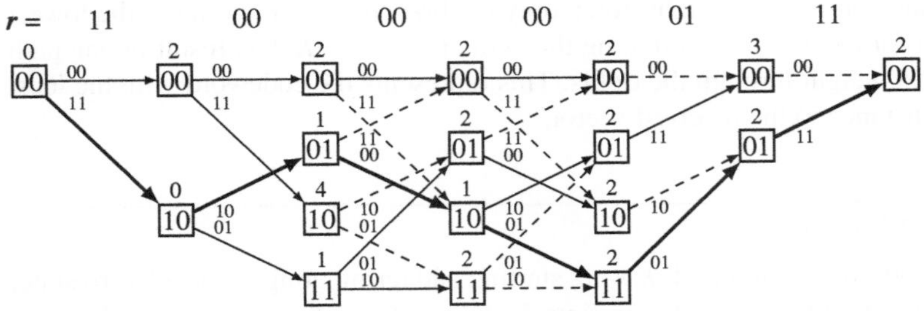

Figure 7.9 The trellis used to decode $\boldsymbol{r}$ = 11 00 00 00 01 11 for the (7, 5) code.

Assume that two errors occur during the transmission; in the third and eight bit, so the received vector is

$$\boldsymbol{r} = 11\ 00\ 00\ 00\ 01\ 11 \tag{7.89}$$

According to (7.86), an MD decoder should be able to correct these two errors. The trellis in Figure 7.9 is used for decoding. For the first and second steps in the algorithm, there are no competing paths entering the states and the metric of the paths is derived as the Hamming distance between the received path and the considered path. In the third step, there are two branches entering each state. Starting with state 00, one path is coming from state 00 at the previous level and one from 01. The path from 00 has a cumulative metric of $\mu_{00} = 2$ and an additional branch metric of $d_H(00, 00) = 0$, which gives in total 2. The second path has a cumulative metric of $\mu_{01} = 1$ at time 2 and an addition of $d_H(11, 00) = 2$ which gives a total metric of 3. Hence, the first path has the least metric and therefore the second should be discarded. This is marked in the figure with a dashed line. Similarly, the state 01 at time 3 has two entering paths with total metric of $4 + 1 = 5$ and $1 + 1 = 2$, where the second is the surviving path and the first should be discarded. Continuing in the same way for the remaining states at time three, there will be exactly one path entering each state, yielding the minimum cumulative metric. Since the Hamming distance is used as metric, the paths entering each state represent the paths with minimum Hamming distance compared to the received vector up to this time.

The fourth step in the trellis follows similarly. Then, at step five the input sequence is known to be a zero, appended to drive the encoder back to the all-zero state. Then there are only the possible states 00 and 01. For state 00, the two entering paths both have the metric 3, and then the survivor should be chosen randomly, using, e.g., coin flipping. In this example, the lower path from 01 is chosen. In the last step in the trellis, there is only one state left, 00. The surviving path entering this state corresponds to the path through the trellis with least metric. Following it back to the starting state gives the closest code vector as

$$\hat{\boldsymbol{y}} = 11\ 10\ 00\ 01\ 01\ 11 \tag{7.90}$$

which gives the most likely information vector

$$\hat{x} = 1011\ (00) \tag{7.91}$$

The two inserted errors have been corrected by the algorithm, as was anticipated from the free distance of 5. If there are more errors, the outcome depends on their distribution. If they are far apart in a long vector, the code will probably be able to correct them, but if they are closely together in a burst there is a higher risk that the decoder will give an erroneous answer.

7.2.2 $\mathcal{D}$-Transform Representation

In the beginning of this section, it was seen that the code sequence equals the information sequences convolved with the impulse response of the circuit. This convolution can be expressed as the multiplication by an (infinite) matrix. As in many applications, a convolution in the time domain is easily viewed as a multiplication in a transform domain. Since there is a finite number of amplitude levels, without order, the normal discrete time transforms for real or complex sequences, such as the discrete Fourier transform or the $\mathcal{Z}$-transform, cannot be used. Instead it is common to define a new transform, often named the $\mathcal{D}$-transform.[3]

Definition 7.6 Consider a sequence

$$\boldsymbol{x} = x_0 x_1 x_2 x_3 \dots \tag{7.92}$$

with or without starting and/or ending time. Then the $\mathcal{D}$-transform of the sequence is

$$x(D) = x_0 + x_1 D + x_2 D^2 + x_3 D^3 + \cdots = \sum_{i=-\infty}^{\infty} x_i D^i \tag{7.93}$$

□

In the definition, the sum is taken from $i = -\infty$ but it is often assumed that the sequences are causal, i.e. starting at time 0. This can be solved by arguing that the sequence is zero up to time 0. The variable D works as a position marker in the sense that the coefficient before D^k describes what is happening at time instant k. Next two important properties of the $\mathcal{D}$-transform will be derived. First, a convolution at the time domain equals a multiplication in the transform domain, and, second, the transform representation of periodic sequences.

Considering a sequence $\boldsymbol{x}$ that is fed to a linear circuit with impulse response $\boldsymbol{g}$, then the output sequence is given by the convolution $\boldsymbol{y} = \boldsymbol{x} * \boldsymbol{g}$. The symbol at time i

[3] In a strict mathematical meaning, it is doubtful that it should be called a transform. The variable D does not have a mathematical meaning as for the frequency in the Fourier transform or a complex number as in the $\mathcal{Z}$-transform. But for our purpose, considering sequences of elements from a finite field, the usage is very similar.

in $\boldsymbol{y}$ can then be expressed as

$$y_i = \sum_j x_j g_{i-j} \tag{7.94}$$

Hence, the D-transform of $\boldsymbol{y}$ becomes

$$\begin{aligned} y(D) &= \sum_i y_i D^i = \sum_i \sum_j x_j g_{i-j} D^i \\ &= \sum_j \sum_m x_j g_m D^{j+m} \\ &= \left(\sum_j x_j D^j\right)\left(\sum_j g_m D^m\right) = x(D)g(D) \end{aligned} \tag{7.95}$$

where in the third equality the summation order is interchanged and the variable change $m = i - j$ is applied. This shows that a convolution in the time domain equals a multiplication in the D-domain.

A periodic sequence can be written as

$$[x_0 x_1 \dots x_{n-1}]^\infty = x_0 x_1 \dots x_{n-1} x_0 x_1 \dots x_{n-1} \dots \tag{7.96}$$

where $x_0 x_1 \dots x_{n-1}$ is the periodically repeated sequence and n the period. To derive the D-transform, first consider a sequence with period n and only one 1,

$$\begin{aligned} [10\dots 0]^\infty &\rightarrow 1 + D^n + D^{2n} + \cdots \\ &= \frac{(1 + D^n)(1 + D^n + D^{2n} + \cdots)}{1 + D^n} = \frac{1}{1 + D^n} \end{aligned} \tag{7.97}$$

In the last equality, there is also term D^M in the numerator, where M tends to infinity. Since M denotes the time instant, the term is vanishing in infinite time and will not affect the derivations.

By similar derivations, if the 1 is in position i, the D-transform is

$$[0\dots 010\dots 0]^\infty \rightarrow \frac{D^i}{1 + D^n} \tag{7.98}$$

Altogether, the D-transform of a general periodic sequence is

$$\begin{aligned} [x_0 x_1 x_2 \dots x_{n-1}]^\infty &\rightarrow x_0 \frac{1}{1 + D^n} + x_1 \frac{D}{1 + D^n} + \cdots + x_{n-1} \frac{D^{n-1}}{1 + D^n} \\ &= \frac{x_0 + x_1 D + x_2 D^2 + \cdots + x_{n-1} D^{n-1}}{1 + D^n} \end{aligned} \tag{7.99}$$

That is, a periodical sequence is represented by a rational function in the D-transform, and vice versa. The next theorem summarizes the properties of the D-transform.

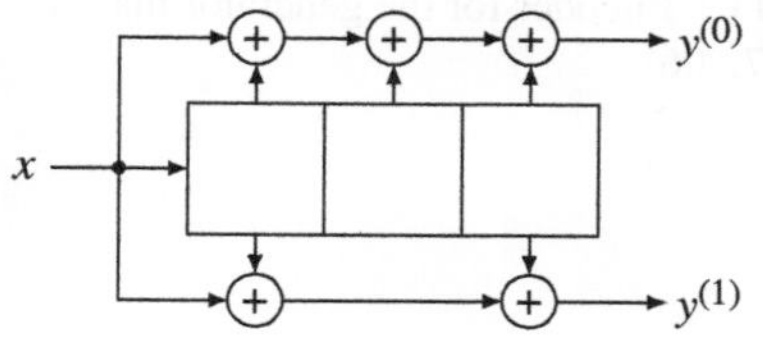

Figure 7.10 Circuit for the (17, 15) encoder.

Theorem 7.16 For the D-transform, the following properties hold:

$$\boldsymbol{x} * \boldsymbol{g} \xrightarrow{D} x(D)g(D) \tag{7.100}$$

$$[x_0 x_1 \ldots x_{n-1}]^{\infty} \xrightarrow{D} \frac{x_0 + x_1 D + \cdots + x_{n-1} D^{n-1}}{1 + D^n} \tag{7.101}$$

□

According to (7.74), the code sequence $\boldsymbol{y}$ is formed by a convolution with $\boldsymbol{g} = g_0 g_1 \cdots g_m$, which can be written by the D-transform as the *generator matrix*,

$$G(D) = g_0 + G_1 D + \cdots + g_m D^m \tag{7.102}$$

The code sequence is formed as

$$y(D) = x(D)G(D) \tag{7.103}$$

The (7, 5) encoder is characterized by $g_0 = (1\ 1)$, $g_1 = (1\ 0)$, and $g_2 = (1\ 1)$, and the generator matrix becomes

$$G(D) = (1\ 1) + (1\ 0)D + (1\ 1)D^2 = (1 + D + D^2 \quad 1 + D^2) \tag{7.104}$$

There are several other encoders than the here described (7, 5) encoder. As an example, the generator matrix

$$G(D) = (1 + D + D^2 + D^3 \quad 1 + D + D^3) \tag{7.105}$$

describes an encoder with memory $m = 3$ and free distance $d_{\text{free}} = 6$. The vectors for the coefficients of the polynomials are (1111) and (1101), which in octal representations are 17_8 and 15_8. The encoder is therefore mentioned as the (17, 15) encoder. In Figure 7.10, the encoder circuit is shown.

The encoder circuit can also have more than one input sequence as for the encoder circuit in Figure 7.11. The number of inputs for the encoder equals the number of rows in the generator matrix. That is, each input corresponds to a row in the matrix and each output a column. In this case, the generator matrix becomes

$$G(D) = \begin{pmatrix} 1 + D & D & 1 \\ D^2 & 1 & 1 + D + D^2 \end{pmatrix} \tag{7.106}$$

Furthermore, if the circuit contains feedback the entries in the generator matrix becomes rational functions, such as

$$G(D) = \begin{pmatrix} \dfrac{1 + D + D^2}{1 + D^2} & 1 \end{pmatrix} \tag{7.107}$$

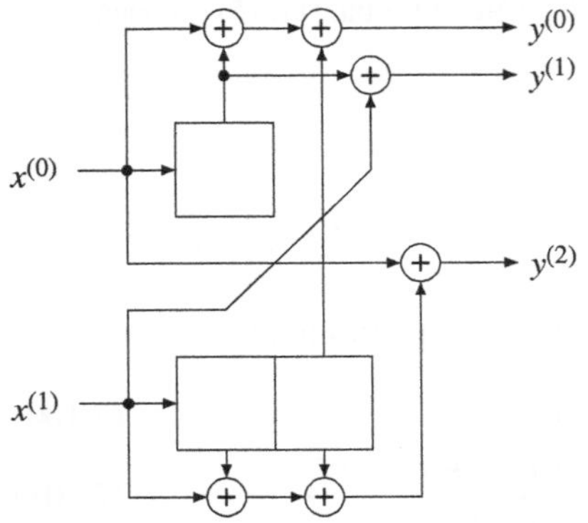

Figure 7.11 Encoder for the generator matrix $G(D)$ in (7.106).

and

$$G(D) = \begin{pmatrix} \frac{1}{1+D+D^2} & \frac{D}{1+D^3} & \frac{1}{1+D^3} \\ \frac{D^2}{1+D^3} & \frac{1}{1+D^3} & \frac{1}{1+D} \end{pmatrix} \tag{7.108}$$

From system theory, the first generator matrix can be realized as in Figure 7.12.

In Figure 7.13, the resulting bit error probability after the decoder is shown for a Hamming code, described in the previous section, and three convolutional codes with memory 2, 6, and 11. The memory 2 encoder is the described above with the generator matrix

$$G(D) = (1 + D + D^2 \quad 1 + D^2) \tag{7.109}$$

Since the rates for the Hamming code and the convolutional codes are not the same, it would not be fare to compare them for the same crossover probability in the BSC. Therefore, the bit error probability is plotted against the signal-to-noise ratio E_b/N_0, where E_b is the energy per information bit and N_0 the Gaussian noise parameter (see Chapter 9). When using a binary antipodal signalling, i.e. binary phase shift keying or BPSK, and hard decision at the receiver, the channel is modeled as a BSC with crossover probability

$$\varepsilon = Q\left(\sqrt{2\frac{E_b}{N_0}R}\right) \tag{7.110}$$

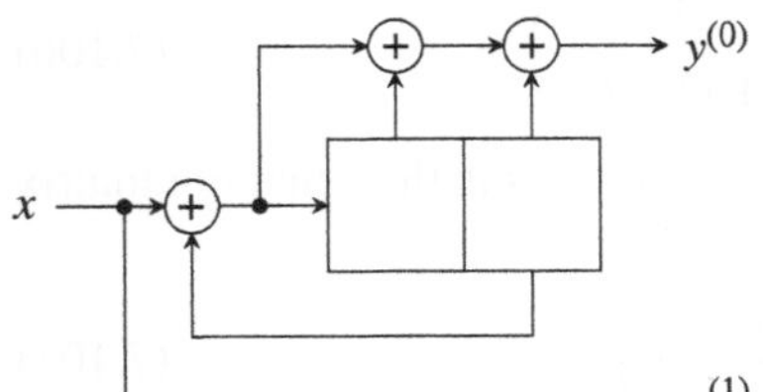

Figure 7.12 Encoder for the generator matrix $G(D)$ in (7.107).

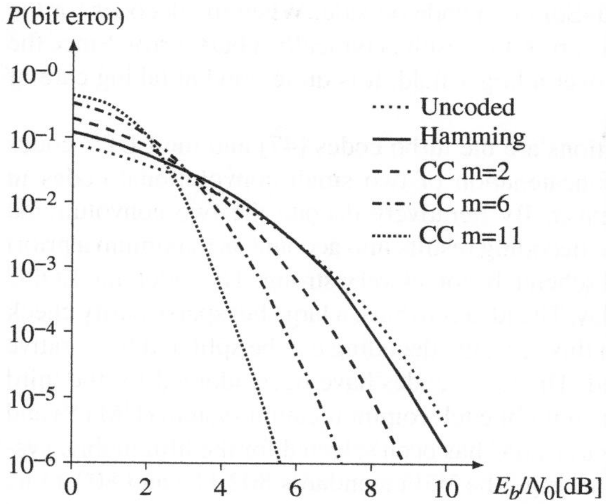

Figure 7.13 The resulting bit error probability as a function of the signal-to-noise ratio E_b/N_0 in decibels, for the (16,7) Hamming code and three different convolutional codes (CC) with rate $R = 1/2$ and memory 2, 6, and 11, respectively. For comparison, the uncoded case is also plotted. In the simulations, the BSC was applied.

where $Q(\cdot)$ is an error function

$$Q(x) = \int_x^\infty \frac{1}{\sqrt{2\pi}} e^{-z^2/2} dz \tag{7.111}$$

i.e. the probability that the outcome of a normalized Gaussian random variable exceeds x. The results in Figure 7.13 are based on hard decision, i.e., a BSC with crossover probability according to (7.110). It is possible to improve the results if the received values are fed directly to the decoder instead of making hard decision for the signal alternatives.

The purpose of the bit error rate plots in Figure 7.13 is to show how the coding schemes considered in this chapter works. The described codes, both block and convolutional, are of low complexity as well as performance, but if the complexity is increased the performance will also improve. This can be seen by the increased performance of the convolutional codes, where the decoding complexity increases exponentially with the encoder memory. The comparison in the figure with the Hamming code is not really fair since it is a very simple code. There are both block codes and convolutional codes that perform much better than shown here.

One widely used class of block codes is the class of Reed–Solomon codes [45]. This is a block code derived over a higher order field, often with alphabet size $q = 2^8$ implemented as bytes, see e.g., [83]. The generated codes are known to be minimum distance separable, MDS, i.e. they fulfill the Singleton bound (see Theorem 7.6), with equality. That means an (n, k) Reed–Solomon code has the minimum distance $d_{\min} = n - k + 1$. There are implementations in various communication systems, such as CD and DVD standards, ADSL and VDSL copper access, QR codes, and data storage.

Even more efficient systems can be obtained by combing several low complexity codes into one. A very well-spread combination is convolutional codes in series with the Reed–Solomon block code [46]. The idea is to have a convolutional codes closest to the channel and a Reed–Solomon code outside. When the decoder for the inner convolutional code makes an error, the result is typically a burst error. Since the Reed–Solomon code is working over a larger field, it is quite good at taking care of such burst errors.

Other larger code constructions are the turbo codes [47] and the LDPC codes [48]. The former is a parallel concatenation of two small convolutional codes in combination with a large interleaver. By iteratively decode the two convolutional codes, and taking the intermediate decoding results into account as maximum a priori information, the resulting overall scheme becomes very strong. The latter, the LDPC codes, are the most promising today. The idea is to have a large but sparse parity check matrix to define the code. Also in this case, the decoding can be split and by iterative decoding the results are improved. The turbo codes have been adopted for the third and fourth mobile systems universal mobile telecommunication system (UMTS) and long-term evolution (LTE), whereas LDPC has been selected for the fifth mobile system. They are also implemented in, e.g., the WiFi standards 802.11n and 802.11ac, as well as 10 Gbps Ethernet (10Gbase-T) and the G.hn standard, which is a standard for all wired in-home communication, such as powerline communication and coax and telephone lines.

7.2.3 Bounds on Convolutional Codes

In this part, two famous bounds on the free distance will be covered. First, it is Heller's upper bound [43], that is based on the Plotkin bound in Theorem 7.7. For this, view a convolutional code with rate $R = b/c$, that is, the encoder circuit has b inputs and c outputs, and memory m. Then consider an input sequence of length l. In total, there are 2^{bl} such sequences, and the corresponding code sequences are of length $c(l+m)$. By using these sequences, a block code can be formed, where the codeword length is $n = c(l+m)$ and $M = 2^{bl}$, i.e. it is an $\big(c(l+m), bl\big)$ code. Applying Plotkin's bound to this gives an upper bound on the minimum distance

$$d_{\min,l} \leq \frac{nM}{2(M-1)} = \frac{c(l+m)2^{bl}}{2(2^{bl}-1)} = \frac{c(l+m)}{2(1-2^{-bl})} \tag{7.112}$$

Minimizing over all possible lengths for the information sequence gives Heller's bound.

Theorem 7.17 The free distance for a convolutional code with rate $R = b/c$ with memory m satisfies

$$d_{\text{free}} \leq \min_{l \geq 1} \frac{c(l+m)}{2(1-2^{-bl})} \tag{7.113}$$

□

it is interesting to consider the case when the memory of the encoder grows, $m \to \infty$. This will mean that the codeword lengths will also grow, as well as the free distance.

Therefor, the *relative free distance* can be assigned as

$$\delta = \lim_{m\to\infty} \frac{d_{\text{free}}}{mc} \tag{7.114}$$

Inserting this in Heller's bound gives the following theorem, as a counterpart of Theorem 7.12.

Theorem 7.18 The relative free distance $\delta = \frac{d_{\text{free}}}{mc}$ for a convolutional code with the rate $R = b/c$ with encoder memory m satisfies

$$\delta \leq \frac{1}{2}, \quad m \to \infty \tag{7.115}$$

□

As for the block coding case, there are lower bounds on the relative free distance. One first example is to use the Gilbert–Varshamov bound from Theorem 7.14. Thus, there exists a code with rate $R = b/c$ such that the relative free distance is bounded by

$$\delta \geq h^{-1}(1-R) \tag{7.116}$$

Due to Costello [49], a stronger lower bound on time-varying convolutional codes can be stated. The proof of this lies outside the scope of this text, so here it is not stated. For a more thorough treatment, refer to [49, 34, 35].

Theorem 7.19 (Costello) For any rate $R = b/c$, there exists a (time-varying) convolutional code such that

$$\delta \geq \frac{R(1-2^{R-1})}{h(2^{R-1}) + R - 1} \tag{7.117}$$

□

7.3 ERROR-DETECTING CODES

In the previous sections of this chapter, the aim has been to correct errors occurred on the channel. In many situations, the aim of the decoder is instead to detect errors. Then, it is up to the higher layer communication protocols to take care of the result, e.g., by requesting retransmission of a packet. In most communication systems, there are both an error-correcting code and an error-detecting code. The idea is to have the error-correcting code closest to the physical layer. Most errors occurred on the channel should be caught by this and corrected. However, as was seen earlier, no matter how strong the code is there will always be error patterns where the decoder makes the wrong decision, resulting in decoding errors. These errors should be caught by an error-detecting scheme. Often these two coding schemes are located on different logical layers in the communication model. Typically the error-correcting scheme is located close to the channel, i.e. on OSI layer 1, whereas the

u_{11}	$\cdots$	u_{1m}	h_1
$\vdots$		$\vdots$	$\vdots$
u_{n1}	$\cdots$	u_{nm}	h_n
v_1	$\cdots$	v_m	r

Figure 7.14 Horizontal and vertical parity check.

error detection is included in the communication protocol on higher layers. There are, for example, error detection codes both in Ethernet protocol on layer 2, Internet protocol (IP) on layer 3, and transmission control protocol (TCP) on layer 4. The main part of this section will be devoted to CRC codes, typically used in layer 2 protocols.

The simplest version of an error-detecting code is the parity check code. Given a binary vector of length n, the idea is to add one more bit, which is the modulo 2 sum of the bits in the vector. Then there will always be an even number of ones in the vector, hence this scheme is called even parity check.

Example 7.14 Assume a vector of length 7 according to

$$\boldsymbol{u} = 1001010 \tag{7.118}$$

Then the parity bit is derived as the modulo 2 sum, $v = 1$, and the codeword is the concatenation,

$$\boldsymbol{x} = 10010101 \tag{7.119}$$

In this way, the codeword is identified as a byte with even weight.

If a codewords with a parity bit is transmitted over a channel and a single error is introduced, the received word will contain an odd number of ones. This way the receiver can directly see if there has been an error. With the decoding rule that a codeword is accepted if it has an even number of ones, all error patterns with odd weight will be detected. However, all even weight error patterns will pass undetected. The described error detection scheme was used in the original ASCII table, where 128 characters were encoded as bytes, each with even weight. This is equivalent to 7 bits describing 128 characters, and then a parity bit is appended.

An easy way to improve the error detection capabilities is to concatenate two parity check codes. Consider a binary matrix of size $n \times m$, and at each row a parity bit, h_i, is appended and to each column a parity bit, v_j, is appended (see Figure 7.14). This scheme is sometimes called *horizontal and vertical parity check*. The extra parity bit added, in the figure denoted r, can be derived either as a parity bit for the horizontal checks, h_i, or the vertical checks, v_j. The horizontal checks are derived as $h_i = \sum_j u_{ij}$

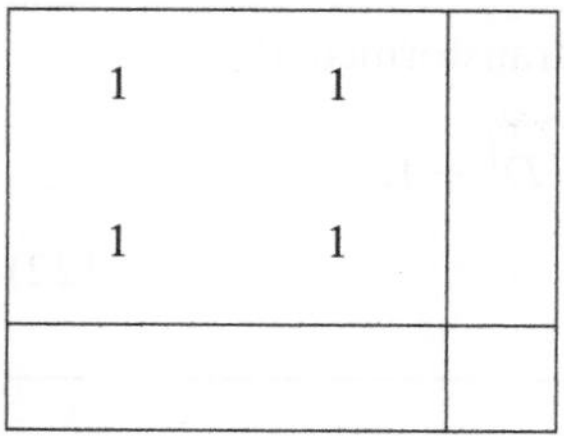

Figure 7.15 An undetected error event of weight four.

and the vertical checks as $v_j = \sum_i u_{ij}$. Then, as a start, let the common parity bit be derived over the horizontal checks, by a change of summation order it can be derived over the vertical checks,

$$r = \sum_i h_i = \sum_i \sum_j u_{ij} = \sum_j \sum_i u_{ij} = \sum_j v_j \tag{7.120}$$

where all summations are performed modulo 2.

A single error in this structure will naturally be detected since it is based on parity check codes. The error events that will not be detected are the cases when there are an even number of errors in both directions, i.e. in columns and rows. The least weight for such event is when four errors are located on the corners of a rectangle (see Figure 7.15). These errors can also occur in the check bits, which is the reason the extra parity bit in the lower right corner is needed.

7.3.1 CRC Codes

To generalize the ideas of the parity bit, more than one bit can be added to a vector. For example in the Hamming code in Example 7.8, three different parity bits are derived over a different set of information bits. An alternative way to generalize the parity bit ideas is given by a CRC code, described in this section. They have become well known and are used in several applications, e.g., Ethernet. One of the reasons is that they can be implemented very efficiently in hardware, and thus, does not add much computational or hardware overhead. The description given here for CRC codes follows partly [50]. The derivations for the codes are based on vectors viewed as polynomials in the variable D, where the exponent of D is the place marker for the coefficient.[4] The vector $\boldsymbol{a} = a_{L-1}a_{L-2} \dots a_1 a_0$ is represented by the polynomial

$$a(D) = \sum_{i=0}^{L} a_i D^i = a_{L-1}D^{L-1} + a_{L-2}D^{L-2} + \cdots + a_1 D + a_0 \tag{7.121}$$

[4] Compared to the notation of the D used for convolutional codes, there is a slight difference since here it is a marker for a position in a vector while for convolutional codes it is a marker for a time instant.

Example 7.15 For example, the vector $\boldsymbol{a} = 1000011$ is transformed as

$$a(D) = 1D^6 + 0D^5 + 0D^4 + 0D^3 + 0D^2 + 1D^1 + 1D^0$$
$$= D^6 + D + 1 \tag{7.122}$$

Let the data to be transmitted consist of a length k binary vector and represent it by the degree $k-1$ polynomial

$$d(D) = d_{k-1}D^{k-1} + d_{k-2}D^{k-2} + \dots + d_1 D + d_0 \tag{7.123}$$

To add redundant bits so the total length of the codeword is n, $n-k$ bits should be appended. These redundant bits, the CRC bits, can be represented by a degree $n-k-1$ polynomial

$$r(D) = r_{n-k-1}D^{n-k-1} + \dots + r_1 D + r_0 \tag{7.124}$$

and the codeword polynomial can be written as

$$c(D) = d(D)D^{n-k} + r(D)$$
$$= d_{k-1}D^{n-1} + \dots + d_0 D^{n-k} + r_{n-k-1}D^{n-k-1} + \dots + r_1 D + r_0 \tag{7.125}$$

To derive the CRC polynomial, $r(D)$, a degree $n-k$ *generator polynomial* is used

$$g(D) = D^{n-k} + g_{n-k-1}D^{n-k-1} + \dots + g_1 D + 1 \tag{7.126}$$

It is a binary polynomial with degree $n-k$ where the highest and lowest coefficients are nonzero, i.e. $g_{n-k} = 1$ and $g_0 = 1$. Then the CRC polynomial is derived as[5]

$$r(D) = R_{g(D)}\big(d(D)D^{n-k}\big) \tag{7.127}$$

The polynomial division is performed in the same manner as normal polynomial division, except that all coefficients are binary and modulo 2 arithmetic is used. The procedure is shown in the following example.

Example 7.16 Assume the data word $\boldsymbol{d} = 1001$ and three CRC bits should be added. In that case $k = 4$ and $n = 7$. The data word is represented as a polynomial as $d(D) = D^3 + 1$. Find a degree three generator polynomial, say $g(D) = D^3 + D + 1$.

[5] The notation $R_a(b)$ means the reminder from the division b/a. That is, if a and b are polynomials, then d and r can (uniquely) be found such that $b = a \cdot d + r$ where $\deg(r) < \deg(a)$. In some texts, it is denoted with the modulo operator as $\mathrm{mod}(b, a)$

Later in this section, it will be considered how to choose these polynomials. Performing the division $d(D)D^3/g(D)$,

$$\begin{array}{r|l}
 & D^3 + D \\ \hline
D^3 + D + 1 & D^6 + \qquad\quad D^3 \\
 & D^6 + \quad D^4 + D^3 \\ \cline{2-2}
 & \qquad\quad D^4 \\
 & \qquad\quad D^4 + \quad D^2 + D \\ \cline{2-2}
 & \qquad\qquad\qquad D^2 + D
\end{array}$$

gives that

$$\frac{d(D)D^3}{g(D)} = \frac{D^6 + D^3}{D^3 + D + 1} = D^3 + D + \frac{D^2 + D}{D^3 + D + 1} \tag{7.128}$$

Hence, the CRC polynomial is

$$r(D) = R_{D^3+D+1}\big(D^6 + D^3\big) = D^2 + D \tag{7.129}$$

and the codeword polynomial

$$c(D) = D^6 + D^3 + D^2 + D \tag{7.130}$$

The codeword rewritten as a binary vector becomes $\boldsymbol{c} = 1001110$.

To see how the receiver side can use this codeword to detect errors, first a couple of properties are noted. Let $z(D)$ denote the quotient in the division $d(D)D^{n-k}/g(D)$. In the previous example, $z(D) = D^3 + D$. Then, the data polynomial can be written as

$$d(D)D^{n-k} = g(D)z(D) + r(D) \tag{7.131}$$

Equivalently, in modulo 2 arithmetic, the codeword polynomial becomes

$$c(D) = d(D)D^{n-k} + r(D) = g(D)z(D) \tag{7.132}$$

That is, all codeword polynomials are divisible by the generator polynomial $g(D)$, and all polynomials (with degree less than n) that are divisible by $g(D)$ are polynomials for codewords. Stated as a theorem the following is obtained.[6]

Theorem 7.20 A polynomial $c(D)$ with $\deg(c(D)) < n$ is a codeword if and only if $g(D)|c(D)$. □

If $c(D)$ is transmitted over a channel and there occur errors, they can be represented by an addition of the polynomial $e(D)$, and the received polynomial is $y(D) = c(D) + e(D)$.

[6] The notation $a|b$ means the division b/a has a zero reminder, i.e. that b is a multiple of a. For integers, for example $3|12$ but $3 \nmid 10$.

Example 7.17 According to the previous example the codeword transmitted over the channel is $\boldsymbol{c} = 1001110$. Assuming the channel introduces an error in the third bit, the received vector is $\boldsymbol{y} = 1011110$. The error vector can be seen as $\boldsymbol{e} = 0010000$ and the

$$y = c \oplus e = 1001110 + 0010000 = 1011110 \tag{7.133}$$

where the addition is performed positionwise. Since modulo 2 arithmetic is used the function $a + 1$ will always invert the bit a. Expressed in polynomial form the error polynomial is $e(D) = D^4$, and the received polynomial

$$y(D) = c(D) + e(D) = D^6 + D^4 + D^3 + D^2 + D \tag{7.134}$$

where the addition is performed over the binary field, i.e. as modulo 2 addition for the coefficients.

For error detection, the receiver can use the fact that $g(D)$ is a factor of each transmitted codeword. Similar to the parity check example, where the receiver detects an error if the received vector has an odd number of ones, in this case the receiver will detect an error if $g(D)$ is not a factor. As a test the syndrome is derived as the reminder of the division $c(D)/g(D)$,

$$\begin{aligned} s(D) &= R_{g(D)}(y(D)) = R_{g(D)}(c(D) + e(D)) \\ &= R_{g(D)}(R_{g(D)}(c(D)) + R_{g(D)}(e(D))) = R_{g(D)}(e(D)) \end{aligned} \tag{7.135}$$

Notice that the syndrome is directly a function of the error since $R_{g(D)}\big(c(D)\big) = 0$. If there are no errors in the transmission, the error polynomial is $e(D) = 0$ and the syndrome $s(D) = 0$. So the criteria for assuming error-free transmission at the receiver side is that $s(D) = 0$. In the case when there are errors, and $e(D) \neq 0$, these will be detected if $s(D) \neq 0$.

Example 7.18 In the previous example, the received vector $\boldsymbol{y} = 1011110$ was considered. The corresponding polynomial representation is $y(D) = D^6 + D^4 + D^3 + D^2 + D$. With the generator polynomial $g(D) = D^3 + D + 1$, the division

$$\frac{y(D)}{g(D)} = \frac{D^6 + D^4 + D^3 + D^2 + D}{D^3 + D + 1} = D^3 + \frac{D^2 + D}{D^3 + D + 1} \tag{7.136}$$

gives the syndrome $s(D) = R_{g(D)}(y(D)) = D^2 + D$. Since this is nonzero, $\boldsymbol{y}$ is not a codeword and an error has been detected.

In some cases, when $e(D) = g(D)p(D)$ for some nonzero polynomial $p(D)$, the syndrome will also be zero and the error will not be detected. That is, an error event will not be detected if it is a codeword. For this purpose, a short investigation on the error detection capability is conducted. For a deeper analysis, refer to [50].

TABLE 7.1 A list of primitive polynomials up to degree 17.

$p(D)$	$p(D)$
D^2+D+1	$D^{10}+D^3+1$
D^3+D+1	$D^{11}+D^2+1$
D^4+D+1	$D^{12}+D^6+D^4+D+1$
D^5+D^2+1	$D^{13}+D^4+D^3+D+1$
D^6+D+1	$D^{14}+D^{10}+D^6+D+1$
D^7+D^3+1	$D^{15}+D+1$
$D^8+D^4+D^3+D^2+1$	$D^{16}+D^{12}+D^9+D^7+1$
D^9+D^4+1	$D^{17}+D^3+1$

First, assume that a single error in position i has occurred, i.e. $e(D)=D^i$. Since $g(D)$ has at least two nonzero coefficients so will $g(D)p(D)$, and it cannot be on the form D^i. Hence, all single errors will be detected by the scheme. In fact, by using the generator polynomial $g(D)=1+D$ is equivalent to adding a parity check bit, which is used to detect one error.

If there are two errors during the transmission, say in position i and j, where $i<j$, the error polynomial is

$$e(D)=D^j+D^i=D^i\left(D^{j-i}+1\right) \tag{7.137}$$

This error will not be detected in the case when $g(D)$ divides $D^{j-i}+1$. From algebra, it is known that if $\deg(g(D))=L$ the least K such that $g(D)|D^K+1$ does not exceed 2^L-1. Furthermore, it is always possible to find a polynomial for which there is equality, i.e. where $K=2^L-1$ is the least integer such that $g(D)|D^K+1$. These polynomials are called *primitive polynomials*. So, by using a primitive polynomial $p(D)$ of degree $\deg(p(D))=L$, all errors on the form $D^{j-i}+1$ will be detected as long as $j-i<2^L-1$. By choosing the codeword length n such that $n-1<2^L-1$, or equivalently by choosing $\deg(p(D))=L>\log n$, there is no combination of i and j such that $p(D)$ divides $D^{j-i}+1$, and therefore all double errors will be detected. In Table 7.1, primitive polynomials up to degree 17 is listed. For a more extensive list, refer to, e.g., [35].

To see how the system can cope with three errors, notice that if a binary polynomial is multiplied with $D+1$ it will contain an even number of nonzero coefficients.[7] Thus, if the generator polynomial contains the factor $D+1$, all polynomials on the form $g(D)z(D)$ will have an even number of nonzero coefficients and all occurrences of an odd number of errors will be detected.

So, to chose a generator polynomial, assume that the length of the data frame is k bits. Then to be able to detect double errors, a primitive polynomial $p(D)$ of degree L is chosen. To get the generator polynomial, this should be multiplied with $D+1$, $g(D)=p(D)(1+D)$. That is, the degree of the generating polynomial is $L+1$ and the

[7] For binary polynomials, it can be found that $(D+1)(D^\alpha+D^{\alpha-1}+\cdots+D^{\alpha-\beta})=D^{\alpha+1}+D^{\alpha-\beta}$ where $\alpha,\beta\in\mathbb{Z}^+$. Generalizing this leads to $(D+1)q(D)$ that will have an even number of nonzero coefficients.

length of the codeword is $n = k + L + 1$. The primitive polynomial should therefore be chosen such that

$$k + L < 2^L - 1 \tag{7.138}$$

To summarize the error detection capabilities, if the generator polynomial is chosen correctly, it is always possible to detect single, double, and triple errors. Apart from this, all errors with odd weight can be detected.

Some well-known CRC generator polynomials are

$$g(D) = D^8 + D^7 + D^6 + D^4 + D^2 + 1 \qquad \text{(CRC-8)}$$

$$g(D) = D^8 + D^2 + D + 1 \qquad \text{(CRC-8 CCITT)}$$

$$g(D) = D^{10} + D^9 + D^5 + D^4 + D^2 + 1 \qquad \text{(CRC-10)}$$

$$g(D) = D^{16} + D^{15} + D^2 + 1 \qquad \text{(CRC-16)}$$

$$g(D) = D^{16} + D^{12} + D^5 + 1 \qquad \text{(CRC-16 CCITT)}$$

$$g(D) = D^{32} + D^{26} + D^{23} + D^{22} + D^{16} + D^{12} + D^{11} + D^{10} + D^8 + D^7 + D^5 + D^4 + D^2 + D + 1 \qquad \text{(CRC-32)}$$

$$g(D) = D^{64} + D^4 + D^3 + D + 1 \qquad \text{(CRC-64)}$$

PROBLEMS

7.1 In a coding scheme, three information bits $\boldsymbol{u} = (u_0, u_1, u_2)$ are appended with three parity bits according to

$$v_0 = u_1 + u_2$$
$$v_1 = u_0 + u_2$$
$$v_2 = u_0 + u_1$$

Hence, an information word $\boldsymbol{u} = (u_0, u_1, u_2)$ is encoded to the codeword $\boldsymbol{x} = (u_0, u_1, u_2, v_0, v_1, v_2)$.

(a) What is the code rate R?

(b) Find a generator matrix G.

(c) What is the minimum distance, $d_{\min}$, of the code?

(d) Find a parity check matrix H, such that $GH^T = 0$.

(e) Construct a syndrome table for decoding.

(f) Make an example where a three bit vector is encoded, transmitted over a channel and decoded.

7.2 Show that if $d_{\min} \geq \lambda + \gamma + 1$ for a linear code, it is capable of correcting λ errors and simultaneously detecting γ errors, where $\gamma > \lambda$.

7.3 One way to extend the code $\mathcal{B}$ is to add one more bit such that the codeword has even Hamming weight, i.e.

$$\mathcal{B}_E = \{(y_1 \dots y_n y_{n+1}) | (y_1 \dots y_n) \in \mathcal{B} \text{ and } y_1 + \cdots + y_n + y_{n+1} = 0 \text{ (mod 2)}\}$$

(a) Show that if $\mathcal{B}$ is a linear code, so is $\mathcal{B}_E$. If you instead extend the code with a bit such that the number of ones is odd, will the code still be linear?

(b) Let H be the parity check matrix for the code $\mathcal{B}$ and show that

$$H_E = \begin{pmatrix} & & & 0 \\ & H & & \vdots \\ & & & 0 \\ 1 & \cdots & 1 & 1 \end{pmatrix}$$

is the parity check matrix for the extended code $\mathcal{B}_E$.

(c) What can you say about the minimum distance for the extended code?

7.4 In the early days of computers, the ASCII table consisted of seven bit vectors where an extra parity bit was appended such that the vector always had even number of ones. This was an easy way to detect errors in, e.g., punch cards. What is the parity check matrix for this code?

7.5 Plot, using, e.g., MATLAB, the resulting bit error rate as a function of E_b/N_0 when using binary repetition codes of rate $R = 1/3$, $R = 1/5$, and $R = 1/7$. Compare with the uncoded case. Notice that E_b is the energy per information bit, i.e. for a rate $R = 1/N$ the energy per transmitted bit is E_b/N. The noise parameter is naturally independent of the code rate.

7.6 Verify that the free distance for the code generated by the generator matrix generator matrix

$$G(D) = \begin{pmatrix} 1 + D + D^2 & 1 + D^2 \end{pmatrix}$$

is $d_{\text{free}} = 5$. Decode the received sequence

$$\boldsymbol{r} = 01\ 11\ 00\ 01\ 11\ 00\ 01\ 00\ 10$$

7.7 A convolutional code is formed from the generator matrix

$$G(D) = (1 + D \quad 1 + D + D^2)$$

(a) Derive the free distance d_{free}.

(b) Decode the received sequence

$$\boldsymbol{r} = 01\ 11\ 00\ 01\ 11\ 00\ 01\ 00\ 10$$

Assume that the encoder is started and ended in the all-zero state.

7.8 Repeat Problem 7.7 for the generator matrix

$$G(D) = (1 + D + D^2 + D^3 \quad 1 + D + D^3)$$

7.9 For the generator matrix in Problem 7.6, show that the generator matrix

$$G_s(D) = \begin{pmatrix} \frac{1 + D + D^2}{1 + D^2} & 1 \end{pmatrix}$$

will give the same code as $G(D)$.

Data	CRC

Figure 7.16 Six data bits and four bits CRC.

7.10 Suppose a 4-bit CRC with generator polynomial $g(x) = x^4 + x^3 + 1$ has been used. Which, if any, of the following three messages will be accepted by the receiver?

(a) 11010111

(b) 10101101101

(c) 10001110111

7.11 Consider a data frame with six bits where a four bit CRC is added at the end, see Figure 7.16.

To calculate the CRC bit,s the following generator polynomial is used:

$$g(x) = (x + 1)(x^3 + x + 1) = x^4 + x^3 + x^2 + 1$$

(a) Will the encoding scheme be able to detect all

- single errors?
- double errors?
- triple errors?
- quadruple errors?

(b) Assume the data vector $\boldsymbol{d} = 010111$ should be transmitted. Find the CRC bits for the frame. Then, introduce an error pattern that is detectable and show how the detection works.

CHAPTER 8

INFORMATION MEASURES FOR CONTINUOUS VARIABLES

THE CHANNELS CONSIDERED in the previous chapters are discrete valued and memoryless. These channel models are very widespread and can represent the real-world signaling in many cases. On that abstraction level, they typically represent correct transmission, errors, and erasures. However, in many cases the modeling is done on another abstraction level, closer to the actual signals and transmission. In those models, the signals can be viewed as continuous valued random variables, and the noise on the channel is often considered to be additive white noise. In this chapter, real-valued random variables are considered and the information measures adopted for this case. In the next chapter, this will be used to find the capacity for time discrete channels like the additive white Gaussian noise channel.

8.1 DIFFERENTIAL ENTROPY AND MUTUAL INFORMATION

For a discrete valued random variable X, the entropy is defined as the expected value of the self-information, $H(X) = E\big[-\log p(X)\big]$. Using a similar function for continuous variables results in the following definition. To keep the definitions apart, it is common to name this value the *differential entropy.*[1]

Definition 8.1 Let X be a real continuous random variable with probability density function $f(x)$. The *differential entropy* is

$$H(X) = E\big[-\log f(X)\big] = -\int_{\mathbb{R}} f(x)\log f(x)dx \tag{8.1}$$

where the convention $0 \log 0 = 0$ is used. □

[1] For simplicity, it is assumed that the variables are "real-valued, $x \in \mathbb{R}$," hence the notation $\int_{\mathbb{R}} dx = \int_{-\infty}^{\infty} dx$. For variables defined on other ranges, e.g., $x \in S$, the integration should be $\int_S dx$. The definition can also include multidimensional variables or vectors, as in $\int_{\mathbb{R}^n} dx$.

Information and Communication Theory, First Edition. Stefan Höst.

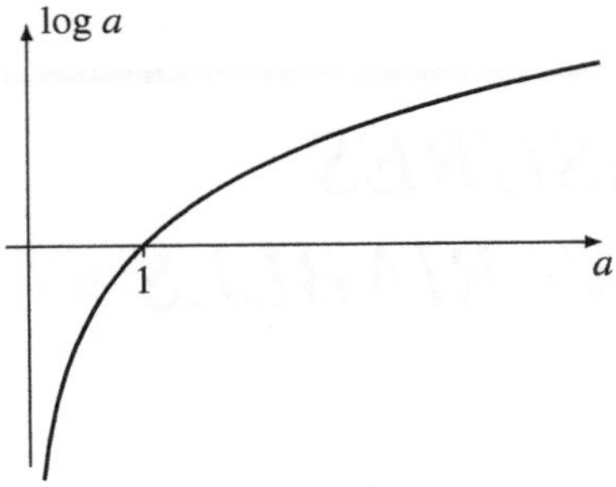

Figure 8.1 The log function.

In the literature, there are other notations as well, among others $H(f)$ is used to denote the differential entropy of the density function $f(x)$. This is in some cases quite handy and will occasionally be used in this text, similar to the use of $H(p_1, \dots, p_N)$ for the discrete case.

Example 8.1 For a uniformly distributed random variable $X \sim U(a)$, $a > 0$, the density function is given by

$$f(x) = \begin{cases} \frac{1}{a}, & 0 \leq x \leq a \\ 0, & \text{otherwise} \end{cases} \tag{8.2}$$

The differential entropy becomes

$$H(X) = -\int_0^a \frac{1}{a} \log \frac{1}{a} dx = \log a \int_0^a \frac{1}{a} dx = \log a \tag{8.3}$$

As can be seen in Figure 8.1, the log function is negative when the argument is less than 1. Hence, for a uniform distribution with $a < 1$ the entropy is negative. If $a = 1$ the entropy is zero, and if $a > 1$ the entropy is positive. The fact that the entropy can now be negative means that the interpretation of entropy as the uncertainty can no longer be motivated.

In Appendix A, the most common distributions are listed together with their mean, variance, and entropy. In this listing, the entropy is derived over the natural base, since this makes the integration slightly more natural. Then the unit is *nats* instead of bits. By elementary logarithm laws, the translation between base 2 and base e can be found as

$$H(X) = E\big[-\log f(X)\big] = \frac{E\big[-\ln f(X)\big]}{\ln 2} = \frac{H_e(X)}{\ln 2} \tag{8.4}$$

It can be useful to notice that translation with a constant c of a random variable does not affect the entropy,

$$\begin{aligned} H(X + c) &= -\int_{\mathbb{R}} f(x - c) \log f(x - c) dx \\ &= -\int_{\mathbb{R}} f(z) \log f(z) dz = H(X) \end{aligned} \tag{8.5}$$

where the variable change $z = x - c$ was used. If a random variable X with density function $f(x)$ is scaled by a constant α the density functions is $f_{\alpha X}(x) = \frac{1}{\alpha}f\left(\frac{x}{\alpha}\right)$. Then the entropy becomes

$$\begin{aligned}
H(\alpha X) &= -\int_{\mathbb{R}} \frac{1}{\alpha}f\left(\tfrac{x}{\alpha}\right) \log \frac{1}{\alpha}f\left(\tfrac{x}{\alpha}\right)dx \\
&= \log \alpha \int_{\mathbb{R}} \frac{1}{\alpha}f\left(\tfrac{x}{\alpha}\right)dx - \int_{\mathbb{R}} f\left(\tfrac{x}{\alpha}\right) \log f\left(\tfrac{x}{\alpha}\right)\frac{dx}{\alpha} \\
&= \log \alpha - \int_{\mathbb{R}} f(z) \log f(z)dz \\
&= H(X) + \log \alpha
\end{aligned} \tag{8.6}$$

The above derivations can be summarized in the following theorem.

Theorem 8.1 Consider the continuous random variable X and form a new random variable $Y = \alpha X + c$, where α and c are real-valued constants. Then

$$H(Y) = H(\alpha X + c) = H(X) + \log \alpha \tag{8.7}$$

□

In the next example, the differential entropy for a Gaussian variable is derived by first scaling it to a normalized Gaussian variable.

Example 8.2 Let X be a Gaussian (normal) distributed random variable, $X \sim N(\mu, \sigma)$. To make derivations easier, consider first a normalized variable $Y = \frac{X-\mu}{\sigma}$, where $Y \sim N(0, 1)$, with the density function

$$f(y) = \frac{1}{\sqrt{2\pi}}e^{-y^2/2} \tag{8.8}$$

The entropy of Y can be derived as

$$\begin{aligned}
H(Y) &= -\int_{\mathbb{R}} f(y) \log f(y)dy = -\int_{\mathbb{R}} f(y) \log \frac{1}{\sqrt{2\pi}}e^{-y^2/2}dy \\
&= -\int_{\mathbb{R}} f(y) \log \frac{1}{\sqrt{2\pi}}dy - \int_{\mathbb{R}} f(y) \log e^{-y^2/2}dy \\
&= \frac{1}{2}\log(2\pi)\int_{\mathbb{R}} f(y)dy + \frac{1}{2}\log(e)\int_{\mathbb{R}} y^2 f(y)dy \\
&= \frac{1}{2}\log(2\pi e)
\end{aligned} \tag{8.9}$$

According to Theorem 8.1, the entropy for X can be derived as

$$\begin{aligned}
H(X) &= H(\sigma Y + \mu) = H(Y) + \log(\sigma) \\
&= \frac{1}{2}\log(2\pi e) + \frac{1}{2}\log(\sigma^2) = \frac{1}{2}\log(2\pi e\sigma^2)
\end{aligned} \tag{8.10}$$

which is the listed function in Appendix A, except that the natural base in the logarithm in the appendix. It should also be noted that for a Gaussian distribution the entropy can be negative. It will be zero for $\log(2\pi e\sigma^2) = 0$, or equivalently, $\sigma^2 = \frac{1}{2\pi e}$. That is,

$$\sigma^2 > \frac{1}{2\pi e} \Rightarrow H(X) > 0 \tag{8.11}$$

$$\sigma^2 = \frac{1}{2\pi e} \Rightarrow H(X) = 0 \tag{8.12}$$

$$\sigma^2 < \frac{1}{2\pi e} \Rightarrow H(X) < 0 \tag{8.13}$$

As for the discrete case, random vectors can be viewed as multidimensional random variables. $(X_1, X_2, \ldots, X_n)$. The entropy can still be defined as the expectation of the logarithmic density function,

$$H(X_1, \ldots, X_n) = E\big[-\log f(X_1, \ldots, X_n)\big] \tag{8.14}$$

Especially for the two-dimensional case, the joint differential entropy is defined below.

Definition 8.2 The joint differential entropy for a two-dimensional random variable (X, Y) with density function $f(x, y)$ is

$$H(X, Y) = E\big[-\log f(X, Y)\big] = -\int_{\mathbb{R}^2} f(x, y) \log f(x, y) dxdy \tag{8.15}$$

□

Example 8.3 The two-dimensional continuous random variable (X, Y) has the density function

$$f(x, y) = \begin{cases} A, & x > 0, \quad y > 0, \quad ax + by < ab \\ 0, & \text{otherwise} \end{cases} \tag{8.16}$$

That is, it is uniformly distributed over the gray area as shown in Figure 8.2.

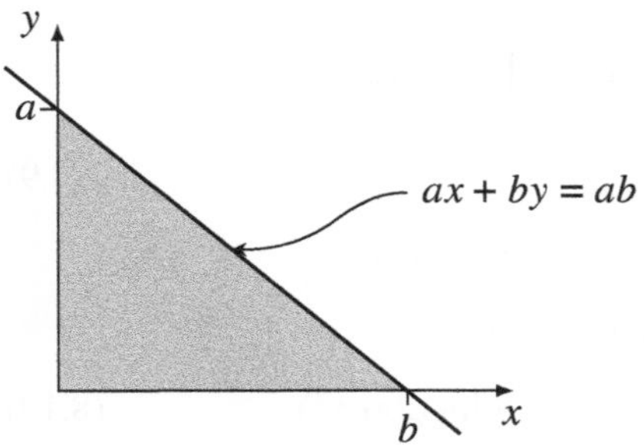

Figure 8.2 The area where $f(x, y) = A$.

First the amplitude of the density function, A, must be determined. Using that

$$1 = \int_{\triangle} A dx dy = A \int_{\triangle} 1 dx dy = A\frac{ab}{2} \tag{8.17}$$

gives $A = \frac{2}{ab}$. Then the joint entropy becomes

$$\begin{aligned} H(X, Y) &= -\int_{\triangle} A \log A dx dy \\ &= -\log A \int_{\triangle} A dx dy \\ &= \log \frac{1}{A} = \log \frac{ab}{2} = \log ab - 1 \end{aligned} \tag{8.18}$$

Next, the mutual information is defined for continuous variables as a straightforward generalization of the discrete case.

Definition 8.3 The *mutual information* for a pair of continuous random variables (X, Y) with joint probability density function $f(x, y)$ is

$$I(X;Y) = E\Big[\log \frac{f(X,Y)}{f(X)f(Y)}\Big] = \int_{\mathbb{R}^2} f(x,y) \log \frac{f(x,y)}{f(x)f(y)} dx dy \tag{8.19}$$

□

From the definition, it can be directly concluded that the mutual information is symmetric, i.e. $I(X;Y) = I(Y;X)$. By breaking up the logarithm in a sum, the function can be rewritten using the entropy functions,

$$\begin{aligned} I(X;Y) &= E\Big[\log \frac{f(X,Y)}{f(X)f(Y)}\Big] \\ &= E\big[\log f(X,Y) - \log f(X) - \log f(Y)\big] \\ &= E\big[-\log f(X)\big] + E\big[-\log f(Y)\big] - E\big[-\log f(X,Y)\big] \\ &= H(X) + H(Y) - H(X,Y) \end{aligned} \tag{8.20}$$

Similar to the discrete case, the conditional entropy is defined as follows.

Definition 8.4 The conditional differential entropy is the differential entropy for the random variable X conditioned on the random variable Y and is written as

$$H(X|Y) = E\big[-\log f(X|Y)\big] = -\int_{\mathbb{R}^2} f(x,y) \log f(x|y) dx dy \tag{8.21}$$

where $f(x, y)$ is the joint density function and $f(x|y)$ is the conditional density function.

□

As a counterpart to (3.33), the conditional differential entropy can be derived as

$$\begin{aligned}
H(X|Y) &= -\int_{\mathbb{R}^2} f(x,y)\log f(x|y)dxdy \\
&= -\int_{\mathbb{R}^2} f(x|y)f(y)\log f(x|y)dxdy \\
&= \int_{\mathbb{R}} \left(-\int_{\mathbb{R}} f(x|y)\log f(x|y)dx\right) f(y)dy \\
&= \int_{\mathbb{R}} H(X|Y=y)f(y)dy
\end{aligned} \tag{8.22}$$

where $H(X|Y=y) = -\int_{\mathbb{R}} f(x|y)\log f(x|y)dx$ is the differential entropy conditioned on the event $Y = y$.

The joint entropy can be written as

$$\begin{aligned}
H(X,Y) &= E\big[-\log f(X,Y)\big] \\
&= E\big[-\log f(X|Y)\big] + E\big[-\log f(Y)\big] \\
&= H(X|Y) + H(Y)
\end{aligned} \tag{8.23}$$

or, similarly,

$$H(X,Y) = H(Y|X) + H(X) \tag{8.24}$$

Combining the above gives the following theorem for the mutual information.

Theorem 8.2 Let X and Y be two continuous random variables. Then the mutual information can be derived as

$$\begin{aligned}
I(X;Y) &= H(X) - H(X|Y) \\
&= H(Y) - H(Y|X) \\
&= H(X) + H(Y) - H(X,Y)
\end{aligned} \tag{8.25}$$

□

Example 8.4 To derive the mutual information between the two variables in Example 8.3, the entropy for the individual variables X and Y is needed. Starting with X, the density function is

$$\begin{aligned}
f(x) &= \int_{\mathbb{R}} f(x,y)dy = \int_{y=0}^{-\frac{a}{b}x+a} \frac{2}{ab}dy \\
&= \frac{2}{ab}\left(-\frac{a}{b}x + a\right) = -\frac{2}{b^2}x + \frac{2}{b}
\end{aligned} \tag{8.26}$$

This is a triangular distribution starting at $f(0) = \frac{2}{b}$ and decreasing linearly to $f(b) = 0$ (see Figure 8.3).

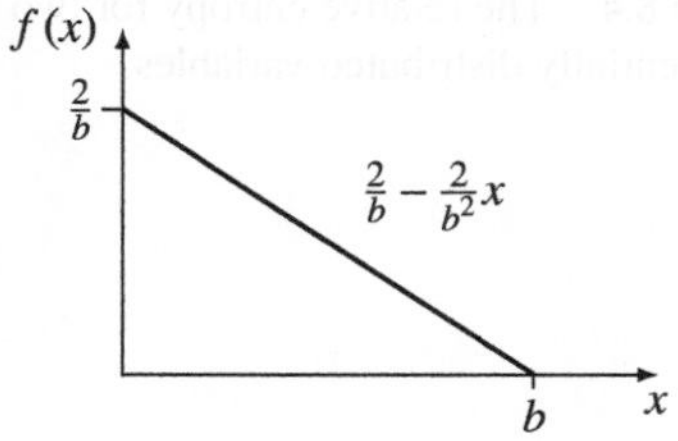

Figure 8.3 The density function of X.

To derive the entropy use the variable change $z = -\frac{2}{b^2}x + \frac{2}{b}$ and integration by parts to get

$$\begin{aligned}
H(X) &= -\int_0^b \Big(-\frac{2}{b^2}x + \frac{2}{b}\Big)\log\Big(-\frac{2}{b^2}x + \frac{2}{b}\Big)dx \\
&= -\frac{b^2}{2\ln 2}\int_0^{2/b} z\ln z dz \\
&= -\frac{b^2}{2\ln 2}\Big(\Big[\frac{z^2}{2}\ln z - \frac{z^2}{4}\Big]_0^{2/b}\Big) \\
&= -\frac{b^2}{2\ln 2}\Big(\frac{2}{b^2}\ln\frac{2}{b} - \frac{1}{b^2}\Big) \\
&= \log\frac{b}{2} + \frac{1}{2\ln 2} = \log b\sqrt{e} - 1
\end{aligned} \tag{8.27}$$

Similarly, the entropy of Y is $H(Y) = \log a\sqrt{e} - 1$. Then, the mutual information between X and Y becomes

$$\begin{aligned}
I(X;Y) &= H(X) + H(Y) - H(X,Y) \\
&= \log b\sqrt{e} - 1 + \log a\sqrt{e} - 1 - \log ab + 1 \\
&= \log e - 1 = \log\frac{e}{2}
\end{aligned} \tag{8.28}$$

There are two things to notice by the previous example. First, the mutual information, in this case, is not dependent on the constants a and b. Second, which is more important, is that the mutual information is a positive number. To see that it is not only in this example where the mutual information is nonegative, the relative entropy is generalized for the continuous case.

Definition 8.5 The *relative entropy* for a pair of continuous random variables with probability density functions $f(x)$ and $g(y)$ is

$$D(f||g) = E_f\Big[\log\frac{f(X)}{g(X)}\Big] = \int_{\mathbb{R}} f(x)\log\frac{f(x)}{g(x)}dx \tag{8.29}$$

□

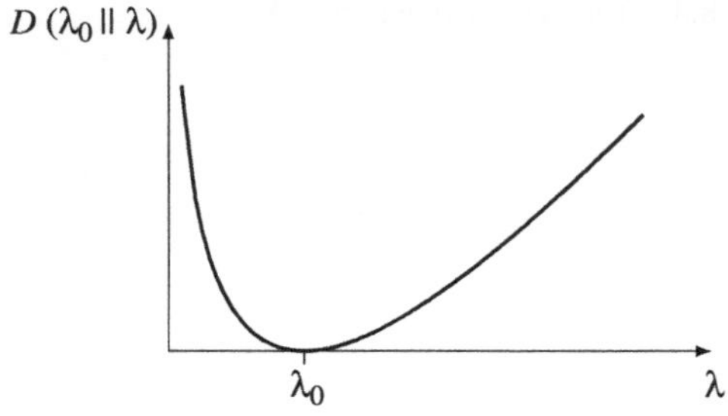

Figure 8.4 The relative entropy for two exponentially distributed variables.

Example 8.5 Consider a random variable that is exponentially distributed, Exp(λ_0). The density function is

$$f(x) = \lambda_0 e^{-\lambda_0 x}, \qquad x \geq 0 \tag{8.30}$$

and the mean is $E[X] = \frac{1}{\lambda_0}$. The relative entropy between this distribution and another exponential distribution Exp(λ) is

$$\begin{aligned} D(\lambda_0||\lambda) &= E_{\lambda_0}\left[\log \frac{\lambda_0 e^{-\lambda_0 X}}{\lambda e^{-\lambda X}}\right] \\ &= E_{\lambda_0}\left[\log \frac{\lambda_0}{\lambda} - (\lambda_0 - \lambda)X \log e\right] \\ &= \log \frac{\lambda_0}{\lambda} - (\lambda_0 - \lambda)E_{\lambda_0}[X] \log e \\ &= \left(\frac{\lambda}{\lambda_0} - 1\right) \log e - \log \frac{\lambda}{\lambda_0} \end{aligned} \tag{8.31}$$

In Figure 8.4, the function is shown. Since

$$\frac{\partial}{\partial \lambda} D(\lambda_0||\lambda) = \frac{1}{\ln 2}\left(\frac{1}{\lambda_0} - \frac{1}{\lambda}\right) \tag{8.32}$$

it has a minimum at $\lambda = \lambda_0$, where $D(\lambda_0||\lambda) = 0$.

As seen in the previous example, the relative entropy between two exponential functions is a nonnegative function. To generalize the result and see that the relative entropy is nonnegative for the continuous case, first denote the support for $f(x)$ and $g(x)$ by $\mathcal{S}_f$ and $\mathcal{S}_g$, respectively. The support is the interval where the function is strictly positive. When both functions are zero, use the convention that $0 \log \frac{0}{0} = 0$ and when $f(x) = 0$ but $g(x) \neq 0$ that $0 \log 0 = 0$. The problem comes when $f(x) \neq 0$ and $g(x) = 0$, as the function tends to infinity. However, right now the purpose is to show that the function is nonnegative, which is of course true for infinity. Therefore, without loss of generality, consider the case when $S_f \subseteq S_g$. Then, by using the IT-inequality,

$$\begin{aligned}
D(f||g) &= \int_{S_f} f(x) \log \frac{f(x)}{g(x)} dx \\
&= -\int_{S_f} f(x) \log \frac{g(x)}{f(x)} dx \\
&\geq -\int_{S_f} f(x) \Big(\frac{g(x)}{f(x)} - 1 \Big) \log e \, dx \\
&= \Big(\int_{S_f} f(x) dx - \int_{S_f} g(x) dx \Big) \log e \\
&\geq (1-1) \log e = 0
\end{aligned} \tag{8.33}$$

where there is equality if and only if $f(x) = g(x)$ for all x. The result can be stated as a theorem.

Theorem 8.3 The relative entropy for continuous random distributions is a non-negative function,

$$D(f||g) \geq 0 \tag{8.34}$$

with equality if and only if $f(x) = g(x)$, for all x. □

The mutual information can be expressed with the relative entropy as

$$I(X;Y) = D(f(X,Y)||f(X)f(Y)) \tag{8.35}$$

which is a nonnegative function.

Corollary 8.1 The mutual information for continuous random variables is the non-negative function,

$$I(X;Y) \geq 0 \tag{8.36}$$

with equality if and only if X and Y are independent. □

From $I(X;Y) = H(X) - H(X|Y)$ the following corollary can be derived.

Corollary 8.2 The differential entropy will not increase by considering side information,

$$H(X|Y) \leq H(X) \tag{8.37}$$

with equality if and only if X and Y are independent. □

The latter corollary can be generalized by the chain rule for probabilities,

$$f(x_1, \dots, x_n) = \prod_{i=1}^{n} f(x_i|x_1, \dots, x_{i-1}) \tag{8.38}$$

Hence, the n-dimensional differential entropy can be written as

$$\begin{aligned}
H(X_1,\dots,X_n) &= E\big[-\log f(X_1,\dots,X_n)\big]\\
&= E\Big[-\log \prod_{i=1}^{n} f(X_i|X_1,\dots,X_{i-1})\Big]\\
&= \sum_{i=1}^{n} E\big[-\log f(X_i|X_1,\dots,X_{i-1})\big]\\
&= \sum_{i=1}^{n} H(X_i|X_1,\dots,X_{i-1})
\end{aligned} \tag{8.39}$$

Using $H(X_i|X_1,\dots,X_{i-1}) \leq H(X_i)$, it also means the differential entropy for the vector is not more than the sum of the differential entropies for the individual variables,

$$H(X_1,\dots,X_n) \leq \sum_{i=1}^{n} H(X_i) \tag{8.40}$$

8.1.1 Relation between Discrete and Continuous Information Measures

In the previous section, it was seen that the definition for the differential entropy is not consistent with the interpretation of the entropy as uncertainty of the random variable. One way to understand this is to discretize a continuous density function to obtain a discrete variable. Given the continuous random variable X with density function $f(x)$, define a discrete random variable X^{Δ}, where the probability for the outcome x_k^{Δ} is

$$p(x_k^{\Delta}) = \int_{k\Delta}^{(k+1)\Delta} f(x)dx = \Delta f(x_k) \tag{8.41}$$

The existence of such x_k in the interval $k\Delta \leq x_k \leq (k+1)\Delta$ in the second equality is guaranteed by the mean value theorem in integral calculus (see Figure 8.5). The entropy, or uncertainty, of this discrete variable is

$$\begin{aligned}
H(X^{\Delta}) &= -\sum_k p(x_k^{\Delta}) \log p(x_k^{\Delta})\\
&= -\sum_k \Delta f(x_k) \log \Delta f(x_k)\\
&= -\sum_k \Delta\big(f(x_k) \log f(x_k)\big) - \Big(\sum_k \Delta f(x_k)\Big) \log \Delta
\end{aligned} \tag{8.42}$$

The first term in (8.42) is a Riemann sum with a limit value as the differential entropy,

$$-\sum_k \Delta\big(f(x_k) \log f(x_k)\big) \rightarrow -\int_{\mathbb{R}} f(x) \log f(x)dx = H(X), \quad \Delta \rightarrow 0 \tag{8.43}$$

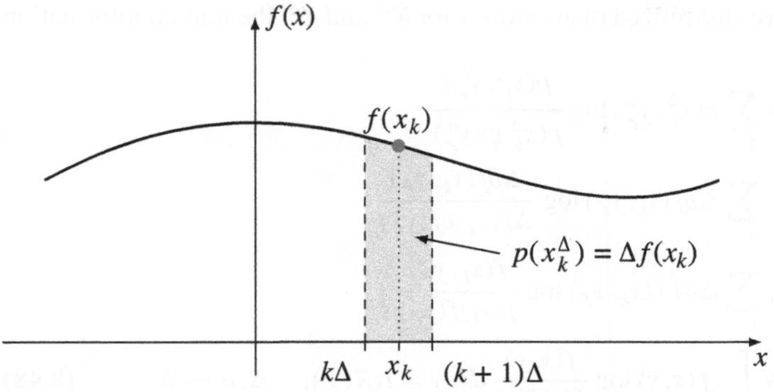

Figure 8.5 Creation of a discrete random variable from a continuous.

Similarly, the first part of the second term is $\sum_k \Delta f(x_k) \to \int_{\mathbb{R}} f(x)dx = 1$, as $\Delta \to \infty$. However, that means the second term becomes $-\log \Delta \to \infty$ as $\Delta \to 0$. As long as the differential entropy is finite, the uncertainty $H(X^{\Delta})$ does not converge as $\Delta \to 0$. Actually, this is reasonable, since for most distributions the number of outcomes for X^{Δ} grows to infinity as $\Delta \to 0$, and then the uncertainty of the outcome also goes to infinity. This divergence is the reason that the interpretation of uncertainty in the discrete case cannot be used for continuous variables. There are simply too many values for it to be reasonable to talk about the uncertainty of specific outcomes.

For the mutual information, however, it can be seen from the same type of derivation that the interpretation as a measure of the information obtained about one variable by observing another still holds. Consider two continuous random variables, X and Y, with joint density function $f(x, y)$ and marginals $f(x)$ and $f(y)$. Define the two discrete random variables X^{Δ} and Y^{δ} with joint probability $p(x_k^{\Delta}, y_{\ell}^{\delta}) = \Delta\delta f(x_k, y_{\ell})$. From the Riemann sums

$$f(x_k) = \sum_{\ell} \delta f(x_k, y_{\ell}) \to \int_{\mathbb{R}} f(x_k, y)dy,\ \delta \to 0 \tag{8.44}$$

$$f(y_{\ell}) = \sum_{k} \Delta f(x_k, y_{\ell}) \to \int_{\mathbb{R}} f(x, y_{\ell})dx,\ \Delta \to 0 \tag{8.45}$$

the marginals can be defined as $p(x_k^{\Delta}) = \Delta f(x_k)$ and $p(y_{\ell}^{\delta}) = \delta f(y_{\ell})$. Then, as both δ and Δ approach zero

$$\sum_k \Delta f(x_k) = \sum_{k,\ell} \delta\Delta f(x_k, y_{\ell}) \to \int_{\mathbb{R}^2} f(x, y)dxdy = 1 \tag{8.46}$$

$$\sum_{\ell} \delta f(x_k) = \sum_{k,\ell} \delta\Delta f(x_k, y_{\ell}) \to \int_{\mathbb{R}^2} f(x, y)dxdy = 1 \tag{8.47}$$

With the above-discretized distributions for X^{Δ} and Y^{δ}, the mutual information can be derived as

$$\begin{aligned}
I(X^{\Delta}, Y^{\delta}) &= \sum_{k,\ell} p(x_k^{\Delta}, y_\ell^{\delta}) \log \frac{p(x_k^{\Delta}, y_\ell^{\delta})}{p(x_k^{\Delta})p(y_\ell^{\delta})} \\
&= \sum_{k,\ell} \Delta\delta f(x_k, y_\ell) \log \frac{\Delta\delta f(x_k, y_\ell)}{\Delta f(x_k)\delta f(y_\ell)} \\
&= \sum_{k,\ell} \Delta\delta \Big(f(x_k, y_\ell) \log \frac{f(x_k, y_\ell)}{f(x_k)f(y_\ell)} \Big) \\
&\to \int_{\mathbb{R}^2} f(x,y) \log \frac{f(x,y)}{f(x)f(y)} dxdy = I(X;Y), \quad \Delta, \delta \to 0 \qquad (8.48)
\end{aligned}$$

where the limit value as Δ and δ approaches zero, individually, follows from the Riemann integration formula. The conclusion from the above derivation is that the properties for the mutual information in the discrete case are inherited to the continuous case. Especially, this means that the interpretation of mutual information is still valid for continuous variables. It also means that there is no problem to consider the information exchange between one discrete and one continuous variable. Similarly, the properties for the relative entropy, $D(p||q)$, for discrete distributions are still valid for continuous distributions.

8.2 GAUSSIAN DISTRIBUTION

In many applications, the Gaussian distribution plays an important role and information theory is not an exception. In this section, it is seen that the entropy is maximized for the Gaussian distribution, over all distributions for a given mean and variance.[2] In the next chapter, this will give the means to calculate the capacity for a case when the noise is Gaussian distributed.

In Example 8.2, the differential entropy for the Gaussian distribution, $N(\mu, \sigma)$, was derived as

$$\begin{aligned}
H(X) &= E\Big[-\log \frac{1}{\sqrt{2\pi\sigma^2}} e^{-(X-\mu)^2/2\sigma^2}\Big] \\
&= \frac{1}{2}\log(2\pi\sigma^2) + \frac{1}{2}\log(e)\frac{E\big[(X-\mu)^2\big]}{\sigma^2} = \frac{1}{2}\log(2\pi e\sigma^2) \qquad (8.49)
\end{aligned}$$

To derive the mutual information between two Gaussian variables, the density function for the two-dimensional case is needed. In the next section, the n-dimensional case for the Gaussian distribution will be treated a bit more thoroughly, but at this point the two-dimensional case is sufficient. The density function for a pair of Gaussian variables $(X, Y) \in N(0, \Lambda)$ with zero mean and covariance matrix

$$\Lambda = \begin{pmatrix} E[X^2] & E[XY] \\ E[XY] & E[Y^2] \end{pmatrix} \qquad (8.50)$$

[2] Here, only real valued Gaussian variables are considered. For treatment of complex valued variables, the reader is referred to [75].

is defined as

$$f(x,y) = \frac{1}{2\pi\sqrt{|\Lambda|}} e^{-\frac{1}{2}(x \ \ y)\Lambda^{-1}\binom{x}{y}} \tag{8.51}$$

The joint entropy can be derived as

$$\begin{aligned}
H(X,Y) &= E\Big[\log \frac{1}{2\pi\sqrt{|\Lambda|}} e^{-\frac{1}{2}(X \ \ Y)\Lambda^{-1}\binom{X}{Y}}\Big] \\
&= E\Big[\frac{1}{2}\log(2\pi)^2|\Lambda| + \frac{1}{2}\begin{pmatrix}X & Y\end{pmatrix}\Lambda^{-1}\begin{pmatrix}X\\ Y\end{pmatrix}\log e\Big] \\
&= \frac{1}{2}\log(2\pi)^2|\Lambda| + \frac{1}{2}\log e^2\Big] \\
&= \frac{1}{2}\log(2\pi e)^2|\Lambda|
\end{aligned} \tag{8.52}$$

where it is used that $E\big[(X \ Y)\Lambda^{-1}\binom{X}{Y}\big] = 2$. This fact is not directly obvious, and to obtain the result for the two-dimensional case, start with the covariance matrix and its inverse as

$$\Lambda = \begin{pmatrix}\sigma_x^2 & c_{xy}\\ c_{xy} & \sigma_y^2\end{pmatrix} \text{ and } \Lambda^{-1} = \begin{pmatrix}\lambda_{11} & \lambda_{12}\\ \lambda_{21} & \lambda_{22}\end{pmatrix} \tag{8.53}$$

where $c_{xy} = E[XY]$ is the covariance. Their product should be equal to the unit matrix,

$$\Lambda\Lambda^{-1} = \begin{pmatrix}\sigma_x^2\lambda_{11} + c_{xy}\lambda_{21} & \sigma_x^2\lambda_{12} + c_{xy}\lambda_{22}\\ c_{xy}\lambda_{11} + \sigma_y^2\lambda_{21} & c_{xy}\lambda_{12} + \sigma_y^2\lambda_{22}\end{pmatrix} = \begin{pmatrix}1 & 0\\ 0 & 1\end{pmatrix} \tag{8.54}$$

By identifying $\sigma_x^2\lambda_{11} + c_{xy}\lambda_{21} = 1$ and $c_{xy}\lambda_{12} + \sigma_y^2\lambda_{22} = 1$, the desired result can be achieved as

$$\begin{aligned}
E\Big[\begin{pmatrix}X & Y\end{pmatrix}\Lambda^{-1}\begin{pmatrix}X\\ Y\end{pmatrix}\Big] &= E\big[X^2\lambda_{11} + XY\lambda_{21} + XY\lambda_{12} + Y^2\lambda_{22}\big] \\
&= E\big[X^2\big]\lambda_{11} + E\big[XY\big]\lambda_{21} + E\big[XY\big]\lambda_{12} + E\big[Y^2\big]\lambda_{22} \\
&= \sigma_x^2\lambda_{11} + c_{xy}\lambda_{21} + c_{xy}\lambda_{12} + \sigma_y^2\lambda_{22} = 2
\end{aligned} \tag{8.55}$$

To derive the mutual information between the two Gaussian variables X and Y, it should first be noted that by the Cauchy–Schwarz inequality

$$\big|E[XY]\big|^2 \le \big|E[X^2]\big| \cdot \big|E[Y^2]\big| \tag{8.56}$$

That is, the covariance is upper bounded by the product of the standard deviations and can be written as

$$c_{xy} = E\big[XY\big] = \rho\sqrt{E\big[X^2\big]E\big[Y^2\big]} = \rho\sigma_x\sigma_y \tag{8.57}$$

where $|\rho| \leq 1$. Then, the determinant of the covariance matrix for the two-dimensional case is

$$|\Lambda| = \begin{vmatrix} \sigma_x^2 & \rho\sigma_x\sigma_y \\ \rho\sigma_x\sigma_y & \sigma_y^2 \end{vmatrix} = \sigma_x^2\sigma_y^2 - \rho^2\sigma_x^2\sigma_y^2 = \sigma_x^2\sigma_y^2\big(1-\rho^2\big) \tag{8.58}$$

and the joint entropy

$$H(X,Y) = \frac{1}{2}\log\big((2\pi e)^2\sigma_x^2\sigma_y^2(1-\rho^2)\big) \tag{8.59}$$

The mutual information becomes

$$\begin{aligned} I(X;Y) &= H(X) + H(Y) - H(X,Y) \\ &= \frac{1}{2}\log(2\pi e\sigma_x^2) + \frac{1}{2}\log(2\pi e\sigma_y^2) - \frac{1}{2}\log\big((2\pi e)^2\sigma_x^2\sigma_y^2(1-\rho^2)\big) \\ &= -\frac{1}{2}\log(1-\rho^2) \end{aligned} \tag{8.60}$$

The third function to derive is the relative entropy. For this, consider two Gaussian distributions with equal mean but different variances, $N(\mu,\sigma_0^2)$ and $N(\mu,\sigma^2)$. Then

$$\begin{aligned} D(\sigma_0||\sigma) &= E_{f_0}\left[-\log\frac{\frac{1}{\sqrt{2\pi\sigma_0^2}}e^{-(X-\mu)^2/2\sigma_0^2}}{\frac{1}{\sqrt{2\pi\sigma^2}}e^{-(X-\mu)^2/2\sigma^2}}\right] \\ &= E_{f_0}\left[\log\frac{\sigma_0}{\sigma} + \frac{1}{2}\Big(\frac{(X-\mu)^2}{\sigma_0^2} - \frac{(X-\mu)^2}{\sigma^2}\Big)\log e\right] \\ &= \log\frac{\sigma_0}{\sigma} + \frac{1}{2}\Big(\frac{E_{f_0}\left[(X-\mu)^2\right]}{\sigma_0^2} - \frac{E_{f_0}\left[(X-\mu)^2\right]}{\sigma^2}\Big)\log e \\ &= \log\frac{\sigma_0}{\sigma} + \Big(1-\frac{\sigma_0^2}{\sigma^2}\Big)\log\sqrt{e} \end{aligned} \tag{8.61}$$

An important result for the Gaussian distribution is that it maximizes the entropy for a given mean and variance. This will be used in the next chapter when the capacity for a channel with Gaussian noise is derived. As a start the next lemma is shown. It states that the averaging distribution in the entropy formula is not of importance, as long as the mean and variance are not changed.

Lemma 8.1 Let $g(x)$ be a density function for a Gaussian distribution, $N(\mu,\sigma)$, with mean μ and variance σ^2. If $f(x)$ is an arbitrary distribution with the same mean and variance, then

$$\int_{\mathbb{R}} f(x)\log g(x)dx = \int_{\mathbb{R}} g(x)\log g(x)dx \tag{8.62}$$

□

The lemma can be shown by the following derivation:

$$\begin{aligned}
-E_f\big[\log g(X)\big] &= -E_f\Big[\log \frac{1}{\sqrt{2\pi\sigma^2}} e^{-(X-\mu)^2/2\sigma^2}\Big] \\
&= \frac{1}{2}\log(2\pi\sigma^2) + E_f\Big[\frac{(X-\mu)^2}{2\sigma^2}\log e\Big] \\
&= \frac{1}{2}\log(2\pi\sigma^2) + \frac{E_f\big[(X-\mu)^2\big]}{2\sigma^2}\log e \\
&= \frac{1}{2}\log(2\pi\sigma^2) + \frac{1}{2}\log e \\
&= \frac{1}{2}\log(2\pi e\sigma^2) = -E_g\big[\log g(X)\big]
\end{aligned} \tag{8.63}$$

which completes the proof of the lemma.

By comparing the entropies for a Gaussian distribution, $N(\mu, \sigma)$, with an arbitrary distribution with the same mean and variance, the following derivation is obtained:

$$\begin{aligned}
H_g(X) - H_f(X) &= -\int_{\mathbb{R}} g(x)\log g(x)dx + \int_{\mathbb{R}} f(x)\log f(x)dx \\
&= -\int_{\mathbb{R}} f(x)\log g(x)dx + \int_{\mathbb{R}} f(x)\log f(x)dx \\
&= \int_{\mathbb{R}} f(x)\log\frac{f(x)}{g(x)}dx = D(f||g) \geq 0
\end{aligned} \tag{8.64}$$

with equality if and only if $f(x) = g(x)$ for all x. Stated differently, it is seen that if $g(x)$ is the density function for a Gaussian distribution and $f(x)$ the density function for any other distribution with the same mean and variance,

$$H_g(X) \geq H_f(X) \tag{8.65}$$

The result is stated in the next theorem.

Theorem 8.4 The Gaussian distribution maximizes the differential entropy over all distributions with mean μ and variance σ^2. □

8.2.1 Multidimensional Gaussian Distribution

In this section, the Gaussian distribution has been treated with extra care. Here the theory is expanded to the n-dimensional case. As a first step, the Gaussian distribution will be defined for an n-dimensional random vector. Through the density function, the differential entropy function is derived.

A random n-dimensional column vector $\boldsymbol{X} = (X_1, \dots, X_n)^T$, where ^{T} denotes the matrix transpose, is said to be Gaussian distributed if every linear combination of its entries forms a scalar Gaussian variable, i.e. if $\boldsymbol{a}^T\boldsymbol{X} = \sum_i a_i X_i \sim N(\mu, \sigma)$ for every real-valued vector $\boldsymbol{a} = (a_1, \dots, a_N)^T$. Since any linear combination of Gaussian variables is again Gaussian, the way to achieve this is to consider the case where each entrance in $\boldsymbol{X}$ is Gaussian with mean μ_i and variance σ_i^2, i.e. $X_i \in N(\mu_i, \sigma_i)$. The mean of the vector $\boldsymbol{X}$ is

$$\boldsymbol{\mu} = E[\boldsymbol{X}] = (\mu_1, \ldots, \mu_n)^T \tag{8.66}$$

and the covariance matrix

$$\Lambda_X = E[(\boldsymbol{X} - \boldsymbol{\mu})(\boldsymbol{X} - \boldsymbol{\mu})^T] = \Big(E[(X_i - \mu_i)(X_j - \mu_j)]\Big)_{i,j=1,\ldots,n} \tag{8.67}$$

Clearly, the diagonal elements of Λ_X contain the variances of $\boldsymbol{X}$. The Gaussian distribution is denoted by $\boldsymbol{X} \sim N(\boldsymbol{\mu}, \Lambda_X)$.[3]

To find the density function of the distribution, consider a general scaling and translation of a random variable $\boldsymbol{X}$. Let $\boldsymbol{X}$ be an n-dimensional random variable according to an n-dimensional distribution with mean $\boldsymbol{\mu}$ and covariance Λ_X. If A is a square matrix of full rank and $\boldsymbol{a}$ an n-dimensional column vector, a new random vector $\boldsymbol{Y} = A\boldsymbol{X} + \boldsymbol{a}$ is formed. The mean and covariance of $\boldsymbol{Y}$ are

$$E[\boldsymbol{Y}] = E[A\boldsymbol{X} + \boldsymbol{a}] = AE[\boldsymbol{X}] + \boldsymbol{a} = A\boldsymbol{\mu} + \boldsymbol{a} \tag{8.68}$$

$$\begin{aligned}
\Lambda_Y &= E[(\boldsymbol{Y} - E[\boldsymbol{Y}])(\boldsymbol{Y} - E[\boldsymbol{Y}])^T] \\
&= E[A\boldsymbol{X} + \boldsymbol{a} - A\boldsymbol{\mu} - \boldsymbol{a})(A\boldsymbol{X} + \boldsymbol{a} - A\boldsymbol{\mu} - \boldsymbol{a})^T] \\
&= E[(A(\boldsymbol{X} - \boldsymbol{\mu}))(A(\boldsymbol{X} - \boldsymbol{\mu}))^T] \\
&= E[A(\boldsymbol{X} - \boldsymbol{\mu})(\boldsymbol{X} - \boldsymbol{\mu})^T A^T] \\
&= AE[(\boldsymbol{X} - \boldsymbol{\mu})(\boldsymbol{X} - \boldsymbol{\mu})^T]A^T = A\Lambda_X A^T
\end{aligned} \tag{8.69}$$

The idea is to transform the Gaussian vector $\boldsymbol{X}$ into a normalized Gaussian vector. In the case when X is a one-dimensional random variable, this is done with $Y = \frac{X-\mu}{\sigma}$. To see how the corresponding equation looks for the n-dimensional case, some definitions and results from matrix theory are needed. For a more thorough treatment of this topic, refer to, e.g., [8]. Most of the results here will be given without proofs.

First, the covariance matrix is characterized to see how the square root of its inverse can be derived.

Definition 8.6 A real matrix A is *symmetric*[4] if it is symmetric along the diagonal, $A^T = A$. □

If the matrix A is symmetric and has an inverse, the unity matrix can be used to get $I = AA^{-1} = A^T A^{-1} = (A^{-T}A)^T = A^{-T}A$, where $^{-T}$ denotes the transpose of the inverse. Then, $A^{-1} = IA^{-1} = A^{-T}AA^{-1} = A^{-T}$. Hence, the inverse of a symmetric matrix is again symmetric. From its definition, it is directly seen that the covariance matrix is symmetric, since $E[(X_i - \mu_i)(X_j - \mu_j)] = E[(X_j - \mu_j)(X_i - \mu_i)]$.

[3] In this text, it is assumed that Λ_X has full rank. In the case of lower rank, the dimensionality of the vector can be decreased.

[4] A complex matrix A is *Hermitian* if $A^* = A$, where * denotes complex conjugate and transpose. For a real matrix, it is equivalent to being symmetric, i.e., $A^T = A$.

In the one-dimensional case, the variance is nonnegative. For matrices, this corresponds to that the covariance matrix is positive semidefinite.

Definition 8.7 A real matrix A is *positive definite* if $\boldsymbol{a}^T A\boldsymbol{a} > 0$, for all vectors $\boldsymbol{a} \neq \mathbf{0}$. □

Definition 8.8 A real matrix A is *positive semidefinite*, or nonzero definite, if $\boldsymbol{a}^T A\boldsymbol{a} \geq 0$, for all vectors $\boldsymbol{a} \neq \mathbf{0}$. □

Consider the covariance matrix Λ_X and a real-valued column vector $\boldsymbol{a} \neq \mathbf{0}$. Then

$$\begin{aligned} \boldsymbol{a}^T \Lambda_X \boldsymbol{a} &= \boldsymbol{a}^T E\big[(\boldsymbol{X}-\boldsymbol{\mu})(\boldsymbol{X}-\boldsymbol{\mu})^T\big]\boldsymbol{a} \\ &= E\big[\boldsymbol{a}^T(\boldsymbol{X}-\boldsymbol{\mu})(\boldsymbol{X}-\boldsymbol{\mu})^T\boldsymbol{a}\big] \\ &= E\big[(\boldsymbol{a}^T\boldsymbol{X}-\boldsymbol{a}^T\boldsymbol{\mu})(\boldsymbol{a}^T\boldsymbol{X}-\boldsymbol{a}^T\boldsymbol{\mu})^T)\big] = V\big[\boldsymbol{a}^T\boldsymbol{X}\big] \geq 0 \end{aligned} \tag{8.70}$$

since the variance of a one-dimensional random variable is nonnegative. To conclude, the following theorem is obtained.

Theorem 8.5 Given an n-dimensional random vector $\boldsymbol{X} = (X_1, \dots, X_n)^T$ with mean $E\big[\boldsymbol{X}\big] = (\mu_1, \dots, \mu_n)^T$, the covariance matrix $\Lambda_X = E\big[(\boldsymbol{X}-\boldsymbol{\mu})(\boldsymbol{X}-\boldsymbol{\mu})^T\big]$ is symmetric and positive semidefinite. □

In, e.g., [8], it can be found that for every symmetric positive semidefinite matrix A, there exists a unique symmetric positive semidefinite matrix $A^{1/2}$ such that

$$\left(A^{1/2}\right)^2 = A \tag{8.71}$$

This matrix $A^{1/2}$ is the equivalence of the scalar square root function. Furthermore, it can be shown that the inverse of the square root is equivalent to the square root of the inverse,

$$\left(A^{1/2}\right)^{-1} = \left(A^{-1}\right)^{1/2} \tag{8.72}$$

often denoted by $A^{-1/2}$. The determinant of $A^{-1/2}$ equals the inverse of the square root of the determinant,

$$|A^{-1/2}| = |A|^{-1/2} = \frac{1}{\sqrt{|A|}} \tag{8.73}$$

With this at hand, consider an n-dimensional Gaussian vector, $\boldsymbol{X} \sim \mathrm{N}(\boldsymbol{\mu}, \Lambda_X)$, and let

$$\boldsymbol{Y} = \Lambda_X^{-1/2}(\boldsymbol{X}-\boldsymbol{\mu}) \tag{8.74}$$

The mean and covariance can be derived as

$$E\big[\boldsymbol{Y}\big] = E\big[\Lambda_X^{-1/2}\boldsymbol{X} - \Lambda_X^{-1/2}\boldsymbol{\mu}\big] = \Lambda_X^{-1/2}E\big[\boldsymbol{X}\big] - \Lambda_X^{-1/2}\boldsymbol{\mu} = 0 \tag{8.75}$$

and

$$\Lambda_Y = \Lambda_X^{-1/2}\Lambda_X\Lambda_X^{-1/2} = \Lambda_X^{-1/2}\Lambda_X^{1/2}\Lambda_X^{1/2}\Lambda_X^{-1/2} = I \tag{8.76}$$

Hence, $\boldsymbol{Y} \sim N(\mathbf{0}, I)$ is normalized Gaussian distributed with zero mean and covariance I. Since Λ_X is assumed to have full rank, $|\Lambda_X| > 0$, there exists a density function that is uniquely determined by the mean and covariance. To find this, use that the entries of $\boldsymbol{Y}$ are independent and write the density function as

$$f_Y(y) = \prod_{i=1}^{n} \frac{1}{\sqrt{2\pi}} e^{-\frac{1}{2}y_i^2} = \frac{1}{(2\pi)^{n/2}} e^{-\frac{1}{2}\sum_i y_i^2} = \frac{1}{(2\pi)^{n/2}} e^{-\frac{1}{2}y^T y} \tag{8.77}$$

The entropy for this vector follows from the independency as

$$H(\boldsymbol{Y}) = \sum_{i=1}^{n} H(Y_i) = n\frac{1}{2}\log(2\pi e) = \frac{1}{2}\log(2\pi e)^n \tag{8.78}$$

To calculate the entropy for the vector $\boldsymbol{X} \sim N(\boldsymbol{\mu}, \Lambda_X)$, first consider the density function. Assume a general n-dimensional random vector $\boldsymbol{Z}$ with density function $f_Z(z)$, and let A be an $n \times n$ nonsingular matrix and $\boldsymbol{a}$ be an n-dimensional static vector. Then, form $\boldsymbol{X} = A\boldsymbol{Z} + \boldsymbol{a}$, which leads to that $\boldsymbol{Z} = A^{-1}(\boldsymbol{X} - \boldsymbol{a})$ and $d\boldsymbol{x} = |A|dz$, where $|A|$ is the Jacobian for the variable change. Thus the density function for $\boldsymbol{X}$ can then be written as

$$f_X(\boldsymbol{x}) = \frac{1}{|A|} f_Z\big(A^{-1}(\boldsymbol{x} - \boldsymbol{a})\big) \tag{8.79}$$

which gives the entropy as

$$\begin{aligned}
H(\boldsymbol{X}) &= -\int_{\mathbb{R}^n} f_X(\boldsymbol{x}) \log f_X(\boldsymbol{x}) d\boldsymbol{x} \\
&= -\int_{\mathbb{R}^n} \frac{1}{|A|} f_Z\big(A^{-1}(\boldsymbol{x} - \boldsymbol{a})\big) \log \frac{1}{|A|} f_Z\big(A^{-1}(\boldsymbol{x} - \boldsymbol{a})\big) d\boldsymbol{x} \\
&= -\int_{\mathbb{R}^n} f_Z(z) \log \frac{1}{|A|} f_Z(z) dz \\
&= -\int_{\mathbb{R}^n} f_Z(z) \log f_Z(z) dz + \log |A| \int_{\mathbb{R}^n} f_Z(z) dz \\
&= H(\boldsymbol{Z}) + \log |A|
\end{aligned} \tag{8.80}$$

Hence, the following result can be stated, similar to the one-dimensional case.

Theorem 8.6 Let $\boldsymbol{Z}$ is an n-dimensional random vector with entropy $H(\boldsymbol{Z})$. If A is an $n \times n$ nonsingular matrix and $\boldsymbol{a}$ an n-dimensional static vector, then $\boldsymbol{X} = A\boldsymbol{Z} + \boldsymbol{a}$ has the entropy

$$H(\boldsymbol{X}) = H(\boldsymbol{Z}) + \log |A| \tag{8.81}$$

□

To get back from the normalized Gaussian vector $\boldsymbol{Y}$ to $\boldsymbol{X} \sim N(\boldsymbol{\mu}, \Lambda_X)$, use the function

$$\boldsymbol{X} = \Lambda_X^{1/2} \boldsymbol{Y} + \boldsymbol{\mu} \tag{8.82}$$

The above theorem states that the entropy for the vector $\boldsymbol{X}$ is

$$\begin{aligned} H(\boldsymbol{X}) &= \frac{1}{2}\log(2\pi e)^n + \log|\Lambda_X|^{1/2} \\ &= \frac{1}{2}\log(2\pi e)^n|\Lambda_X| = \frac{1}{2}\log|2\pi e\Lambda_X| \end{aligned} \tag{8.83}$$

Theorem 8.7 Let $\boldsymbol{X} = (X_1, \dots, X_n)^T$ be an n-dimensional Gaussian vector with mean $\boldsymbol{\mu} = (\mu_1, \dots, \mu_n)^T$ and covariance matrix $\Lambda_X = E\big[(\boldsymbol{X}-\boldsymbol{\mu})(\boldsymbol{X}-\boldsymbol{\mu})^T\big]$, i.e., $\boldsymbol{X} \sim \mathrm{N}(\boldsymbol{\mu}, \Lambda_X)$. Then the differential entropy of the vector is

$$H(\boldsymbol{X}) = \frac{1}{2}\log|2\pi e\Lambda_X| \tag{8.84}$$

□

An alternative way to show the above theorem is to first derive the density function for $\boldsymbol{X}$ and then use this to derive the entropy. Since this derivation will be reused later, it is also shown here. So, again use the variable change $\boldsymbol{X} = \Lambda_X^{1/2}\boldsymbol{Y} + \boldsymbol{\mu}$ and (8.79) to get the density function for an n-dimensional Gaussian distribution

$$\begin{aligned} f_X(\boldsymbol{x}) &= \frac{1}{\sqrt{|\Lambda_X|}}\frac{1}{(2\pi)^{n/2}}e^{-\frac{1}{2}(\Lambda_X^{-1/2}(\boldsymbol{x}-\boldsymbol{\mu}))^T(\Lambda_X^{-1/2}(\boldsymbol{x}-\boldsymbol{\mu}))} \\ &= \frac{1}{\sqrt{|2\pi\Lambda_X|}}e^{-\frac{1}{2}(\boldsymbol{x}-\boldsymbol{\mu})^T\Lambda_X^{-1}(\boldsymbol{x}-\boldsymbol{\mu})} \end{aligned} \tag{8.85}$$

Before progressing toward the entropy, the argument in the exponent needs some extra attention. Assume a random variable $\boldsymbol{X}$ (not necessarily Gaussian) with mean $E[\boldsymbol{X}] = \boldsymbol{\mu}$ and covariance matrix $\Lambda_X = E\big[(\boldsymbol{X}-\boldsymbol{\mu})(\boldsymbol{X}-\boldsymbol{\mu})^T\big]$, and form $\boldsymbol{Y} = \Lambda_X^{-1/2}(\boldsymbol{X}-\boldsymbol{\mu})$ to get a normalized version with $E[\boldsymbol{Y}] = \boldsymbol{0}$ and $\Lambda_Y = I$. Then

$$\begin{aligned} E\big[(\boldsymbol{X}-\boldsymbol{\mu})^T\Lambda_X^{-1}(\boldsymbol{X}-\boldsymbol{\mu})\big] &= E\big[(\boldsymbol{X}-\boldsymbol{\mu})^T\Lambda_X^{-1/2}\Lambda_X^{-1/2}(\boldsymbol{X}-\boldsymbol{\mu})\big] \\ &= E\big[\boldsymbol{Y}^T\boldsymbol{Y}\big] = E\Big[\sum_{i=1}^{n} Y_i^2\Big] = \sum_{i=1}^{n} 1 = n \end{aligned} \tag{8.86}$$

If $\boldsymbol{X}$ is Gaussian with $\boldsymbol{X} \sim \mathrm{N}(\boldsymbol{\mu}, \Lambda_X)$, then $\boldsymbol{Y} = \Lambda_X^{-1/2}(\boldsymbol{X}-\boldsymbol{\mu})$ is normalized Gaussian, $\boldsymbol{Y} \sim \mathrm{N}(\boldsymbol{0}, I)$, and so is each of the entries, $Y_i \sim \mathrm{N}(0, 1)$. Since

$$Z = (\boldsymbol{X}-\boldsymbol{\mu})^T\Lambda_X^{-1}(\boldsymbol{X}-\boldsymbol{\mu}) = \sum_{i=1}^{n} Y_i^2 \sim \chi^2(n) \tag{8.87}$$

this also gives the mean of a chi-square distributed random variable, $E[Z] = n$.

The entropy for the Gaussian distribution can now be derived using the density function above as

$$\begin{aligned}
H(X) &= E_f\left[-\log \frac{1}{\sqrt{|2\pi\Lambda_X|}} e^{-\frac{1}{2}(X-\mu)^T\Lambda_X^{-1}(X-\mu)}\right] \\
&= E_f\left[\frac{1}{2}\log|2\pi\Lambda_X| + \frac{1}{2}(X-\mu)^T\Lambda_X^{-1}(X-\mu)\log e\right] \\
&= \frac{1}{2}\log|2\pi\Lambda_X| + \frac{1}{2}n\log e \\
&= \frac{1}{2}\log\big(e^n|2\pi\Lambda_X|\big) = \frac{1}{2}\log|2\pi e\Lambda_X| \qquad (8.88)
\end{aligned}$$

Looking back at Lemma 8.1 and Theorem 8.4, the corresponding result for the n-dimensional case can be derived. Starting with the lemma, assume that $g(\boldsymbol{x})$ is a density function for a normal distribution, $N(\boldsymbol{\mu}, \Lambda_X)$, and that $f(\boldsymbol{x})$ is an arbitrary density function with the same mean $\boldsymbol{\mu}$ and covariance matrix Λ_X. Then, the expectation of $-\log g(\boldsymbol{X})$ with respect to $g(\boldsymbol{x})$ and $f(\boldsymbol{x})$, respectively, are equal. This can be seen from the exact same derivation as above when $f(\boldsymbol{x})$ is non-Gaussian. Hence, the following lemma, corresponding to Lemma 8.1, can be stated.

Lemma 8.2 Let $g(\boldsymbol{x})$ be an n-dimensional Gaussian distribution, $N(\boldsymbol{\mu}, \Lambda_X)$, with mean $\boldsymbol{\mu}$ and covariance matrix Λ_X. If $f(\boldsymbol{x})$ is an arbitrary distribution with the same mean and covariance matrix, then

$$E_f\big[-\log g(\boldsymbol{X})\big] = E_g\big[-\log g(\boldsymbol{X})\big] \qquad (8.89)$$

□

To see that the Gaussian distribution maximizes the entropy consider

$$\begin{aligned}
H_g(\boldsymbol{X}) - H_f(\boldsymbol{X}) &= E_g\big[-\log g(\boldsymbol{X})\big] - E_f\big[-\log f(\boldsymbol{X})\big] \\
&= E_f\big[-\log g(\boldsymbol{X})\big] - E_f\big[-\log f(\boldsymbol{X})\big] \\
&= E_f\left[\log\frac{f(\boldsymbol{X})}{g(\boldsymbol{X})}\right] = D(f||g) \geq 0 \qquad (8.90)
\end{aligned}$$

Theorem 8.8 The n-dimensional Gaussian distribution maximizes the differential entropy over all n-dimensional distributions with mean $\boldsymbol{\mu}$ and covariance matrix Λ_X.

□

PROBLEMS

8.1 Derive the differential entropy for the following distributions:

(a) Rectangular distribution: $f(x) = \frac{1}{b-a}, \quad a \leq x \leq b$.

(b) Normal distribution: $f(x) = \frac{1}{\sqrt{2\pi\sigma^2}}e^{-\frac{(x-\mu)^2}{2\sigma^2}}, \quad -\infty \leq x \leq \infty$.

(c) Exponential distribution: $f(x) = \lambda e^{-\lambda x}, \quad x \geq 0$.

(d) Laplace distribution: $f(x) = \frac{1}{2}\lambda e^{-\lambda|x|}, \quad -\infty \leq x \leq \infty$.

8.2 The joint distribution on X and Y is given by

$$f(x,y) = \alpha^2 e^{-(x+y)}$$

for $x \geq 0$ and $y \geq 0$. (Compare with Problem 3.10)

(a) Determine α.

(b) Derive $P(X < 4, Y < 4)$.

(c) Derive the joint entropy.

(d) Derive the conditional entropy $H(X|Y)$.

8.3 Repeat Problem 8.2 for

$$f(x,y) = \alpha^2 2^{-(x+y)}$$

8.4 In wireless communication, the attenuation due to a shadowing object can be modeled as a log-normal random variable, $X \sim \text{logN}(\mu, \sigma)$. If the logarithm of a random variable X is normal distributed, i.e., $Y = \ln X \sim N(\mu, \sigma)$, then X is said to be log-normal distributed. Notice that $X \in [0, \infty]$ and $Y \in [-\infty, \infty]$.

(a) Use the probability

$$P(X < a) = \int_0^a f_X(x)dx$$

to show that the density function is

$$f_X(x) = \frac{1}{\sqrt{2\pi\sigma^2}} e^{-\frac{(\ln x - \mu)^2}{2\sigma^2}}$$

(b) Use the density function in (a) to find

$$\begin{aligned} E[X] &= e^{\mu + \frac{\sigma^2}{2}} \\ E[X^2] &= e^{2\mu + 2\sigma^2} \\ V[X] &= e^{2\mu + \sigma^2}(e^{\sigma^2} - 1) \end{aligned}$$

(c) Show that the entropy is

$$H(X) = \frac{1}{2}\log 2\pi e\sigma^2 + \frac{\mu}{\ln 2}$$

8.5 Let X and Y be two independent equally distributed random variables and form $Z = X + Y$. Derive the mutual information $I(X; Z)$ if

(a) X and Y are Gaussian with zero mean and unit variance, i.e., $X, Y \sim N(0, 1)$.

(b) X and Y are uniformly distributed between $-\frac{1}{2}$ and $\frac{1}{2}$, i.e., $X, Y \sim U(-\frac{1}{2}, \frac{1}{2})$.

Hint: $\int t \ln t \, dt = \frac{t^2}{2}\ln t - \frac{t^2}{4}$.

8.6 Show that for a continuous random variable X

$$E\big[(X - \alpha)^2\big] \geq \frac{1}{2\pi e} 2^{2H(X)}$$

for any constant α.

Hint: Use that $E\big[(X - \alpha)^2\big]$ is minimized for $\alpha = E\big[X\big]$.

8.7 The vector $X_1, X_2, \dots, X_n$ consists of n i.i.d. Gaussian distributed random variables. Their sum is denoted by $Y = \sum_i X_i$. Derive the information $I(X_k; Y)$ for the case when

(a) $X_i \sim N(0, 1)$.

(b) $X_i \sim N(m_i, \sigma_i)$.

8.8 For a one-dimensional discrete random variable over a finite interval the uniform distribution maximizes the entropy. In this problem it will be shown that this is a more general rule. Consider a finite region, $\mathcal{R}$, in N dimensions.

(a) Assume that $\boldsymbol{X} = X_1, \dots, X_N$ is discrete valued N-dimensional random vector with probability function $p(\boldsymbol{x})$ such that

$$\sum_{\boldsymbol{x}\in\mathcal{R}} p(\boldsymbol{x}) = 1$$
$$p(\boldsymbol{x}) = 0, \boldsymbol{x} \notin \mathcal{R}$$

where $\mathcal{R}$ has a finite number of outcomes, $\sum_{\boldsymbol{x}\in\mathcal{R}} 1 = k$. Show that the uniform distribution maximizes the entropy over all such distributions.

(b) Assume that $\boldsymbol{X} = X_1, \dots, X_N$ is a continuous valued N-dimensional random vector with density function $f(\boldsymbol{x})$ such that

$$\int_{\mathcal{R}} f(\boldsymbol{x}) d\boldsymbol{x} = 1$$
$$f(\boldsymbol{x}) = 0, \boldsymbol{x} \notin \mathcal{R}$$

where $\mathcal{R}$ has finite volume, $\int_{\mathcal{R}} 1 d\boldsymbol{x} = V$. Show that the uniform distribution maximizes the differential entropy over all such distributions.

8.9 A two-dimensional uniform distribution is defined over the shaded area shown in Figure 8.6. Derive

(a) $H(X, Y)$.

(b) $H(X)$ and $H(Y)$.

(c) $I(X; Y)$.

(d) $H(X|Y)$ and $H(Y|X)$.

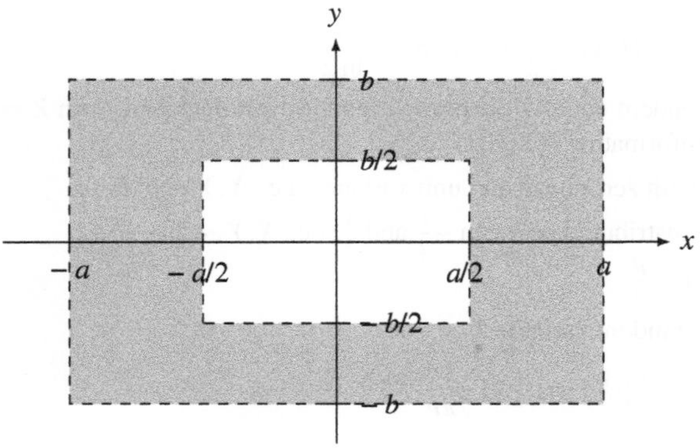

Figure 8.6 A two-dimension uniform distribution.

8.10 If $h(x)$ is the density function for a normal distribution, $N(\mu, \sigma)$, and $f(x)$ any distribution with the same mean and variance, show that

$$D(f(x)||h(x)) = H_h(X) - H_f(X)$$

8.11 The two independent Gaussian random variables X_1 and X_2 are distributed according to $N(\mu_1, \sigma_1)$ and $N(\mu_2, \sigma_2)$. Construct a new random variable $X = X_1 + X_2$.

(a) What is the distribution of X?

(b) Derive the differential entropy of X.

8.12 Let X_1 and X_2 be two Gaussian random variables, $X_1 \sim N(\mu_1, \sigma_1)$ and $X_1 \sim N(\mu_1, \sigma_1)$, with density functions $f_1(x)$ and $f_2(x)$.

(a) Derive $D(f_1||f_2)$.

(b) Let $X = X_1 + X_2$ with density function $f(x)$. Derive $D(f||f_2)$.

8.13 A random length, X, is uniformly distributed between 1 and 2 m. Derive the differential entropy if the length is considered distributed according to the two cases below.

(a) The length varies between 1 and 2 m, i.e.

$$f(x) = \begin{cases} 1, & 1 \le x \le 2 \\ 0, & \text{otherwise} \end{cases}$$

(b) The length varies between 100 and 200 cm, i.e.

$$f(x) = \begin{cases} 0.01, & 100 \le x \le 200 \\ 0, & \text{otherwise} \end{cases}$$

(c) Explain the difference.

8.14 **(a)** Let $X \sim N(\mathbf{0}, \Lambda)$ be an n-dimensional Gaussian vector with zero mean and covariance matrix $\Lambda = E[\boldsymbol{X}\boldsymbol{X}^T]$. Use the chain rule for differential entropy to show that

$$H(X) \le \frac{1}{2}\log\Big((2\pi e)^n \prod_i \sigma_i^2\Big)$$

where $\sigma_i^2 = E[X_i^2]$ is the variance for the ith variable.

(b) Use the result in (a) to show Hadamad's inequality for covariance matrices, i.e. that

$$|\Lambda| \le \prod_i \sigma_i^2$$

CHAPTER 9

GAUSSIAN CHANNEL

IN COMMUNICATION THEORY, it is often assumed that the transmitted signals are distorted by some noise. The most common noise to assume is additive Gaussian noise, the so-called additive white Gaussian noise channel (AWGN). Even though the noise in reality often is more complex, this model is very efficient when simulating, for example, background noise or amplifier noise. The model can be complemented by, e.g., colored noise, impulse noise, or other typical noise models. In this chapter, the AWGN channel will be studied in more detail. A fundamental limit will be derived for the signal-to-noise ratio (SNR) specifying when it is not possible to achieve reliable communication.

9.1 GAUSSIAN CHANNEL

In a communication system, data are often represented in a binary form. However, binary digits are elements in a discrete world, and, in all cases of transmission they need to be represented in a continuous form. It can be as a magnetization on a hard disk, charging of a transistor in a flash memory, or signals in a cable or an antenna. The noise added on the channel is typically modeled as Gaussian (i.e., normal distributed) and represents, for example, the background noise, amplifier noise in the transceivers, and signals from other communication systems picked up in the frequency bands.

A signal in a digital communication system can be represented by a continuous random variable. This value can be decomposed into two parts added together

$$Y = X + Z \tag{9.1}$$

where X is the information carrier component, Z noise component, and Y the received variable. The average power allocated by the variable X is defined as the second moment, $E[X^2]$. The Gaussian channel can then be defined as follows.

Definition 9.1 A *Gaussian channel* is a time-discrete channel with input X and output $Y = X + Z$, where Z models the noise and is normal distributed, $Z \sim N(0, \sqrt{N})$. The communication signaling is limited by a power constraint on the transmitter side,

$$E[X^2] \leq P \tag{9.2}$$

□

Information and Communication Theory, First Edition. Stefan Höst.

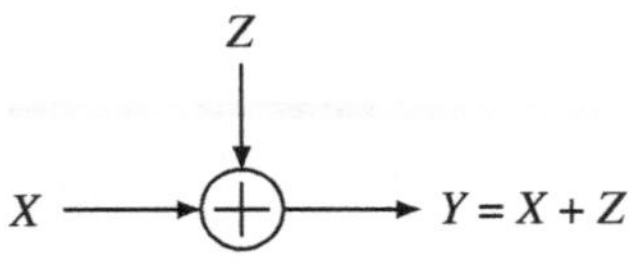

Figure 9.1 A Gaussian channel.

The Gaussian channel can be viewed as a block diagram as shown in Figure 9.1.

Without the power constraint in the definition, it would be possible to transmit as much information as desired in a single-channel use. With the power constraint, the system is more realistic and other means are needed than increasing the power to get a higher information throughput over the channel.[1]

The interpretation of the mutual information as the amount of information obtained about one variable by observing another is still valid for continuous variables. In a communication system where the signal X is transmitted and Y received, the aim for the receiver is to gain as much information as possible about X by observing Y. This is the transmitted information for each channel use, or transmission. To maximize the throughput over the channel, the information should be maximized with respect to the distribution of X and the transmitter power constraint.

Definition 9.2 The *information channel capacity* for a Gaussian channel is

$$C = \max_{\substack{f(x) \\ E[X^2]=P}} I(X;Y) \tag{9.3}$$

□

In the continuation of the text, the notation $f(x), P$ will be used representing $f(x), E[X^2] = P$. As before, when calculating the capacity the mutual information can be expressed as

$$I(X;Y) = H(Y) - H(Y|X) \tag{9.4}$$

The second term is

$$H(Y|X) = H(X+Z|X) = H(Z|X) = H(Z) \tag{9.5}$$

where in the second equality it is used that, conditioned on X, $X + Z$ is a known shift in position of the noise Z which does not change the differential entropy. To motivate the last equality, it is noted that X and Z are independent (the noise on the transmission channel is independent of the transmitted symbol). Therefore, the information over the channel can be viewed as the difference in entropy between the received symbol and the noise,

$$I(X;Y) = H(Y) - H(Z) \tag{9.6}$$

[1] Before Shannon published his work, it was the common knowledge that to get higher throughput, it was necessary to increase the power of the transmitted signals. Shannon showed that there are ways to reach high communication rates at maintained power.

With the noise known to be a normal distributed with zero mean and variance N, the entropy becomes

$$H(Z) = \frac{1}{2}\log(2\pi eN) \tag{9.7}$$

From the previous chapter, it is known that for a given mean and variance, the Gaussian distribution maximizes the entropy. So, maximizing $H(Y)$ over all distributions of X gives

$$\max_{f(x),P} H(Y) = \frac{1}{2}\log(2\pi e\sigma^2) \tag{9.8}$$

where the maximizing distribution is $Y \sim N(0, \sigma)$. Using that the sum of two Gaussian variables is again Gaussian gives that $Y = X + Z$ will be Gaussian if also X is Gaussian, $X \sim N(0, \sqrt{P})$. Then the variance of the received variable is $\sigma^2 = P + N$ and $Y \sim N(0, \sqrt{P + N})$. Hence, the information capacity is given by

$$\begin{aligned} C &= \max_{f(x),P} I(X;Y) \\ &= \max_{f(x),P} H(Y) - H(Z) \\ &= \frac{1}{2}\log(2\pi e(P+N)) - \frac{1}{2}\log(2\pi eN) \\ &= \frac{1}{2}\log\Big(\frac{2\pi e(P+N)}{2\pi eN}\Big) \\ &= \frac{1}{2}\log\Big(1 + \frac{P}{N}\Big) \end{aligned} \tag{9.9}$$

Formulated as a theorem, the following is obtained.

Theorem 9.1 The information channel capacity of a Gaussian channel with transmitted power constraint P and noise variance N is

$$C = \frac{1}{2}\log\Big(1 + \frac{P}{N}\Big) \tag{9.10}$$

□

The terminology *signal-to-noise ratio* is often used for the relation between the signal power and the noise power. In this case, the signal power is P while the noise has the power $E\big[Z^2\big] = N$ and $\mathrm{SNR} = \frac{P}{N}$. Depending on the topic and what type of system is considered, there are many different ways to define the SNR. It is important to be aware of the specific definition used in a text.

9.1.1 Band-Limited Gaussian Channel

Often in communication systems, the signaling is allowed to occupy a certain bandwidth. Therefore, it is interesting to consider signals with a limited bandwidth. When it comes to the derivations, since the channel is assumed to be time discrete, the signal can be assumed to be sampled and for that purpose the sampling theorem is needed. Then each sample can be considered to be transmitted over a Gaussian channel.

A band limited signal is a signal where the frequency content is limited inside a bandwidth W. For example, speech is normally located within the frequency bandwidth 0–4 kHz. By modulating the signal, it can be shifted up in frequency and located in a higher band. Still, it will occupy 4 kHz bandwidth. In this way, it is possible to allocate several bands of 4 kHz after each other, and in principle it is possible to pack one voice signal every 4 kHz in the frequency band.

To transmit, e.g., a voice signal, analog technology can be used to transmit it as it is. But if there is a need for some signal processing, it is often easier to work with the sampled version as digital data. Then it is, e.g., possible to use source coding to reduce the redundancy and channel coding for error protection. Hence, it is possible to achieve much better quality at a lower transmission cost (i.e., bandwidth or power). Sampling the signal means taking the value from the continuous signal every T_s second. This means there are $F_s = \frac{1}{T_s}$ samples every second, which is denoted by the *sampling frequency*. If the continuous time signal $x(t)$ is sampled with frequency F_s, the sample values are given by

$$x[n] = x(nT_s) = x\left(\tfrac{n}{F_s}\right) \tag{9.11}$$

For a band-limited signal with a bandwidth of W, the sampling theorem states that to be able to reconstruct the original signal, $F_s \geq 2W$ must be satisfied. So, a voice signal that is band limited to $W = 4$ kHz should be sampled with at least $F_s = 8$ kHz. The relation between the signals and samples was first investigated by Nyquist in 1928 [3], and further improved by Shannon in [1]. Actually, Nyquist studied the number of pulses that can be transmitted over a certain bandwidth and did not target sampling as such. It is not clear who first published the sampling theorem as we know it today. Shannon was one of them, but there are also other publications.

Theorem 9.2 (Shannon–Nyquist sampling theorem) Let $x(t)$ be a band-limited signal with the maximum frequency content at $f_{\max} \leq W$. If the signal is sampled with $F_s = 2W$ samples per second to form the discrete time sequence $x[n] = x(\frac{n}{2W})$, it can be reconstructed with

$$x(t) = \sum_{n=-\infty}^{\infty} x[n]\,\mathrm{sinc}\left(t - \frac{n}{2W}\right) \tag{9.12}$$

where

$$\mathrm{sinc}(t) = \frac{\sin(2\pi Wt)}{2\pi Wt} \tag{9.13}$$

□

The proof of the sampling theorem is normally included in a basic course in signal processing and lies a bit outside the scope of this text. For the sake of completeness, in Appendix B one version of a proof is given.

The function sinc(t) is 1 for $t = 0$ and 0 for $t = k/2W$, $k \neq 0$ (see Figure 9.2). If the sampling frequency is less than $2W$, the reconstructed signal will be distorted due to aliasing and perfect reconstruction is not possible. It is, however, always possible to

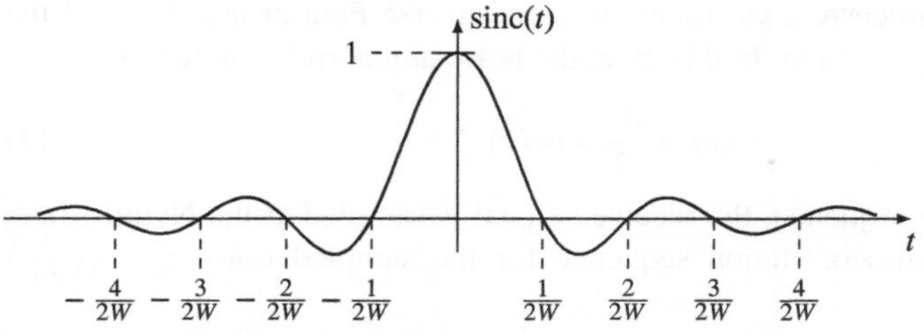

Figure 9.2 The sinc function.

perfectly reconstruct the signal if the sampling frequency is at least $2W$. The sampling frequency $F_s = 2W$ is often called the *Nyquist rate*.

To define a channel model for band-limited signals, assume a base-band signal with highest frequency content $f_{\max} = W$, and hence occupying the bandwidth W. Sampling the signal at the Nyquist rate gives the sampling frequency $F_s = 2W$ and the sampling time, i.e., the time between two samples, $T_s = \frac{1}{F_s} = \frac{1}{2W}$. Then the sampled sequence becomes

$$x[n] = x(nT_s) = x\left(\tfrac{n}{2W}\right) \tag{9.14}$$

Definition 9.3 A band-limited Gaussian channel consists of a band-limited input signal $x(t)$, where $f_{\max} = W$, distorted by additive white noise $\eta(t)$, and filtered by an ideal low-pass filter, as shown in Figure 9.3. □

Since the signal $x(t)$ is limited by the bandwidth W, it passes the ideal filter without changes. The aim of the filter is instead to limit the power of the noise. The meaning of the terminology white noise is that its power spectral density function occupies all frequencies with a constant value, which is normally set to

$$R_\eta(f) = \frac{N_0}{2}, \quad f \in \mathbb{R} \tag{9.15}$$

After filtering the noise, signal is $z(t) = \eta(t) * h(t)$, which is also band limited with power spectral density

$$R_z(f) = \begin{cases} \frac{N_0}{2}, & -W \le f \le W \\ 0, & \text{otherwise} \end{cases} \tag{9.16}$$

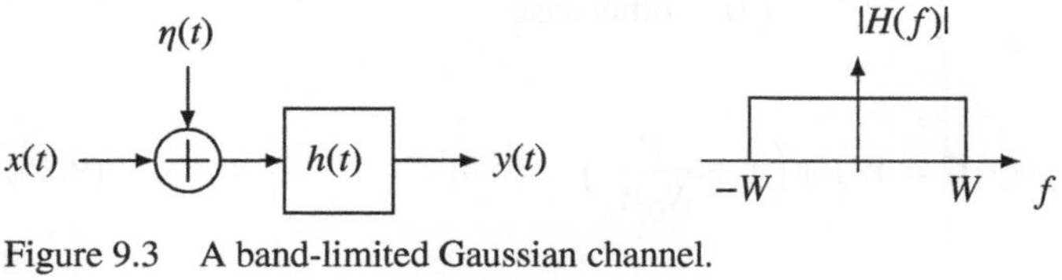

Figure 9.3 A band-limited Gaussian channel.

The corresponding autocorrelation function is the inverse Fourier transform of the power spectral density function. In this case, the noise autocorrelation function is

$$r_z(\tau) = \frac{N_0}{2}\,\mathrm{sinc}(\tau) \tag{9.17}$$

To get a time discrete sequence, the received signal is sampled at the Nyquist rate, $F_s = 2W$. Then the autocorrelation sequence for the sampled noise $z_k = \eta\left(\frac{k}{2W}\right)$ becomes

$$r_z[k] = r_z\left(\tfrac{k}{2W}\right) = \begin{cases} \frac{N_0}{2}, & k = 0 \\ 0, & \text{otherwise} \end{cases} \tag{9.18}$$

This implies the resulting sampled noise is normal distributed with zero mean and variance $N_0/2$,

$$z_k \sim \mathrm{N}\left(0, \sqrt{\tfrac{N_0}{2}}\right) \tag{9.19}$$

Hence, to calculate the achievable bit rate the theory for the Gaussian channel can be used. For each sample transmitted, the capacity is

$$C = \frac{1}{2}\log\left(1 + \frac{2\sigma_x^2}{N_0}\right) \quad \text{[b/sample]} \tag{9.20}$$

The power of the transmitted signal is constraint to P, meaning each transmitted sample has the energy $\sigma_x^2 = \frac{P}{2W}$, which gives the capacity per sample

$$C = \frac{1}{2}\log\left(1 + \frac{P}{N_0 W}\right) \quad \text{[b/sample]} \tag{9.21}$$

With $F_s = 2W$ samples every second, the achievable bit rate becomes

$$C = W\log\left(1 + \frac{P}{N_0 W}\right) \quad \text{[b/s]} \tag{9.22}$$

This important result is formulated as a theorem.

Theorem 9.3 Let $x(t)$ be a band-limited signal with maximum frequency contribution $f_{\max} \le W$, and $\eta(t)$ noise with power spectral density $R_\eta(f) = N_0/2$. The channel

$$y(t) = \Big(x(t) + \eta(t)\Big) * h(t) \tag{9.23}$$

where

$$H(f) = \begin{cases} 1, & f \le |W] \\ 0, & \text{otherwise} \end{cases} \tag{9.24}$$

has the capacity

$$C = W\log\left(1 + \frac{P}{N_0 W}\right) \quad \text{[b/s]} \tag{9.25}$$

□

Example 9.1 A widely used technology for fixed Internet access to the home, is through DSL (digital subscriber line). There are mainly two flavors, asymmetric digital subscriber line (ADSL), with bit rates up to 26 Mb/s, and very high bit rate digital subscriber line (VDSL), that enables bit rates exceeding 100 Mb/s. Recently, a successor, G.fast, has been standardized that will achieve rates in the order of Gb/s. The advantage with DSL technology is that it reuses the old telephone lines to access the households, so it is a relatively cheap technology to roll out. Comparing with optical networks in the access link (fiber to the home, FttH) where a new infrastructure of optical fibers must be installed to each house, this is an economically feasible technology. This, however, is also one of its drawbacks since the old telephone cables have low-quality twists, and the signal may be more distorted than for modern cables.

In both ADSL and VDSL, the data signal coexists with the speech signal for the telephone, located in the band 0–4 kHz. The data signal is positioned from 25 kHz up to 2.2 MHz for ADSL and up to 17 MHz for VDSL.[2] To do capacity calculations on the VDSL band, neglect the speech band and assume that the available bandwidth is $W = 17$ MHz. The signaling level is set by the standardization body ITU-T, and the allowed signal power varies over the utilized frequency band. For simplicity in this example, the signaling power is set to −60 dBm/Hz.[3]

The absolute maximum that is possible to transmit can be found when the noise is as low as possible. The thermal noise, or the Johnson–Nyquist noise, is the noise generated in all electrical circuits. The thermal noise is typically white and at room temperature about −174 dBm/Hz. The power and the noise variance is thus

$$P = 10^{-60/10} \cdot W \quad [\text{mW}] \tag{9.26}$$

$$N_0 = 10^{-174/10} \quad [\text{mW/Hz}] \tag{9.27}$$

Then the capacity for the 17 MHz band is

$$\begin{aligned} C_{-174} &= W \log\left(1 + \frac{P}{N_0 W}\right) \\ &= 17 \times 10^6 \log\left(1 + \frac{10^{-60/10}}{10^{-174/10}}\right) = 644 \text{ Mb/s} \end{aligned} \tag{9.28}$$

While this is a maximum theoretical possible bit rate, in real cables the noise level is higher, somewhere around −145 dBm/Hz. With this noise level, the capacity is

$$C_{-145} = 17 \times 10^6 \log\left(1 + \frac{10^{-60/10}}{10^{-145/10}}\right) = 480 \text{ Mb/s} \tag{9.29}$$

In a real VDSL system, the theoretical bit rate is about 200 Mb/s. However, in practice 150 Mb/s is a more realistic figure even for very short loops. As the distance between the central equipment and the customer increases, the signal is attenuated, especially at high frequencies.

[2] There are several defined band plans in the standard for VDSL2 and include 8, 12, 17, and 30 MHz.In this example, the 17 MHz case is considered.

[3] Often the unit dBm/Hz is used for power spectral density (PSD) levels. This means the power level expressed in mW, normalized with the bandwidth and expressed in dB, i.e., $P_{\text{dBm/Hz}} = 10 \log_{10}(P_{\text{mW/Hz}})$.

9.2 PARALLEL GAUSSIAN CHANNELS

In some cases, there can be several independent parallel Gaussian channels used by the same communication system. Figure 9.4 shows n such independent parallel channels. Each of the channel has a power constraint $P_i = E\left[X_i^2\right]$ and a noise variance N_i. The total power is $P = \sum_i P_i$.

For these n independent Gaussian channels, the mutual information between the transmitted vector $\boldsymbol{X} = X_1, \ldots, X_n$ and the received vector $\boldsymbol{Y} = Y_1, \ldots, Y_n$ can be written as

$$\begin{aligned} I(\boldsymbol{X}, \boldsymbol{Y}) &= I(X_1, \ldots, X_n; Y_1, \ldots, Y_n) \\ &\le \sum_{i=1}^{n} I(X_i; Y_i) \\ &\le \sum_{i=1}^{n} \frac{1}{2} \log\left(1 + \frac{P_i}{N_i}\right) \end{aligned} \tag{9.30}$$

with equality if the variables X_i are independent and Gaussian. Since $\sigma_{X_i}^2 = E[X_i^2] = P_i$, the mutual information can be maximized for the set of P_i by using independent Gaussian variable $X_i \sim N(0, \sqrt{P_i})$. To get the capacity, the above expression should be maximized with respect to P_i and with the additional constraint $\sum_i P_i = P$. From optimization theory (see, e.g., [18]), the Lagrange multiplier method can be applied. The function to maximize is then given by

$$J = \sum_{i=1}^{n} \frac{1}{2} \log\left(1 + \frac{P_i}{N_i}\right) + \lambda\left(\sum_{i=1}^{n} P_i - P\right) \tag{9.31}$$

Setting the derivative equal to zero yields

$$\frac{\partial}{\partial P_i} J = \frac{1}{2 \ln 2} \cdot \frac{\frac{1}{N}}{\left(1 + \frac{P_i}{N}\right)} + \lambda = \frac{1}{2 \ln 2} \cdot \frac{1}{P_i + N_i} + \lambda = 0 \tag{9.32}$$

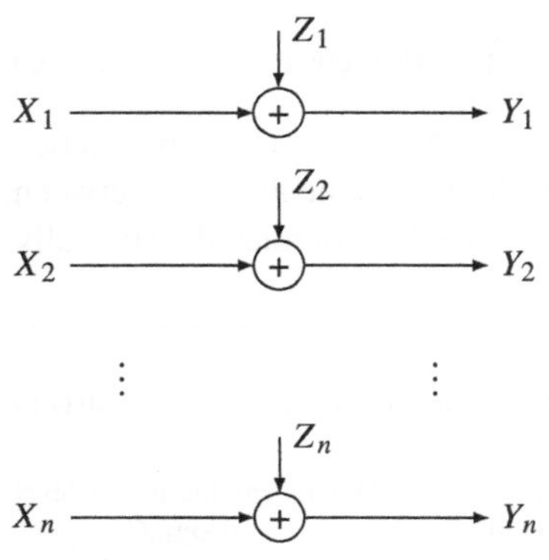

Figure 9.4 A channel with n independent parallel Gaussian subchannels.

or, equivalently,

$$P_i = -\frac{1}{\lambda 2 \ln 2} - N_i = B - N_i \tag{9.33}$$

It is important to notice the constant B is independent of the subchannel and should be determined together with the power constraints for the n channels. Combining (9.33) together with the power constraint gives an equation system for the $n + 1$ variables $(P_1, \ldots, P_n, B)$,

$$\begin{cases} P_i = B - N_i, & \forall i \\ \sum_j P_j = P \end{cases} \tag{9.34}$$

First B can be found from

$$P = \sum_j P_j = \sum_j B - N_j = nB - \sum_j N_j \tag{9.35}$$

and, hence, $B = \frac{1}{n}P + \frac{1}{n}\sum_j N_j$. Then the power for channel i is achieved as $P_i = \frac{1}{n}P + \frac{1}{n}\sum_j N_j - N_i = B - N_i$, which will give an optimal power distribution for the channel. However, there is one more requirement not included above. Solving this equation system might give negative powers for some subchannels, but in reality $P_i \geq 0$. From the Kuhn–Tucker optimization,[4] it can be seen that the negative part can be truncated to zero. That is, by introducing the truncating function

$$(x)^+ = \begin{cases} x, & x \geq 0 \\ 0, & x < 0 \end{cases} \tag{9.36}$$

the condition on the powers for channel i can be expressed as

$$P_i = \left(B - N_i\right)^+ \tag{9.37}$$

to achieve the optimality.

The modification means that some channels may have too much noise and should not be used at all. The derivation is summarized as a theorem.

Theorem 9.4 Given n independent parallel Gaussian channels with noise variances N_i, $i = 1, \ldots, n$, and a restricted total transmit power, $\sum_i P_i = P$. Then, the capacity is given by

$$C = \frac{1}{2}\sum_{i=1}^{n} \log\left(1 + \frac{P_i}{N_i}\right) \tag{9.38}$$

where

$$P_i = (B - N_i)^+, \qquad (x)^+ = \begin{cases} x, & x \geq 0 \\ 0, & x < 0 \end{cases} \tag{9.39}$$

and B is such that $\sum_i P_i = P$. □

[4] The Kuhn–Tucker method can be seen as a generalization of the Lagrange multiplier method, often used in nonlinear optimization. It is also known as Karush–Kuhn–Tucker. For further studies on both Lagrange and Kuhn–Tucker methods, refer to a standard optimization theory text book, e.g., [18].

The above method is often referred to as *water filling*, which can be seen in the next example.

Example 9.2 Assume a system with four independent Gaussian channels with noise variance $N_1 = 2$, $N_2 = 4$, $N_3 = 6$, and $N_4 = 3$. The total power used in transmission is restricted to $P = 6$. The condition $P_i = B - N_i$ is equivalent to

$$B = P_i + N_i \tag{9.40}$$

and summed for all four subchannels

$$\begin{aligned} 4B &= P_1 + N_1 + P_2 + N_2 + P_3 + N_3 + P_4 + N_4 \\ &= \underbrace{P_1 + P_2 + P_3 + P_4}_{P=6} + \underbrace{N_1 + N_2 + N_3 + N_4}_{15} = 21 \end{aligned} \tag{9.41}$$

This gives $B = \frac{21}{4}$. Using this result would require $P_3 = \frac{21}{4} - 6 = -\frac{3}{4}$, which is not possible. The conclusion is that the third subchannel has too much noise and should be used.

In a second attempt to find an optimal distribution of the available power, turn off subchannel 3 and use the other three. Similar as above

$$3B = \underbrace{P_1 + P_2 + P_4}_{P=6} + \underbrace{N_1 + N_2 + N_4}_{9} = 15 \tag{9.42}$$

and $B = \frac{15}{3} = 5$. The power distribution becomes

$$P_1 = 3 \quad P_2 = 1 \quad P_3 = 0 \quad P_4 = 2 \tag{9.43}$$

Since all used subchannels have positive powers, the capacity is

$$\begin{aligned} C &= \frac{1}{2}\log\left(1 + \frac{3}{2}\right) + \frac{1}{2}\log\left(1 + \frac{1}{4}\right) + \frac{1}{2}\log\left(1 + \frac{2}{3}\right) \\ &\approx 1.1904 \text{ bit/channel use} \end{aligned} \tag{9.44}$$

In Figure 9.5 , a graphical interpretation of the power allocation is shown. The noise level for each subchannel acts as surface of a landscape. Then the power is poured in

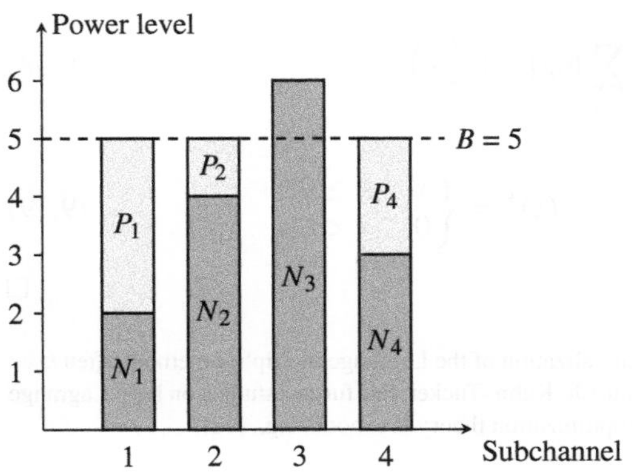

Figure 9.5 Water-filling principle in Example 9.2.

the landscape as water and the surface will stay at the level B. The depth of the power for each subchannel is the amount of power used. If, as in subchannel 3, the water level does not reach above the noise level, this subchannel should not be used. This analogy is the origin of the terminology *water filling*.

9.2.1 Frequency Division

For both mobile and fixed line communication systems, new generations are launched roughly every 10th year, and a new generation normally means a 10-fold of the required data rate. This increase in data rate demand is met by the system design with both a more complex system to operate closer to the capacity, but often also with an increase in frequency bandwidth. One popular method to signal over wide-band channels is OFDM (orthogonal frequency division multiplexing) modulation, used in, e.g., WLAN (802.11), xDSL, DVB-T (digital TV), and LTE (long-term evolution). The bandwidth is divided into several subbands that can be used independently of each other. Typically, a wide-band channel cannot be considered to be constant in neither noise level nor attenuation. In Figure 9.6, a channel with bandwidth W is shown. The noise level has a peak at slightly more than half the band, for example, be due to an external disturbance interfering with the channel. Similarly, the attenuation on the channel increases with frequency, which is typical for a copper-based channel, like a twisted pair copper cable.

To avoid that the signal is damaged by local degradations in SNR, the frequency span can be divided into subchannels of bandwidth W_Δ. The division should be such that the noise level and attenuation can be considered to be constant over the subchannel. Hence, in subchannel i the noise level is $N_{0,i}/2$ and attenuation $|H_i|^2$. Denoting the transmitted power in subchannel i by P_i, the received power is by $|H_i|^2 P_i$. By viewing the channel as independent subchannels, the equivalent channel model consists of n Gaussian channels, as shown in Figure 9.7, where the noise is $Z_i \sim N(0, \sqrt{N_{0,i}/2})$.

Hence, the ith subchannel is a band-limited Gaussian channel with capacity

$$C_i = W_\Delta \log\left(1 + \frac{P_i |H_i|^2}{N_{0,i} W_\Delta}\right) \tag{9.45}$$

and the total capacity is $C = \sum_i C_i$. To maximize the capacity, the distribution of power, $\sum_i P_i = P$, over the subchannels should be optimized. As before, the

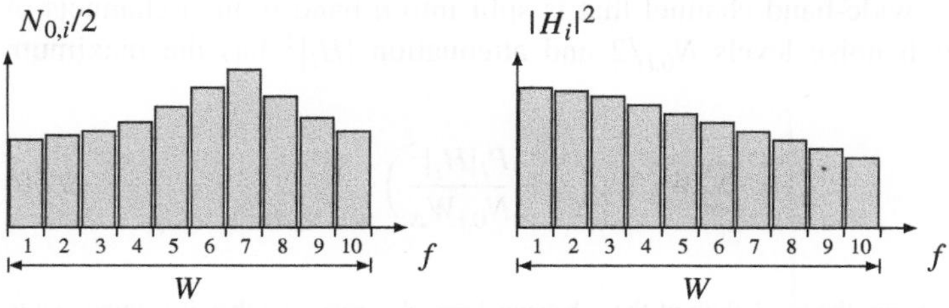

Figure 9.6 A channel with frequency varying noise and attenuation.

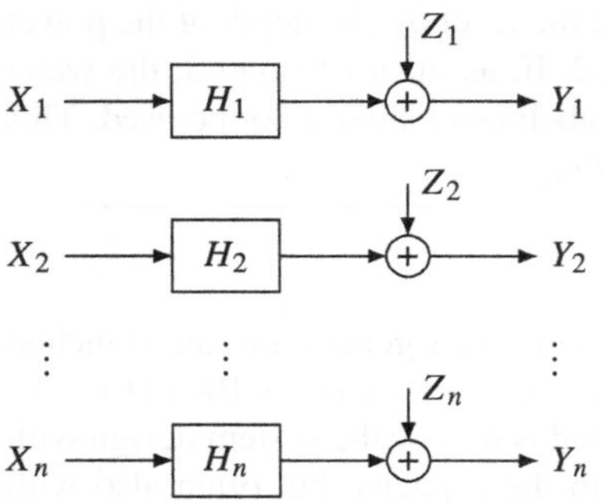

Figure 9.7 A frequency varying wide-band channel viewed as n parallel band-limited Gaussian channels.

Lagrangian multiplier method can be used together with the following optimization function

$$J = \sum_i W_\Delta \log\left(1 + \frac{P_i|H_i|^2}{N_{0,i}W_\Delta}\right) + \lambda\left(\sum_i P_i - P\right) \tag{9.46}$$

Setting the derivative equal to zero yields

$$\frac{\partial}{\partial P_i} J = W_\Delta \frac{\frac{|H_i|^2}{N_{0,i}W_\Delta}}{\ln 2\left(1 + \frac{P_i|H_i|^2}{N_{0,i}W_\Delta}\right)} + \lambda = 0 \tag{9.47}$$

or, equivalently,

$$P_i + \frac{N_{0,i}W_\Delta}{|H_i|^2} = -\frac{1}{\lambda \ln 2} W_\Delta = BW_\Delta \tag{9.48}$$

where $B = -\frac{1}{\lambda \ln 2}$ and BW_Δ is a constant.[5] As before, the power cannot take negative values and the generalization to the nonlinear optimization by Kuhn–Tucker yields

$$P_i = \left(BW_\Delta - \frac{N_{0,i}W_\Delta}{|H_i|^2}\right)^+ \tag{9.49}$$

such that $\sum_i P_i = P$. The result is stated in the following theorem.

Theorem 9.5 A wide-band channel that is split into n band-limited channels of bandwidth W_Δ with noise levels $N_{0,i}/2$ and attenuation $|H_i|^2$ has the maximum capacity

$$C = \sum_i W_\Delta \log\left(1 + \frac{P_i|H_i|^2}{N_{0,i}W_\Delta}\right) \tag{9.50}$$

[5] In a more general scenario, the bandwidth of the subchannel can also vary, and then it is important to separate the constant B with $W_{\Delta,i}$.

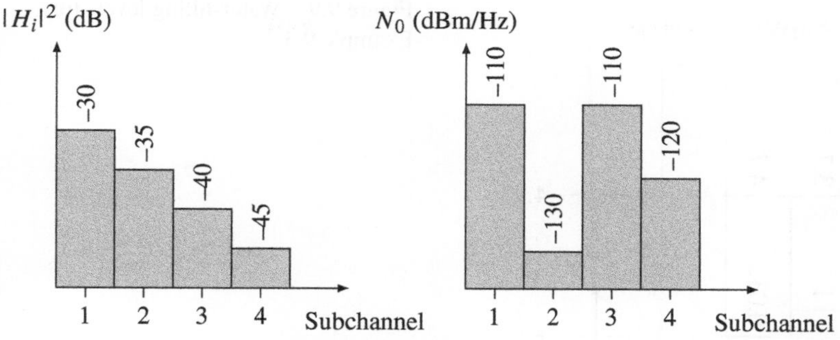

Figure 9.8 Channel parameters for Example 9.3.

where the power levels are derived according to

$$P_i = \left(BW_\Delta - \frac{N_{0,i}W_\Delta}{|H_i|^2}\right)^+ \tag{9.51}$$

$$\sum_i P_i = P \tag{9.52}$$

□

It should be noted that the capacity formula for each subchannel requires that both the noise level and the attenuation can be considered constant within the subchannel bandwidth. That is, the formula can be applied if the channel first is split in small enough subchannels with constant noise and attenuation. As the subchannel bandwidth goes to zero, that means the principles of water filling also apply for continuous (in frequency) channel models.

Example 9.3 A wideband channel with bandwidth $W = 4$ kHz is split into four independent subchannels with $W_\Delta = 1$ kHz. The noise and attenuations are considered constant within the subchannels (see Figure 9.8),

$$|H_i|^2 = \begin{pmatrix} -30 & -35 & -40 & -45 \end{pmatrix} \quad [\text{dB}] \tag{9.53}$$

$$N_0 = \begin{pmatrix} -110 & -130 & -110 & -120 \end{pmatrix} \quad [\text{dBm/Hz}] \tag{9.54}$$

The total allowed power on the channel is $P = -40$ dBm.

To derive the channel capacity for the combined channel, first derive the channel parameters in a linear scale. Since the dB scale is derived as $A_{\text{dB}} = 10\log_{10} A$, the inverse is given by $A = 10^{A_{\text{dB}}/10}$. That gives $P = 10^{-4}$ mW, and

$$|H_i|^2 = \begin{pmatrix} 10^{-3} & 10^{-3.5} & 10^{-4} & 10^{-4.5} \end{pmatrix} \tag{9.55}$$

$$N_0 = \begin{pmatrix} 10^{-11} & 10^{-13} & 10^{-11} & 10^{-12} \end{pmatrix} \quad [\text{mW/Hz}] \tag{9.56}$$

From (9.51) and (9.52), the total power can be written as

$$P = \sum_i BW_\Delta - \frac{N_{0,i}W_\Delta}{|H_i|^2} = 4BW_\Delta - \sum_i \frac{N_{0,i}W_\Delta}{|H_i|^2} \tag{9.57}$$

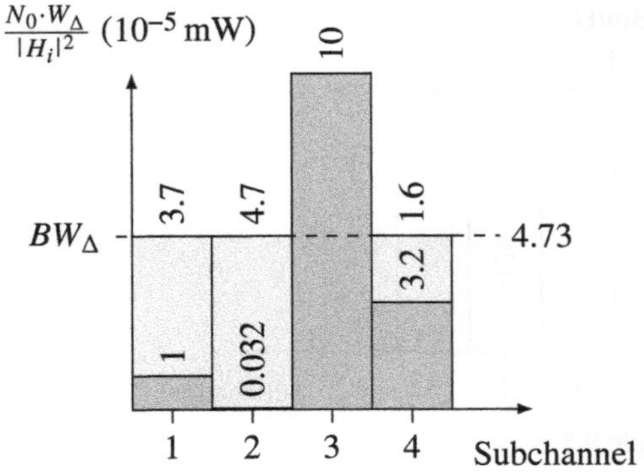

Figure 9.9 Water-filling levels for Example 9.3.

which gives

$$BW_\Delta = \frac{1}{4}\Big(P + \sum_i \frac{N_{0,i} W_\Delta}{|H_i|^2}\Big) = 6.05 \times 10^{-5} \text{ mW} \tag{9.58}$$

The individual subchannel power becomes

$$\begin{aligned} P_i &= BW_\Delta - \frac{N_{0,i} W_\Delta}{|H_i|^2} \\ &= \big(5.0 \times 10^{-5} \quad 6.0 \times 10^{-5} \quad -4.0 \times 10^{-5} \quad 2.9 \times 10^{-5}\big) \quad [\text{mW}] \end{aligned} \tag{9.59}$$

Since the third subchannel has negative power, it should be turned off and the procedure restarted. That is,

$$BW_\Delta = \frac{1}{3}\Big(P + \sum_{i \neq 3} \frac{N_{0,i} W_\Delta}{|H_i|^2}\Big) = 4.73 \times 10^{-5} \text{ mW} \tag{9.60}$$

and the individual subchannel powers as

$$\begin{aligned} P_i &= BW_\Delta - \frac{N_{0,i} W_\Delta}{|H_i|^2} \\ &= \big(3.7 \times 10^{-5} \quad 4.7 \times 10^{-5} \quad 0 \quad 1.6 \times 10^{-5}\big) \quad [\text{mW}] \end{aligned} \tag{9.61}$$

Since all subchannel powers are positive, the capacity can be derived as

$$\begin{aligned} C &= \sum_i W_\Delta \log\Big(1 + \frac{P_i |H_i|^2}{N_{0,i} W_\Delta}\Big) \\ &= 2.24 + 7.23 + 0 + 0.58 = 10 \text{ kb/s} \end{aligned} \tag{9.62}$$

The derived levels of $\frac{N_0 \cdot W_\Delta}{|H_i|^2}$ together with the constant level BW_Δ is shown in Figure 9.9. In the figure, the dark gray area represents $\frac{N_0 \cdot W_\Delta}{|H_i|^2}$ and the light gray areas the derived subchannel power.

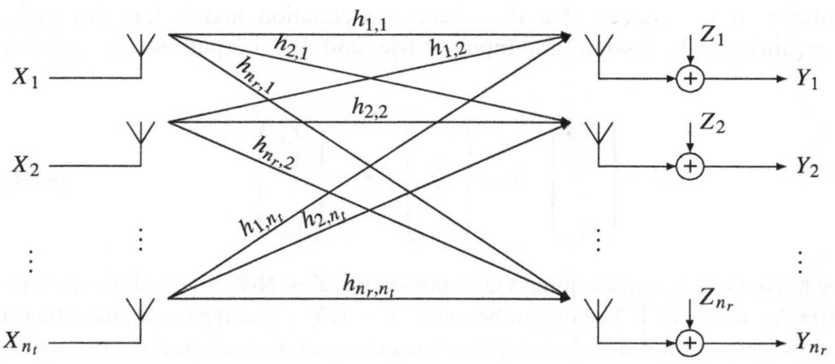

Figure 9.10 The antenna grid in a MIMO setup.

9.2.2 MIMO: The Multidimensional Gaussian Channel

The multiple in, multiple out (MIMO) Gaussian channel refers to a radio communication setup where several transmit and receive antennas are used together. The transmissions over the antennas use the same communication bandwidth, which mean that all receive antennas get contributions from all transmit antennas (see Figure 9.10). For each path, from one transmit antenna to one receive antenna, there is an attenuation factor, and if all transmission paths can be considered independent there are significant gains compared to the single antenna alternative. In many standards today, for example, in later versions of WiFi (802.11) and LTE, there are support for MIMO in the physical link. For the next mobile system, 5G, it will be an essential part of increasing the available bit rates in the system. In this section, the scheme will be considered from an information theoretic view and the capacity for the system derived.

The transmission system shown in Figure 9.10 uses n_t transmit antennas and n_r receive antennas. The attenuation factors between the transmit and received nodes can be given in the $n_r \times n_t$ matrix[6]

$$H = \begin{pmatrix} h_{11} & h_{12} & \cdots & h_{1n_t} \\ h_{21} & h_{22} & \cdots & h_{2n_t} \\ \vdots & \vdots & \ddots & \vdots \\ h_{n_r1} & h_{n_r2} & \cdots & h_{n_rn_t} \end{pmatrix} \tag{9.63}$$

[6] Often, due to the interpretation of the MIMO scheme as a wireless system, the coefficients are assumed to be complex. To maintain consistency with the preceding text, it is here assumed real-valued coefficients. The main difference is the use of transpose, $()^T$, instead of Hermitian transpose, $()^H$, and the factor $\frac{1}{2}$ in the capacity.

For simplicity, it is assumed that the channel attenuation matrix has full rank, $\text{rank}(H) = \min\{n_t, n_r\}$. Assign the input, noise and the output as the column vectors

$$\boldsymbol{X} = \begin{pmatrix} X_1 \\ X_2 \\ \vdots \\ X_{n_t} \end{pmatrix} \quad \boldsymbol{Z} = \begin{pmatrix} Z_1 \\ Z_2 \\ \vdots \\ Z_{n_r} \end{pmatrix} \quad \boldsymbol{Y} = \begin{pmatrix} Y_1 \\ Y_2 \\ \vdots \\ Y_{n_r} \end{pmatrix} \tag{9.64}$$

where the noise is an n_r-dimensional Gaussian vector, $\boldsymbol{Z} \sim N(\boldsymbol{0}, \Lambda_Z)$ with the covariance matrix $\Lambda_Z = E\big[\boldsymbol{Z}\boldsymbol{Z}^T\big]$. The output becomes $\boldsymbol{Y} = H\boldsymbol{X} + \boldsymbol{Z}$ and as such, the MIMO channel constitutes a multidimensional Gaussian channel with attenuation matrix H. As in the one-dimensional case, the total power used at the transmitter side is the limiting factor of the communication link. Constraining it to P, the sum of the transmitted powers must not exceed this. Since the diagonal elements in Λ_X is $P_i = E\big[X_i^2\big]$, the constraint can be written as $\operatorname{tr} \Lambda_X \leq P$. The notation $\operatorname{tr} A$ denotes the trace of the matrix A, which is the sum of the diagonal elements. The trace also equals the sum of the eigenvalues of the matrix.

The first step in deriving the channel capacity for the MIMO channel is to derive the mutual information between the input and output vectors that can be bounded as

$$\begin{aligned} I(\boldsymbol{X};\boldsymbol{Y}) &= H(\boldsymbol{Y}) - H(\boldsymbol{Y}|\boldsymbol{X}) \\ &= H(\boldsymbol{Y}) - H(\boldsymbol{Z}) \\ &= H(\boldsymbol{Y}) - \frac{1}{2}\log|2\pi e\Lambda_Z| \\ &\leq \frac{1}{2}\log|2\pi e\Lambda_Y| - \frac{1}{2}\log|2\pi e\Lambda_Z| \\ &= \frac{1}{2}\log|\Lambda_Y\Lambda_Z^{-1}| \end{aligned} \tag{9.65}$$

where it is used in the second equality that $\boldsymbol{X}$ and $\boldsymbol{Z}$ are independent, in the third equality the differential entropy derived in (8.83) and (8.88), and in the inequality Theorem 8.12 is applied. The bound is fulfilled with equality if and only if $\boldsymbol{Y}$ is Gaussian, i.e., $\boldsymbol{Y} \sim N(\boldsymbol{0}, \Lambda_Y)$. To achieve this distribution, first notice that the sum of two Gaussian variables is again Gaussian. By letting the input vector be Gaussian, $\boldsymbol{X} \sim N(\boldsymbol{0}, \Lambda_X)$, the output $\boldsymbol{Y}$ will also be Gaussian with the covariance matrix

$$\begin{aligned} \Lambda_Y &= E\big[\boldsymbol{Y}\boldsymbol{Y}^T\big] = E\big[(H\boldsymbol{X}+\boldsymbol{Z})(H\boldsymbol{X}+\boldsymbol{Z})^T\big] \\ &= E\big[(H\boldsymbol{X}+\boldsymbol{Z})(\boldsymbol{X}^T H^T + \boldsymbol{Z}^T)\big] \\ &= HE\big[\boldsymbol{X}\boldsymbol{X}^T\big]H^T + E\big[\boldsymbol{Z}\boldsymbol{Z}^T\big] \\ &= H\Lambda_X H^T + \Lambda_Z \end{aligned} \tag{9.66}$$

where $E\big[\boldsymbol{X}\boldsymbol{Z}^T\big] = \boldsymbol{0}$ since $\boldsymbol{X}$ and $\boldsymbol{Z}$ are independent. By maximizing the mutual information using a Gaussian-distributed input vector gives that the general form of the channel capacity can be written as

$$\begin{aligned} C &= \max_{f(\boldsymbol{x}),\mathrm{tr}\,\Lambda_X=P} I(X;Y) \\ &= \max_{\mathrm{tr}\,\Lambda_X=P} \frac{1}{2}\log\left|(H\Lambda_X H^T + \Lambda_Z)\Lambda_Z^{-1}\right| \\ &= \max_{\mathrm{tr}\,\Lambda_X=P} \frac{1}{2}\log\left|I_{n_r} + H\Lambda_X H^T\Lambda_Z^{-1}\right| \end{aligned} \tag{9.67}$$

Since all transmissions in Figure 9.10 work over the same bandwidth, it is reasonable to assume that the noise for the received symbols is independent and identically distributed (i.i.d.). Then the covariance matrix becomes a diagonal matrix with the noise variance N, i.e., $\Lambda_Z = NI_{n_r}$, where I_{n_r} is the $n_r \times n_r$ unit matrix. The capacity in (9.67) becomes

$$C = \max_{\mathrm{tr}\,\Lambda_X=P} \frac{1}{2}\log\left|I_{n_r} + \frac{1}{N}H\Lambda_X H^T\right| \tag{9.68}$$

To get a better understanding of how the distribution of $\boldsymbol{X}$ can be assigned, the channel attenuation matrix can be composed by singular value decomposition (SVD),[7]

$$H = USV^T \tag{9.69}$$

where U and V are orthogonal matrices and S a diagonal (in general nonsquare) matrix with the singular values s_i, $i = 1, \dots, n$ along the diagonal. The fact that U and V are orthogonal means they have unit determinant and that the transpose equals the inverse, i.e.

$$UU^T = U^TU = I_{n_r} \qquad |U| = |U^T| = 1 \tag{9.70}$$

$$VV^T = V^TV = I_{n_t} \qquad |V| = |V^T| = 1 \tag{9.71}$$

The capacity can be rewritten as

$$\begin{aligned} C &= \max_{\mathrm{tr}\,\Lambda_X=P} \frac{1}{2}\log\left|I_{n_r} + \frac{1}{N}(USV^T)\Lambda_X(USV^T)^T\right| \\ &= \max_{\mathrm{tr}\,\Lambda_X=P} \frac{1}{2}\log\left|UU^T + \frac{1}{N}USV^T\Lambda_X VS^TU^T\right| \\ &= \max_{\mathrm{tr}\,\Lambda_X=P} \frac{1}{2}\log|U|\left|I_{n_r} + \frac{1}{N}SV^T\Lambda_X VS^T\right||U^T| \\ &= \max_{\mathrm{tr}\,\Lambda_X=P} \frac{1}{2}\log\left|I_{n_r} + \frac{1}{N}SV^T\Lambda_X VS^T\right| \end{aligned} \tag{9.72}$$

Introducing a basis change of the input vector, such that $\widetilde{\boldsymbol{X}} = V^T\boldsymbol{X}$, gives the covariance matrix $\Lambda_{\widetilde{X}} = E\big[V^T\boldsymbol{X}(V^T\boldsymbol{X})^T\big] = E\big[V^T\boldsymbol{X}\boldsymbol{X}^TV\big] = V^T\Lambda_X V$.

In matrix theory, two matrices A and B, are said to be *similar* if there exists a nonsingular matrix T such that $A = TBT^{-1}$. Furthermore, two similar matrices have the same set of eigenvalues, and hence the same trace and determinant. This means

[7] The SVD can be derived in MATLAB with the command `[U,S,V]=svd(A)`.

that Λ_X and $\Lambda_{\widetilde{X}}$ are similar and since the trace is the sum of the eigenvalues, $\operatorname{tr}\Lambda_X = \operatorname{tr}\Lambda_{\widetilde{X}}$. Thus, the power constraint in the capacity formula can be considered over $\widetilde{X}$ instead of $\boldsymbol{X}$, and the capacity becomes

$$C = \max_{\operatorname{tr}\Lambda_{\widetilde{X}}=P} \frac{1}{2}\log\left|I_{n_r} + \frac{1}{N}S\Lambda_{\widetilde{X}}S^T\right| \tag{9.73}$$

Since $\Lambda_{\widetilde{X}}$ is a covariance matrix, it is positive semidefinite, i.e. $\boldsymbol{a}^T\Lambda_{\widetilde{X}}\boldsymbol{a} \geq 0$ for all vectors $\boldsymbol{a}$. From $\boldsymbol{a}^T S\Lambda_{\widetilde{X}}S^T\boldsymbol{a} = \tilde{\boldsymbol{a}}^T\Lambda_{\widetilde{X}}\tilde{\boldsymbol{a}} \geq 0$, where $\tilde{\boldsymbol{a}} = S^T\boldsymbol{a}$, it is seen that also $S\Lambda_{\widetilde{X}}S^T$ is positive semidefinite.

The Hadamard inequality states that if a matrix A is positive semidefinite, then the determinant is bounded by the product of the diagonal entries, $|A| \leq \prod_i a_{ii}$. Clearly, there is equality if A is a diagonal matrix. Since both I and S in the argument $I_{n_r} + \frac{1}{N}S\Lambda_{\widetilde{X}}S^T$ are diagonal, the minimum of (9.73) is obtained when $\Lambda_{\widetilde{X}}$ is diagonal. By assigning $\Lambda_{\widetilde{X}} = \operatorname{diag}(\widetilde{P}_1, \dots, \widetilde{P}_n)$, the capacity is obtained by

$$C = \max_{\sum_i \widetilde{P}_i=P} \frac{1}{2}\log\prod_{i=1}^{n}\left(1 + \frac{s_i^2}{N}\widetilde{P}_i\right) = \max_{\sum_i \widetilde{P}_i=P} \sum_{i=1}^{n}\frac{1}{2}\log\left(1 + \frac{s_i^2}{N}\widetilde{P}_i\right) \tag{9.74}$$

From the assumption that H has full rank the number of nonzero singular values is $n = \min\{n_t, n_r\}$.

Hence, the MIMO channel is equivalent to a channel with parallel Gaussian channels where the attenuation is given by the singular values of H. As before, to optimize the usage of the channel the power levels $\widetilde{P}_i$ can be found by water filling. The result is summarized in the following theorem.

Theorem 9.6 Given a MIMO channel with n_t transmit antennas, n_r receive antennas and the attenuation matrix H, the capacity is given by

$$C = \sum_{i=1}^{n}\frac{1}{2}\log\left(1 + \frac{s_i^2}{N}\widetilde{P}_i\right) \tag{9.75}$$

where the power levels $\widetilde{P}_i$ is found from the equation system

$$\widetilde{P}_i = \left(B - \frac{N}{s_i^2}\right)^+ \tag{9.76}$$

$$\sum_i \widetilde{P}_i = P \tag{9.77}$$

and s_i are the $n = \min\{n_t, n_r\}$ nonzero singular values in the singular value decomposition $H = USV^T$. □

The optimizing distribution is $\widetilde{\boldsymbol{X}} \sim \mathrm{N}(\boldsymbol{0}, \Lambda_{\widetilde{X}})$ where $\Lambda_{\widetilde{X}} = \operatorname{diag}(\widetilde{P}_1, \dots, \widetilde{P}_{n_t})$. This corresponds to the input distribution $\boldsymbol{X} \sim \mathrm{N}(\boldsymbol{0}, \Lambda_X)$ where $\Lambda_X = V\Lambda_{\widetilde{X}}V^T$.

Example 9.4 A channel with two transmit antennas and three receive antennas is defined by the attenuation matrix

$$H = \begin{pmatrix} 0.6 & 0.35 \\ 0.5 & 0.15 \\ 0.05 & 0.8 \end{pmatrix} \tag{9.78}$$

The transmit power is limited to $P = 4$, and the received additional noise vector is i.i.d. Gaussian with variance $V[Z_i] = N$, i.e., $\boldsymbol{Z} \sim \mathrm{N}(\boldsymbol{0}, NI_3)$, where $N = 2$. That is, when the two-dimensional vector $\boldsymbol{X}$ is transmitted the three-dimensional vector $\boldsymbol{Y}$ is received according to

$$\begin{pmatrix} Y_1 \\ Y_2 \\ Y_3 \end{pmatrix} = \begin{pmatrix} 0.6 & 0.35 \\ 0.5 & 0.15 \\ 0.05 & 0.8 \end{pmatrix} \begin{pmatrix} X_1 \\ X_2 \end{pmatrix} + \begin{pmatrix} Z_1 \\ Z_2 \\ Z_3 \end{pmatrix} \tag{9.79}$$

The svd, $H = USV^T$, of the attenuation matrix gives

$$U = \begin{pmatrix} 0.63 & 0.44 & -0.64 \\ 0.42 & 0.51 & 0.76 \\ 0.65 & -0.74 & 0.14 \end{pmatrix}, \ S = \begin{pmatrix} 1.017 & 0 \\ 0 & 0.602 \\ 0 & 0 \end{pmatrix}, \ V = \begin{pmatrix} 0.61 & 0.79 \\ 0.79 & -0.61 \end{pmatrix} \tag{9.80}$$

which gives the nonzero singular values $s_1 = 1.017$ and $s_2 = 0.602$. The water-filling procedure starts with the power constraint

$$\sum_{i=1}^{2} \left(B - \frac{N}{s_i^2} \right) = 2B + \sum_{i=1}^{2} \frac{N}{s_i^2} = P \tag{9.81}$$

which gives

$$B = \frac{1}{2} \left(P + \sum_{i=1}^{2} \frac{N}{s_i^2} \right) = 5.72 \tag{9.82}$$

The corresponding power distribution over the channels then becomes

$$P_i = B - \frac{N}{s_i^2} = \begin{cases} 3.79, & i = 1 \\ 0.21 & i = 2 \end{cases} \tag{9.83}$$

Since both subchannel powers are positive, they can be used to derive the capacity of the channel

$$C = \sum_{i=1}^{2} \frac{1}{2} \log\left(1 + \frac{s_i^2 P_i}{N} \right) = 0.783 + 0.027 = 0.81 \text{ bit/channel use} \tag{9.84}$$

This capacity is reached for the input covariance matrix

$$\Lambda_{\tilde{X}} = \begin{pmatrix} 3.79 & 0 \\ 0 & 0.21 \end{pmatrix} \tag{9.85}$$

The corresponding covariance matrix for the transmit vector $\boldsymbol{X}$ is

$$\Lambda_X = V\Lambda_{\tilde{X}}V^T = \begin{pmatrix} 1.54 & 1.73 \\ 1.73 & 2.46 \end{pmatrix} \tag{9.86}$$

This gives, of course, the same capacity

$$C = \frac{1}{2}\log\left|I_{n_r} + \frac{1}{N}H\Lambda_X H^T\right| = 0.81 \text{ bit/channel use} \tag{9.87}$$

The resulting capacity for the MIMO channel can be compared with the case when only one-single antenna is used for both transmitter and receiver, and the others are turned off. From the attenuation matrix, it is seen that the maximum capacity is reached when the second transmitter antenna is used in combination with the third receive antenna. This gives

$$C_{32} = \frac{1}{2}\log\left(1 + \frac{h_{32}^2 P}{N}\right) = 0.59 \text{ bit/channel use} \tag{9.88}$$

So, for this example, the gain for the MIMO transmission compared with the best single antenna version of the channel is about 36%.

9.3 FUNDAMENTAL SHANNON LIMIT

One of the most famous results from information theory is the fundamental Shannon limit that sets requirements on the SNR for reliable communication. When deriving the limit, the capacity formula for the band-limited case is considered as the band width grows to infinity. It can also be seen that to reach this limit, it is required that the coding rate goes to zero and with that the codeword length and decoding complexity approach infinity.

Consider a band-limited Gaussian channel with bandwidth W and noise-level $N_0/2$. If the transmitted power constraint is P, the capacity is given by

$$C = W\log\left(1 + \frac{P}{N_0 W}\right) \quad \text{[b/s]} \tag{9.89}$$

The capacity maximizes the throughput for a fixed SNR when using a certain bandwidth. As the bandwidth increases, the noise power will also increase, whereas the signal power will remain, meaning that the capacity increase will diminish with increasing bandwidth. In Figure 9.11, the capacity as a function of the bandwidth is shown. As the bandwidth increases to infinity, the following limit value is reached:

$$\begin{aligned} C_\infty &= \lim_{W\to\infty} W\log\left(1 + \frac{P/N_0}{W}\right) \\ &= \lim_{W\to\infty} \log\left(1 + \frac{P/N_0}{W}\right)^W \\ &= \log e^{P/N_0} = \frac{P/N_0}{\ln 2} \end{aligned} \tag{9.90}$$

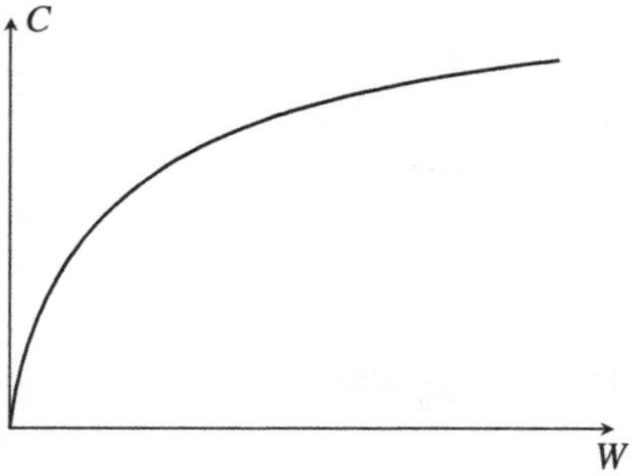

Figure 9.11 Capacity as a function of the bandwidth W for a fixed total signal power P.

Denoting the achieved bit rate as R_b, it is required that this is not more than the capacity, $C_\infty > R_b$. Let E_b be the average transmit energy per information bit. This is a very important parameter since it can be compared between different systems, without having the same number of bits per symbol or even the same coding rate. Hence, a system independent SNR can be used as $\text{SNR} = E_b/N_0$. The energy per information bit can be written as

$$E_b = PT_b = \frac{P}{R_b} \Rightarrow R_b = \frac{P}{E_b} \tag{9.91}$$

Since $C_\infty > R_b$, the ratio between the capacity and the achieved bit rate gives

$$\frac{C_\infty}{R_b} = \frac{\frac{P/N_0}{\ln 2}}{P/E_b} = \frac{E_b/N_0}{\ln 2} > 1 \tag{9.92}$$

Rewriting the above concludes that for reliable communication

$$\frac{E_b}{N_0} > \ln 2 = 0.69 = -1.59\text{ dB} \tag{9.93}$$

The value -1.6 dB is a well-known bound in communication theory and is often referred to as the *fundamental limit*. It constitutes a hard limit for when it is possible to achieve reliable communication. If the SNR is less than this limit, it is not possible to reach error probability that tends to be zero, independent of what system is used.

To see the demands this limit puts on the system, the capacity formula will be rewritten to show a limit on the bandwidth efficiency R_b/W in the used system. Often the bandwidth are a limited resource and then to reach a high data rate the system should have a high bandwidth efficiency. Inserting $P = R_b E_b$ in the capacity formula (9.89), and utilizing that $R_b < C$ the following relation is acquired:

$$C = W\log\Big(1 + \frac{R_b}{W}\cdot\frac{E_b}{N_0}\Big) < W\log\Big(1 + \frac{C}{W}\cdot\frac{E_b}{N_0}\Big) \quad \text{[b/s]} \tag{9.94}$$

That is, a limit on the maximum achievable bandwidth efficiency C/W can be obtained as

$$\frac{C}{W} < \log\Big(1 + \frac{C}{W}\cdot\frac{E_b}{N_0}\Big) \quad \left[\frac{\text{b/s}}{Hz}\right] \tag{9.95}$$

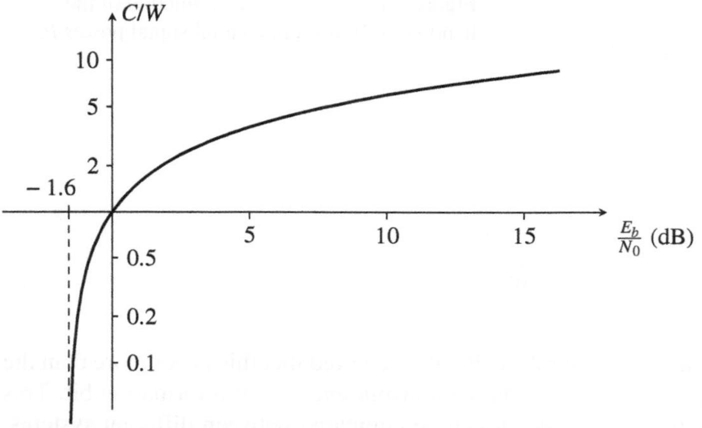

Figure 9.12 Maximum bandwidth utilization C/W as a function of the SNR $\frac{E_b}{N_0}$. Notice the logarithmic scale on the y-axis.

which can be rewritten as

$$\frac{E_b}{N_0} > \frac{2^{C/W} - 1}{C/W} \tag{9.96}$$

As C/W becomes large, the required SNR $\frac{E_b}{N_0} \approx \frac{2^{C/W}}{C/W}$ grows exponentially with the bandwidth efficiency. For a fixed $\frac{E_b}{N_0}$, there is a maximum achievable C/W. On the other hand, if the bandwidth efficiency decreases to zero, the Shannon limit on the SNR is achieved

$$\frac{E_b}{N_0} > \lim_{C/W \to 0} \frac{2^{C/W} - 1}{C/W} = \ln 2 = -1.59 \text{ dB} \tag{9.97}$$

In Figure 9.12, C/W is shown as a function of $\frac{E_b}{N_0}$. To reach the fundamental limit, the bandwidth efficiency has to approach zero.

An alternative way to view the decrease in bandwidth efficiency close to the limit is to consider the coding rate. With an increasing strength of the code, there is an increasing amount of redundancy transmitted. That means the amount of information in each transmission will decrease, which will also decrease the bandwidth efficiency. With a growing portion of redundancy, the code rate will decrease. In fact, next it will be shown that to reach the fundamental limit the code rate has also to approach zero.

First, assume that a codeword consists of N samples and that there are K information bits in it, giving an (N, K) code with a rate

$$R = \frac{K}{N} \tag{9.98}$$

The duration of time for a codeword can then be set to T, and assuming the Nyquist rate, this is coupled to the number of samples through $N = TF_s = 2WT$ sample/codeword. The information bit rate can be derived as the number of information bits in a codeword divided by the duration of the codeword,

$$R_b = \frac{K}{T} = 2W\frac{K}{N} = 2WR \tag{9.99}$$

Similarly, each codeword requires an average energy of KE_b, and the corresponding power is

$$P = \frac{KE_b}{T} \tag{9.100}$$

With this at hand, the SNR in the capacity formula can be written as

$$\frac{P}{N_0 W} = \frac{KE_b}{TN_0 W} = 2WR\frac{E_b}{N_0 W} = 2\frac{E_b}{N_0}R \tag{9.101}$$

Then, since the bit rate is less than the capacity

$$R_b = 2WR < W\log\left(1 + 2\frac{E_b}{N_0}R\right) \tag{9.102}$$

which gives

$$1 + 2\frac{E_b}{N_0}R > 2^{2R} \tag{9.103}$$

or, equivalently,

$$\frac{E_b}{N_0} > \frac{2^{2R} - 1}{2R} \tag{9.104}$$

This gives the requirement on E_b/N_0 for a certain code rate. Using, e.g., a code with rate $R = \frac{1}{2}$ the limit is now shifted to

$$\frac{E_b}{N_0} > 1 = 0 \text{ dB} \tag{9.105}$$

To get a better communication environment, the code rate is decreased. It can only be decreased down to zero, where the bound again hits the fundamental limit.

$$\frac{E_b}{N_0} > \lim_{R\to 0} \frac{2^{2R} - 1}{2R} = \ln 2 = -1.59 \text{ dB} \tag{9.106}$$

Hence, it is seen that to reach the limit -1.6 dB the code rate has to approach zero. In Figure 9.13, the minimum code rate is shown as a function of the SNR.

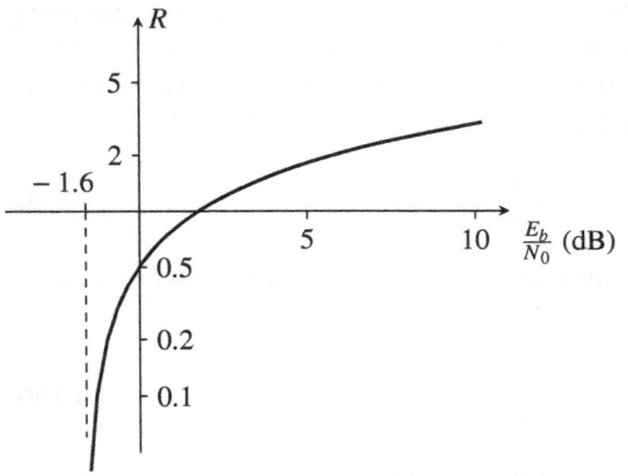

Figure 9.13 Maximum code rate R as a function of the SNR $\frac{E_b}{N_0}$. Notice the logarithmic scale on the y-axis.

PROBLEMS

9.1 An additive channel has input X and output $Y = X + Z$, where the noise is normal distributed with $Z \sim N(0, \sigma)$. The channel has an *output* power constraint $E[Y^2] \leq P$. Derive the channel capacity for the channel.

9.2 A band-limited Gaussian channel has bandwidth $W = 1$ kHz. The transmitted signal power is limited to $P = 10$ W, and the noise on the channel is distributed according to $N(0, \sqrt{N_0/2})$, where $N_0 = 0.01$ W/Hz. What is the channel capacity?

9.3 A random variable X is drawn from a uniform distribution U(1), and transmitted over a channel with additive noise Z, distributed uniformly U(a), where $a \leq 1$. The received random variable is then $Y = X + Z$. Derive the average information obtained about X from the received Y, i.e., $I(X; Y)$.

9.4 A channel consists of two channels, both with attenuation and Gaussian noise. The first channel has the attenuation H_1 and noise distribution $n_1 \sim N(0, \sqrt{N_1/2})$, and the second channel has attenuation H_2 and noise distribution $n_2 \sim N(0, \sqrt{N_2/2})$. The two channels are used in cascade, i.e. a signal X is first transmitted over the first channel and then over the second channel (see Figure 9.14). Assume that both channels work over the same bandwidth W.

(a) Derive an expression for the channel capacity for the cascaded channel.

(b) Denote the signal-to-noise ratio over the cascaded channel as SNR and the two constituent channels as SNR_1 and SNR_2, respectively. Show that

$$\text{SNR} = \frac{\text{SNR}_1 \cdot \text{SNR}_2}{\text{SNR}_1 + \text{SNR}_2}$$

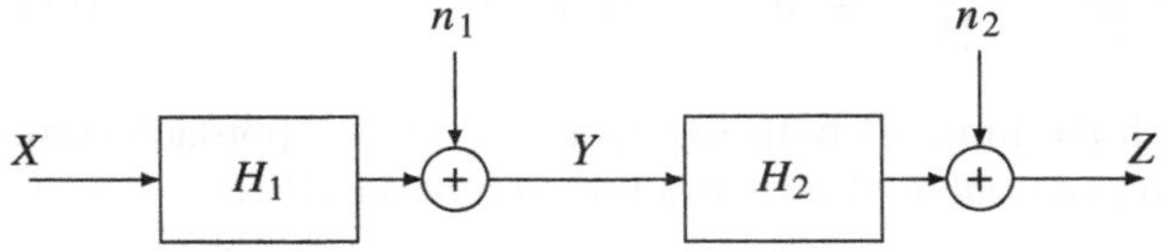

Figure 9.14 A channel consisting of two Gaussian channels.

Notice that the formula is similar to the total resistance of a parallel coupling in electronics design.

9.5 A wide-band channel is split in four independent, parallel, time discrete, additive Gaussian channels. The variance of the noise in the ith channel is $\sigma_i = i^2$, $i = 1, 2, 3, 4$. The total power of the used signals is limited by

$$\sum_{i=1}^{4} P_i \leq 17.$$

Derive the channel capacity.

9.6 A channel consists of six parallel Gaussian channels with the noise levels

$$N = (8, 12, 14, 10, 16, 6)$$

The total allowed power usage in the transmitted signal is $P = 19$.

(a) What is the capacity of the combined channel?

(b) If you must divide the power equally over the six channels, what is the capacity?

(c) If you decide to use only one of the channels, what is the maximum capacity?

9.7 An OFDM channel with five subchannels, each occupying a bandwidth of 10 kHz. Due to regulations, the allowed power level in the utilized band is –60 dBm/Hz. Each of the subchannels has separate attenuation and noise levels according to Figure 9.15. Notice that the attenuation $|G_i|^2$ is given in dB and the noise $N_{0,i}$ in dBm/Hz, where i is the subchannel. Derive the total capacity for the channel.

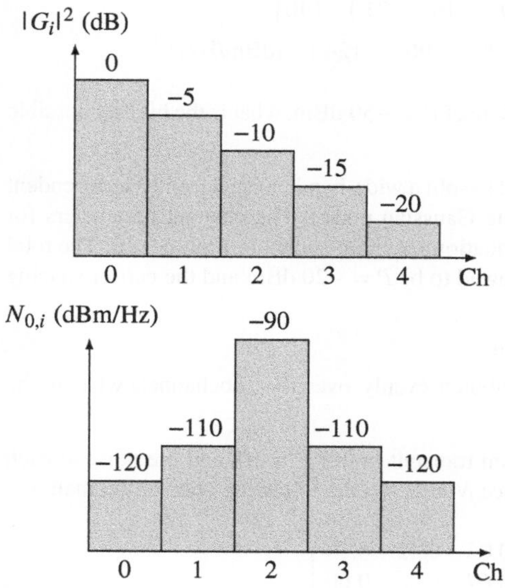

Figure 9.15 Attenuation and noise in the five subchannels.

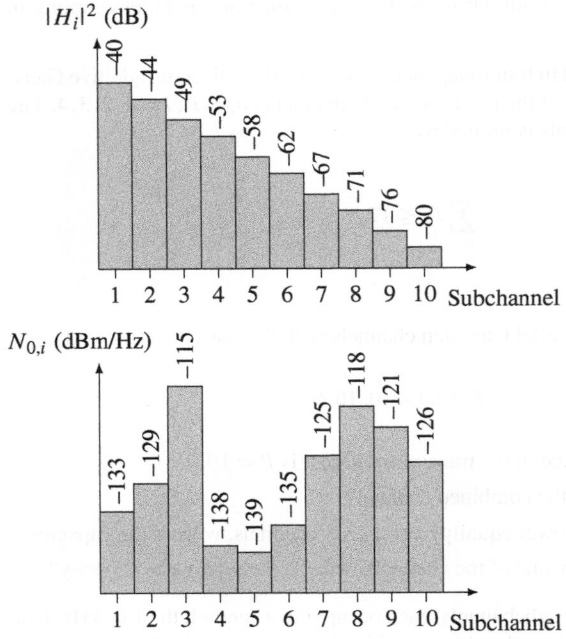

Figure 9.16 Channel parameters for an OFDM communication system.

9.8 A wide-band Gaussian channel is split into four subchannels, each with bandwidth $W_\Delta = 1$ kHz. The attenuations and noise parameters are

$$|H_i|^2 = \begin{pmatrix} -36 & -30 & -16 & -21 \end{pmatrix} \quad [\text{dB}]$$
$$N_0 = \begin{pmatrix} -108 & -129 & -96 & -124 \end{pmatrix} \quad [\text{dBm/Hz}]$$

With the total allowed power on the channel $P = -50$ dBm, what is the highest possible bit rate on the channel?

9.9 An OFDM modulation scheme is used to split a wide-band channel into 10 independent subbands with AWGN (additive white Gaussian noise). The channel parameters for the noise level N_0 and the signal attenuation $|H_i|^2$ are shown in Figure 9.16. The total power in the transmitted signal is allowed to be $P = -20$ dBm and the carrier spacing is $W_\Delta = 10$ kHz.

(a) Derive the capacity for the system.

(b) If the transmitted power is distributed evenly over the subchannels,what is the capacity for the system?

9.10 A 3×3 MIMO system with maximum transmit power $P = 10$, and the noise at each receive antenna Gaussian with variance $N = 1$, has the following attenuation matrix:

$$H = \begin{pmatrix} 0.05 & 0.22 & 0.73 \\ 0.67 & 0.08 & 0.60 \\ 0.98 & 0.36 & 0.45 \end{pmatrix}$$

Derive the channel capacity for the system. What is the capacity achieving distribution on the transmitted vector $\boldsymbol{X}$?

Hint: The singular value decomposition of H is given by

$$U = \begin{pmatrix} -0.36 & 0.88 & 0.30 \\ -0.59 & 0.03 & -0.81 \\ -0.72 & -0.47 & 0.51 \end{pmatrix}, \; S = \begin{pmatrix} 1.51 & 0 & 0 \\ 0 & 0.60 & 0 \\ 0 & 0 & 0.19 \end{pmatrix}, \; V = \begin{pmatrix} -0.74 & -0.65 & -0.16 \\ -0.26 & 0.05 & 0.97 \\ -0.62 & 0.76 & -0.20 \end{pmatrix}$$

9.11 The 5×4 MIMO attenuation matrix

$$H = \begin{pmatrix} 0.76 & 0.40 & 0.61 & 0.90 & 0.42 \\ 0.46 & 0.73 & 0.97 & 0.23 & 0.73 \\ 0.61 & 0.96 & 0.63 & 0.10 & 0.94 \\ 0.36 & 0.88 & 0.78 & 0.83 & 0.30 \end{pmatrix}$$

has the singular value decomposition $H = USV^T$, where

$$\mathrm{diag}(S) = \begin{pmatrix} 2.85 & 0.89 & 0.46 & 0.30 \end{pmatrix}$$

What is the channel capacity for the system if $P = 5$ and $N = 2$?

CHAPTER 10

DISCRETE INPUT GAUSSIAN CHANNEL

IN THE PREVIOUS chapter, it was seen that to reach capacity for a Gaussian channel the input variable should be Gaussian distributed. Normally, it is not possible to use continuous distributions for the transmitted signals and instead discrete values are used. That is, the transmitted variable is discrete while the noise on the channel is Gaussian. Furthermore, in most applications the transmitted signal alternatives are considered to be equally likely. In this chapter, a constraint capacity, in form of the mutual information, for an M-ary pulse amplitude modulated signal will be derived in the case when the (finite number) signal alternatives are equally likely [73]. The loss made by using uniformly distributed inputs will be derived and addressed as the shaping gain. Finally, a parameter called the signal-to-noise ratio (SNR) gap is derived to show how far from the capacity an uncoded system is working. This latter value is derived for a certain obtained bit error rate.

10.1 M-PAM SIGNALING

When transmitting discrete data, the symbols must be represented by signals such that it can be transmitted over a continuous media. It should also be possible for the receiver to decode back to the discrete form even though the signal is distorted on the channel. This process is called modulation and demodulation, and one of the basic modulation scheme is M-ary pulse amplitude modulation (M-PAM). In this text a brief introduction to the subject is given. For a more thorough description refer to e.g., [51, 76, 82, 91]. The number M is the number of signal alternatives, i.e. the number of different signals used in the scheme. Since the transmitted data are often binary, this number will here be assumed to be a power of 2, $M = 2^k$. In an M-PAM scheme, a signal is built from an amplitude and a pulse form, where the amplitude is the information carrier and the pulse form common for all signal alternatives. To minimize the average signal energy, the amplitudes are centered around zero, e.g., the binary case has the amplitudes -1 and 1. If $M = 4$, the amplitudes -3, -1, 1, and 3 are used. In this way, the minimum difference between two amplitude values are always 2. For an arbitrary M, the amplitude values can be described by

$$A_i = M - 1 - 2i, \quad i = 0, 1, 2, \ldots, M - 1 \tag{10.1}$$

Information and Communication Theory, First Edition. Stefan Höst.

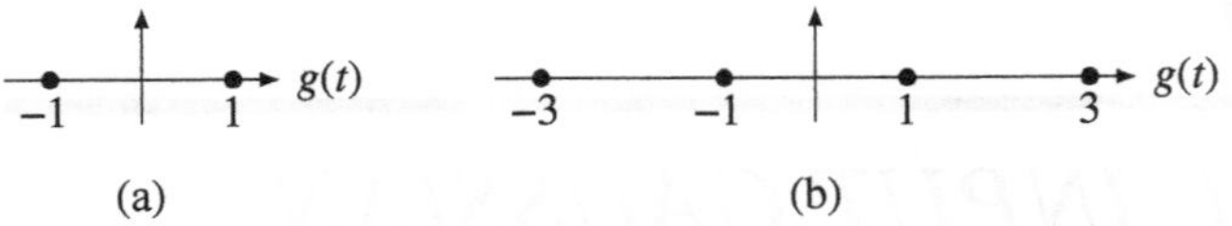

Figure 10.1 Graphical representation of (a) 2-PAM and (b) 4-PAM.

which holds for all positive integer M and not just powers of 2 [51]. Then, to form the signal, the amplitude is applied to a pulse form $g(t)$, meaning that the general form of a signal alternative in M-PAM can be written as

$$s_i(t) = A_i g(t) \tag{10.2}$$

In Figure 10.1, a graphical view of the 2-PAM and 4-PAM signal alternatives is shown.

Assuming an infinite binary information sequence to be transmitted, where tuples of $k = \log M$ bits are mapped to an amplitude A_i, the transmitted signal is

$$s(t) = \sum_{\ell} A_{i_\ell} g(t - \ell T_s) \tag{10.3}$$

where T_s is the signaling interval.

The pulse form $g(t)$ has the energy $\int_{\mathbb{R}} g^2(t)dt = E_g$. By letting A be a random variable for the amplitude level, the symbol energy becomes

$$E_s = E\left[\int_{\mathbb{R}} (Ag(t))^2 dt\right] = E[A^2]\int_{\mathbb{R}} (g(t))^2 dt = E[A^2]E_g \tag{10.4}$$

For equally likely signal alternatives and levels according to an M-PAM constellation,

$$E_s = E[A^2]E_g = \sum_{i=0}^{M-1} \frac{1}{M} A_i^2 E_g = \frac{M^2-1}{3} E_g \tag{10.5}$$

Example 10.1 Considering a 2-PAM signal constellation used to communicate over a channel with additive white Gaussian noise (AWGN). The signals are chosen from the two signal alternatives in Figure 10.1a. For transmission, a mapping between the information bit a and the amplitude is used according to $s_a(t) = s_a \cdot g(t)$, where $s_a = (-1)^a$ and $g(t) = \sqrt{E_g}\phi(t)$, i.e.

$$s_a(t) = \begin{cases} \sqrt{E_g}\phi(t), & a = 0 \\ -\sqrt{E_g}\phi(t), & a = 1 \end{cases} \tag{10.6}$$

The basis function $\phi(t)$ is a scaled version of $g(t)$ such that it has unit energy, $\int_{\mathbb{R}} \phi^2(t)dt = 1$. The energy per transmitted information bit for this constellation is

$$E_b = \sum_{a=0}^{1} \frac{1}{2} \int_{\mathbb{R}} s_a^2(t)dt = \sum_{a=0}^{1} \frac{1}{2} E_g \int_{\mathbb{R}} \phi^2(t)dt = E_g \tag{10.7}$$

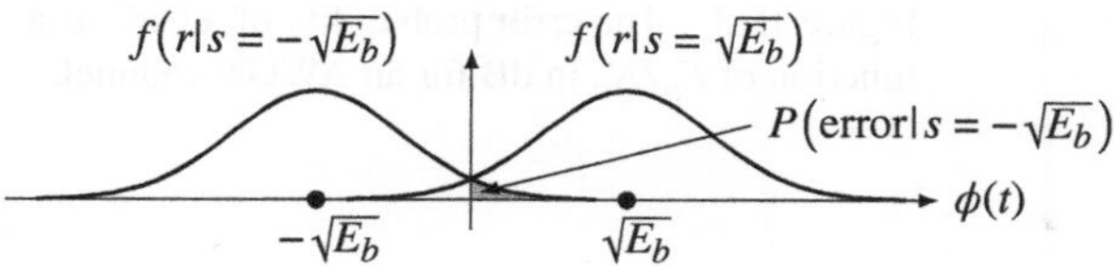

Figure 10.2 The conditional distributions at the receiver side in a 2-PAM transmission over an AWGN channel.

During transmission over an AWGN channel, noise with power spectral density $R_\eta(f) = N_0/2$ is added to the signal. After filtering and maximum likelihood (ML) detection at the receiver side, the received signal can be viewed as the point $r = s + z$ in the signal space, where $s = \pm\sqrt{E_b}$ is the transmitted signal amplitude and $z \sim N(0, \sqrt{N_0/2})$. In Figure 10.2, the probability distributions for the received value conditioned on the transmitted s is shown.

If the two signal alternatives are equally likely, an ML receiver follows a simple decoding rule. If the received value is positive, the estimated transmitted amplitude is $\hat{s} = \sqrt{E_b}$, and if the value is negative the estimated transmitted amplitude is $\hat{s} = -\sqrt{E_b}$. Hence the probability of erroneous estimation, conditioned on the transmitted amplitude $-\sqrt{E_b}$ is the gray-shaded area in the figure, and

$$\begin{aligned} P(\text{error}|s = -\sqrt{E_b}) &= P(r > 0|s = -\sqrt{E_b}) \\ &= P(z > \sqrt{E_b}) \\ &= P\Big(z_{\text{norm}} > \sqrt{\tfrac{E_b}{N_0/2}}\Big) = Q\Big(\sqrt{2\tfrac{E_b}{N_0}}\Big) \end{aligned} \tag{10.8}$$

where $z_{\text{norm}} \sim N(0, 1)$ is a normalized Gaussian variable and

$$Q(x) = \int_x^\infty \frac{1}{\sqrt{2\pi}} e^{-t^2/2} dt \tag{10.9}$$

gives the error probability, $P(z_{\text{norm}} > x)$. Similarly, the error probability conditioned on the transmitted amplitude $\sqrt{E_g}$ gets the same value. The error probability is the probability that a 1 is transmitted and a zero is received and vice versa. That is, the channel can after detection be modeled as a binary symmetric channel (BSC), with the crossover probability equaling to

$$\varepsilon = Q\Big(\sqrt{2\tfrac{E_b}{N_0}}\Big) \tag{10.10}$$

In Figure 10.3, the error probability ε is plotted as a function of the SNR E_b/N_0 in dB. With this mapping, the capacity for the BSC, $C_{BSC} = 1 - h(\varepsilon)$, is plotted in Figure 10.4 as a function of $E - b/N_0$ in dB.

In Chapter 8, it was seen that the interpretation of the mutual information is the same in the discrete and the continuous case, i.e. the amount of information achieved about one variable by observing another. The interpretation still holds if one of the variables is discrete and the other continuous. This is an important fact, since the capacity for a

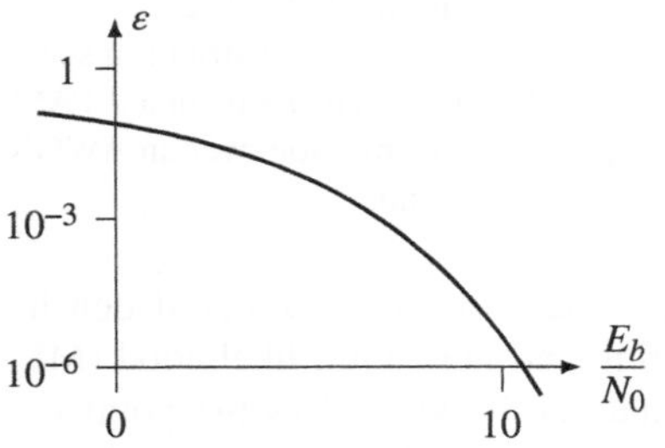

Figure 10.3 The error probability of a BSC as a function of E_b/N_0 in dB for an AWGN channel.

channel is the mutual information maximized over the input distribution. In this chapter, the considered channel model has discrete input signals, and white noise is added on the channel. To derive the capacity for this model, the mutual information should be maximized over all distributions for the input signals. However, often in communication systems the signals are transmitted with equal probabilities. The counterpart of the capacity, a constraint capacity, is the mutual information for the case with discrete, equally likely inputs and white noise added during transmission. The mutual information between discrete and continuous variables can be derived in the same manner as before by $I(X;Y) = H(Y) - H(Y|X)$. Letting $p(x)$ be the distribution for the input symbols X, the conditional entropy can be written as

$$\begin{aligned} H(Y|X) &= -\sum_x \int_{\mathbb{R}} f(x,y) \log f(y|x) dy \\ &= -\sum_x \int_{\mathbb{R}} f(y|x)p(x) \log f(y|x) dy \\ &= \sum_x p(x) \Big(-\int_{\mathbb{R}} f(y|x) \log f(y|x) dy \Big) \\ &= \sum_x p(x) H(Y|X=x) \end{aligned} \tag{10.11}$$

By $f(y) = \sum_x f(y|x)p(x)$, the entropy of Y can be expressed in terms of the input probabilities and the channel density function,

$$\begin{aligned} H(Y) &= -\int_{\mathbb{R}} f(y) \log f(y) dy \\ &= -\int_{\mathbb{R}} \Big(\sum_x f(y|x)p(x) \Big) \log \Big(\sum_x f(y|x)p(x) \Big) dy \end{aligned} \tag{10.12}$$

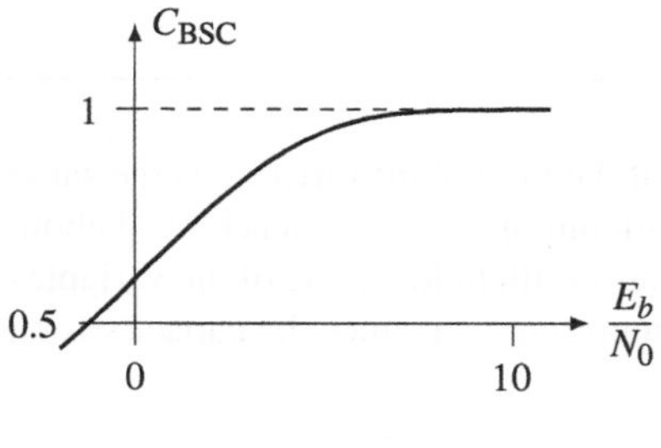

Figure 10.4 The capacity of a BSC as a function of E_b/N_0 in dB for an AWGN channel.

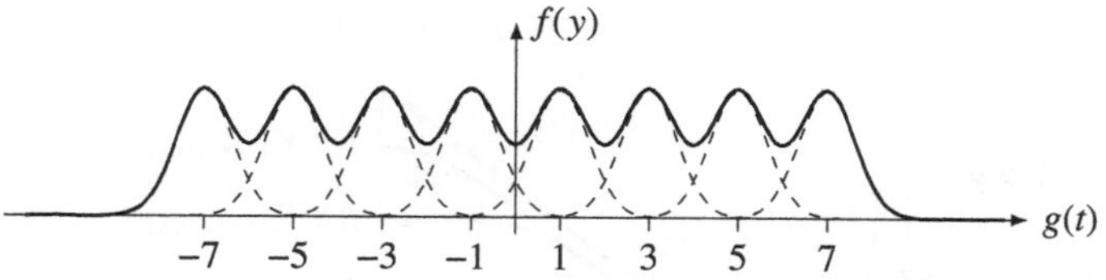

Figure 10.5 The density functions $f(y|x)$ for an 8-PAM signal transmitted over a Gaussian channel are added together to form the resulting $f(y)$.

Even in simple cases, as seen below, it is hard to find closed expressions for $f(y)$, and the derivation for $H(Y)$ falls back to numerical integration. For an M-PAM signal constellation with equally likely signal alternatives, the density of Y becomes

$$f(y) = \frac{1}{M} \sum_x f(y|x) \tag{10.13}$$

Assuming an additive Gaussian noise with zero mean and variance $N_0/2$, the conditional density function is

$$f(y|x) = \frac{1}{\sqrt{\pi N_0}} e^{-(y-x)^2/N_0} \tag{10.14}$$

In Figure 10.5, the resulting density function $f(y)$ for an 8-PAM constellation with equally likely signals is shown. In the figure, the contribution from the eight conditional density functions $f(y|x)$ is shown as dashed graph, whereas the density function for Y is shown as a solid curve. In this example, the noise variance is quite high to show the behavior. In the case of a more moderate noise, the eight peaks corresponding to the signal alternatives will be more separated by deeper valleys.

To get the entropy of Y, the function $-f(y)\log f(y)$ is integrated by numerical methods. The entropy conditional entropy $H(Y|X)$ can be derived from

$$H(Y|X) = \frac{1}{2} \log \pi e N_0 \tag{10.15}$$

since $[Y|X = x] \sim \mathrm{N}(x, \sqrt{N_0/2})$ and $H(Y|X = x) = \frac{1}{2} \log \pi e N_0$.

In Figure 10.6, plots of the mutual information $I(X; Y) = H(Y) - H(Y|X)$ for M-PAM signaling is shown for the case of equiprobable signal alternatives and additive Gaussian noise. The mutual information $I(X; Y)$ is a measure of how much information can be transmitted over the channel for each channel use, i.e. for each signal alternative sent. The plots typically flatten at the maximum transmitted bits for the number of signal alternatives as the channel becomes good. For example, 6 bits can be written as 64 different binary vectors and $I(X; Y)$ for 64-PAM flattens at 6 bits/channel use.

Assuming the Nyquist sampling rate $F_s = 2W$, the signal time is $T_s = 1/2W$. With P as the average power, the energy per signal becomes $E_s = P/2W$. Hence, the SNR in the capacity formula can be derived as

$$\mathrm{SNR} = \frac{P}{WN_0} = 2\frac{E_s}{N_0} \tag{10.16}$$

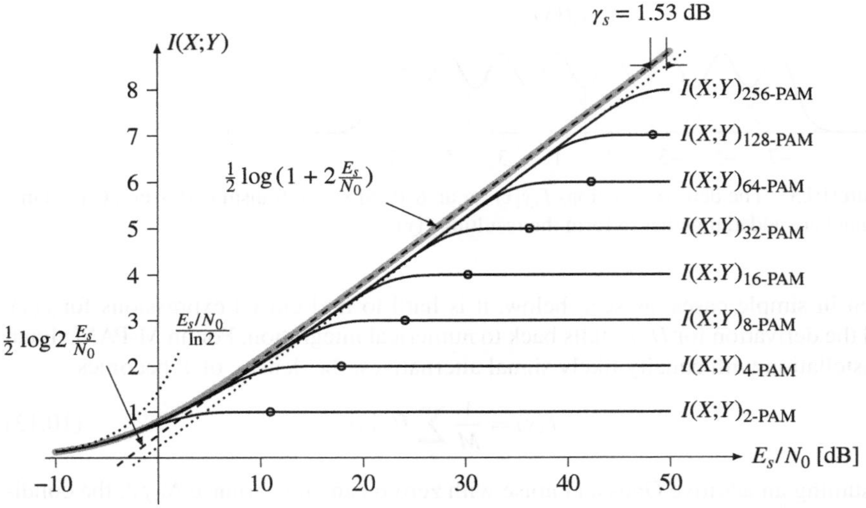

Figure 10.6 Constraint capacity for discrete uniformly distributed signal constellations, like M-PAM, transmitted over an AWGN channel. The gray- shaded line in the figure is the capacity $C = \frac{1}{2}\log(1 + 2\frac{E_s}{N_0})$. The circles on the curves mark the SNR where the uncoded M-PAM signaling gives an error probability of 10^{-6}.

and the expression for the capacity can be written as

$$C = \frac{1}{2}\log\left(1 + \frac{P}{N_0 W}\right)$$
$$= \frac{1}{2}\log\left(1 + 2\frac{E_s}{N_0}\right) \quad \text{[bit/channel use]} \tag{10.17}$$

In Figure 10.6, the capacity is shown as the thick gray line.

For good channels, as the SNR $\frac{E_s}{N_0}$ becomes large, the "+1" in the capacity formula can be neglected. Furthermore, as $\frac{E_s}{N_0}$ becomes small, the following series expansion can be used:

$$\frac{1}{2}\log(1+x) = \frac{1}{2\ln 2}\left(x - \frac{x^2}{2} + \frac{x^3}{3} - \frac{x^4}{4} + \cdots\right) \approx \frac{x}{2\ln 2} \tag{10.18}$$

to estimate the capacity. Hence, the asymptotic behavior of the capacity function is given by

$$C \approx \begin{cases} \frac{1}{2}\log 2\frac{E_s}{N_0}, & \frac{E_s}{N_0} \text{ large} \\ \frac{1}{\ln 2}\frac{E_s}{N_0}, & \frac{E_s}{N_0} \text{ small} \end{cases} \tag{10.19}$$

In Figure 10.6, these two functions are shown as a dashed line and a dotted line, respectively. There is also one more dotted line, located at 1.53 dB to the right of the asymptotic capacity. This shows the asymptotic slope of the achievable bit rate for equiprobable M-PAM signaling. The 1.53-dB gap shows the possible gain by not restricting to equiprobable signal alternatives. This quantity is called the *shaping gain* and will be further elaborated in Section 10.3.

10.2 A NOTE ON DIMENSIONALITY

Figure 10.6 contains a lot of information about how a real system can be expected to behave compared to what the capacity limit promise. The mutual information plotted for different M-PAM constellations in the figure shows how practical systems behave at the best, using equally likely discrete signal alternatives. It also shows the asymptotic loss made by using uniform distribution instead of Gaussian. In the plot, the unit of the vertical axis is often expressed as *bits/channel use per dimension*, where *channel use* means the transmission of one signal alternative. The extra added term *per dimension* can be interpreted in a variety of ways, and it is worth to give a special note on this. In an information theoretical view, the *per dimension* can be any dimension and it is not strictly coupled to the time series of signal alternatives or the dimensionally of the signal constellation.

To get a better understanding of how the capacity relates to the dimensionality of the signal, consider an N-dimensional signal and introduce N orthonormal basis function, $\phi_i(t)$, $i = 1, 2, \dots, N$. The orthonormality requirement means

$$\int_{\mathbb{R}} \phi_i(t)\phi_j(t)dt = \delta_{i-j} = \begin{cases} 1, & i = j, \\ 0, & i \neq j \end{cases} \tag{10.20}$$

The basis functions used represent the span of the signal in different dimensions. However, it does not say how they are differentiated. A PAM signal can be seen as N consecutive signal alternatives separated in time, and then the base pulses $g(t - nT_s)$ and $g(t - kT_s)$ are orthogonal if $n \neq k$ and the pulse duration is T. Another example of dimensionality is the basis functions $\phi_1(t) = \sqrt{2}\cos(2\pi f_c t)$ and $\phi_2(t) = \sqrt{2}\sin(2\pi f_c t)$ often considered for, e.g., quadrature amplitude modulation (QAM) constellations. So, the dimensions here can be seen as the dimensionality of the signal constellation, but it has also a more general interpretation and can, e.g., be seen as separation in time. The signal is constructed from N (real) signal amplitudes, s_i, $i = 1, 2, \dots, N$, as

$$s(t) = \sum_{i=1}^{N} s_i\phi_i(t) \tag{10.21}$$

After transmission over an AWGN channel, the received signal is

$$r(t) = s(t) + \eta(t) \tag{10.22}$$

where $\eta(t)$ is white noise with power density $R_\eta(f) = N_0/2$. The signal can then be represented in dimension i as

$$r_i = \int_{\mathbb{R}} r(t)\phi_i(t)dt = \int_{\mathbb{R}} \big(s(t) + \eta(t)\big)\phi_i(t)dt = s_i + \eta_i \tag{10.23}$$

where $\eta_i = \int_{\mathbb{R}} \eta(t)\phi_i(t)dt$, and it has been used that

$$\begin{aligned}\int_{\mathbb{R}} s(t)\phi_i(t)dt &= \int_{\mathbb{R}} \sum_j s_j\phi_j(t)\phi_i(t)dt \\ &= \sum_j s_j \int_{\mathbb{R}} \phi_j(t)\phi_i(t)dt \\ &= \sum_j s_j\delta_{i-j} = s_i \end{aligned}\tag{10.24}$$

The noise parameter in the received dimension has the following mean and autocorrelation:

$$\begin{aligned} E\big[\eta_i\big] &= E\Big[\int_{\mathbb{R}} \eta(t)\phi_i(t)dt\Big] = \int_{\mathbb{R}} E\big[\eta(t)\big]\phi_i(t)dt = 0 \\ r_\eta(i,j) = E\big[\eta_i\eta_j\big] &= E\Big[\int_{\mathbb{R}} \eta(t)\phi_i(t)dt \int_{\mathbb{R}} \eta(s)\phi_j(s)ds\Big] \\ &= \int_{\mathbb{R}}\int_{\mathbb{R}} E\big[\eta(t)\eta(s)\big]\phi_i(t)\phi_j(s)dtds \\ &= \int_{\mathbb{R}} \frac{N_0}{2}\phi_i(t)\int_{\mathbb{R}} \delta(t-s)\phi_j(s)dsdt \\ &= \frac{N_0}{2}\int_{\mathbb{R}} \phi_i(t)\phi_j(t)dt = \frac{N_0}{2}\delta_{i-j} = \begin{cases}\frac{N_0}{2}, & i=j \\ 0, & i\neq j\end{cases} \end{aligned}\tag{10.25}$$

where it is used that $E\big[\eta(t)\big] = 0$ and $E\big[\eta(t)\eta(s)\big] = \frac{N_0}{2}\delta(t-s)$. This shows the noise component in each dimension is Gaussian with zero mean and variance $\frac{N_0}{2}$, i.e., $\eta_i \sim N(0, \sqrt{N_0/2})$. Hence, the N dimensions are equivalent to N transmissions over a Gaussian channel.

Denoting the total energy in an N-dimensional signal by E_s, the energy per dimension is $E_s^{(\mathrm{dim})} = E_s/N$ and the SNR

$$\mathrm{SNR}_N^{(\mathrm{dim})} = 2\frac{E_s/N}{N_0} = \frac{2}{N}\frac{E_s}{N_0} \tag{10.26}$$

Signaling at the Nyquist sampling rate for a band-limited signal with bandwidth W, the sampling rate is $F_s = 2W$. Thus, a vector with N samples will take the transmission time $T = \frac{N}{2W}$, i.e., $N = 2WT$. Hence, the SNR can be written as

$$\mathrm{SNR}_N^{(\mathrm{dim})} = \frac{E_s}{WTN_0} = \frac{P}{WN_0} \tag{10.27}$$

where in the second equality it is used that $E_s = TP$. Thus, the capacity per dimension is

$$C^{(\text{dim})} = \frac{1}{2}\log\left(1 + \frac{P}{WN_0}\right) \quad \left[\frac{\text{b/ch.u.}}{\text{dim}}\right] \tag{10.28}$$

and the capacity for the N-dimensional signal construction

$$C^{(N)} = \frac{N}{2}\log\left(1 + \frac{P}{WN_0}\right) = WT\log\left(1 + \frac{P}{WN_0}\right) \quad [\text{b}/N\ \text{dim ch.u.}] \tag{10.29}$$

Division by T gives the capacity in b/s as

$$C^{(N)} = W\log\left(1 + \frac{P}{WN_0}\right) \quad [\text{b/s}] \tag{10.30}$$

This means the capacity in bits per second for a band-limited signal is independent of the dimensionality of the signal construction. Especially it is independent of the dimensionality of the signal constellation.

In the derivations, it was seen that each amplitude in a signal constellation is equivalent to a sample in terms of the sampling theorem. In essence, N amplitudes gives N degrees of freedom, which can be translated to N samples. Each sample, or dimension, can transmit $\frac{1}{2}\log(1 + \frac{P}{WN_0})$ bits per channel use. In this aspect, one real amplitude in one dimension in the signal space is regarded as one sample. Hence, from an information theoretic view there is no difference in transmitting N signals from a one-dimensional constellation during time T, or one signal from an N-dimensional constellation in the same time. They both represent an N-dimensional signal.

Even though the above derivations states that the dimensionality of the signal does not matter, one has to be a bit careful. The requirements in the derivations are that the basis functions are orthonormal and that the utilized bandwidth is unchanged. In the above description, M-PAM signals are considered. An M-QAM constellation is essentially formed by using two orthogonal $\sqrt{M}$-PAM constellations (see Figure 10.7, which describes how two 4-PAM constellations are used to form a 16-QAM constellation). In general, such construction can be done using two real signals modulated in terms of a complex signal.

Consider a real base-band signal $s_b(t)$ with the positive bandwidth W (see Figure 10.8a). Since $s_b(t)$ is real, its spectrum is Hermitian symmetric. Denoting the

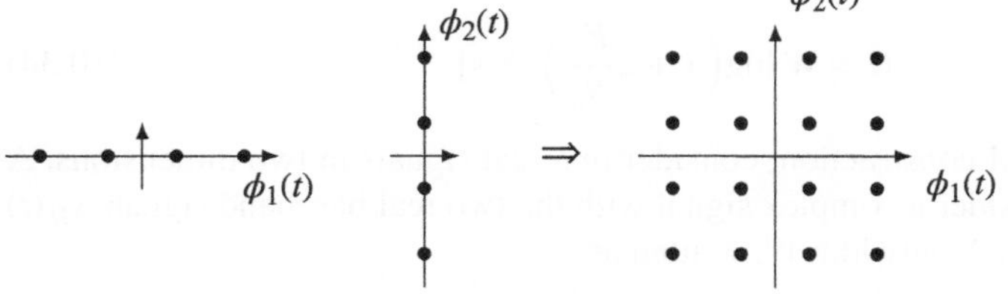

Figure 10.7 Two orthogonal 4-PAM considered as a 16-QAM constellation.

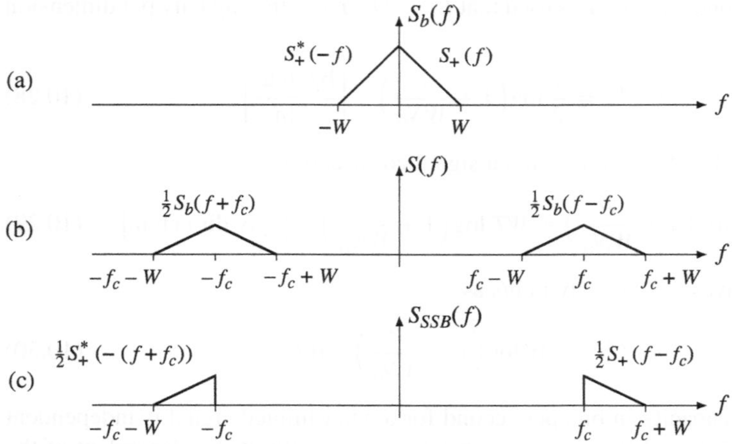

Figure 10.8 Modulation of the signal $s_b(t)$ to a higher carrier frequency f_c using single sideband modulation.

positive frequency side of $S_b(f)$ by $S_+(f)$, the negative side is the complex conjugate mirror image $S_+^*(-f)$.

A frequency-shifted signal centered at carrier frequency f_c is created by

$$s(t) = s_b(t)\cos 2\pi f_c t \tag{10.31}$$

Its Fourier transform is

$$S(f) = \frac{1}{2}S_b(f+f_c) + \frac{1}{2}S_b(f-f_c) \tag{10.32}$$

which is shown in Figure 10.8b. Since the inner half of the signal, i.e. for $f_c - W \leq |f| < f_c$ is a mirror image of the outer half, this can be filtered without losing any information. The procedure is called single sideband modulation and is shown as the function $S_{SSB}(f)$ in Figure 10.8c. This means the effective bandwidth of both the baseband signal $s_b(t)$, and the frequency-shifted version $s_{SSB}(t)$ is W. Hence, the capacity for the system is

$$C = \frac{1}{2}\log\Big(1 + \frac{P}{N_0 W}\Big) = \frac{1}{2}\log\Big(1 + 2\frac{E_s}{N_0}\Big) \text{ [b/ch.u.]} \tag{10.33}$$

or, by using the Nyquist sampling rate $F_s = 2W$,

$$C = W\log\Big(1 + 2\frac{E_s}{N_0}\Big) \text{ [b/s]} \tag{10.34}$$

Now, like in the QAM construction, consider two real signals in two dimensions. A natural way is to consider a complex signal with the two real baseband signals $s_{\mathcal{R}}(t)$ and $s_{\mathcal{I}}(t)$ with positive bandwidth W, written as

$$s_b(t) = s_{\mathcal{R}}(t) + js_{\mathcal{I}}(t) \tag{10.35}$$

Since the signal is complex, there are no longer any symmetries in the frequency domain, and the complete bandwidth $-W \leq f \leq W$ is used for information. However, there is no way to transmit a complex signal directly since the signals are transmitted in a real world. Therefore, the signal is shifted up in frequency using cosine for the real part and sine for the imaginary part as

$$s(t) = \text{Re}\{s_b(t)e^{j2\pi f_c t}\} = \tfrac{1}{2}s_{\mathcal{R}}(t)\cos 2\pi f_c t - \tfrac{1}{2}s_{\mathcal{I}}(t)\sin 2\pi f_c t \tag{10.36}$$

To view the signal in the frequency domain, use $\text{Re}\{x(t)\} = \frac{1}{2}(x(t) + x^*(t))$,

$$\begin{aligned} S(f) = \mathcal{F}\{s(t)\} &= \frac{1}{2}\mathcal{F}\Big\{s_b(t)e^{j2\pi f_c t}\Big\} + \frac{1}{2}\mathcal{F}\Big\{s_b^*(t)e^{-j2\pi f_c t}\Big\} \\ &= \frac{1}{2}S_b(f - f_c) + \frac{1}{2}S_b^*(-(f + f_c)) \end{aligned} \tag{10.37}$$

where the second equality follows from $\mathcal{F}(x^*(t)) = X^*(-f)$. The second term in (10.37), $\frac{1}{2}S_b^*(-(f+f_c))$ is a complex conjugated and mirrored version of $\frac{1}{2}S_b(f-f_c)$ centered around $-f_c$, meaning the negative frequency side of $S(f)$ is a Hermitian reflection of the positive frequency side, as it should for a real sequence. In this case, the whole bandwidth $f_c - W \leq f \leq f_c + W$ contains information and the resulting bandwidth for the modulated signal is $W^{(2)} = 2W$. By assuming the power $P^{(2)}$ used over the signal, the resulting capacity is

$$C^{(2)} = \frac{1}{2}\log\Big(1 + \frac{P^{(2)}}{N_0 W^{(2)}}\Big) \quad \text{[b/ch.u]} \tag{10.38}$$

and, equivalently by using $F_s^{(2)} = 2W^{(2)}$

$$C^{(2)} = W^{(2)}\log\Big(1 + \frac{P^{(2)}}{N_0 W^{(2)}}\Big) \quad \text{[b/s]} \tag{10.39}$$

To compare the two signaling schemes, where one-dimensional or two-dimensional real signals are used, the constants in (10.39) need to be interpreted. Since the bandwidth is doubled in the second scheme, the power consumption will also double, $P^{(2)} = 2P$. Similarly, the energy used in the signaling will be divided over the two dimensions, and the energy in the second signal becomes $E_s^{(2)} = 2E_s$. Hence, the SNR for the second signaling can be expressed as

$$\text{SNR}^{(2)} = \frac{P^{(2)}}{N_0 W^{(2)}} = \frac{2P}{N_0 2W} = \frac{E_s^{(2)}}{N_0} \tag{10.40}$$

and

$$C^{(2)} = 2W\log\Big(1 + \frac{E_s^{(2)}}{N_0}\Big) \text{ b/s} \tag{10.41}$$

This relation also reflects the relation between PAM and QAM signaling since QAM is a two-dimensional version of PAM.

10.3 SHAPING GAIN

The channel capacity depicted as the gray line in Figure 10.6 is the maximum achievable transmission over a Gaussian channel with the SNR measured in E_s/N_0. To reach this limit, the communication system must be optimized in all possible ways. One of many requirements is that the input signal should be chosen according to a continuous Gaussian distribution. In most communication systems, the choice of signal is done according to uniform distribution over a discrete set of signals. In the figure, this asymptotic loss is shown as the gap between the channel capacity and the dotted line. Since this reflects the gain that is possible to achieve by shaping the input distribution from uniform to Gaussian, it is called the *shaping gain* and often denoted γ_s. By viewing the total gain that is possible, compared to the uncoded case, it can be split into two parts, the shaping gain γ_s and coding gain γ_c. Often it is easy to achieve a coding gain of a couple of dB by using some standard channel coding. But to achieve higher gains more complex codes must be used alternative is to consider shaping of the constellation.

The ultimate shaping gain of 1.53 dB denoted in Figure 10.6 denotes the maximum shaping gain [74]. To show this value, consider the case when the SNR, E_s/N_0, becomes large. The interesting part of the plot is then the growth of the mutual information for M-PAM signaling before it flattens due to a finite number of signals. By letting the number of signal alternatives approaching infinity, the distribution of X becomes the continuous rectangular distribution

$$f_u(x) = \frac{1}{2a}, \quad \text{where } -a \le x \le a \tag{10.42}$$

This should be compared to the case of a Gaussian distribution, $f_g(x)$. For the Gaussian case, the average signal energy and the entropy is (see Appendix A)

$$P_g = E\left[X_g^2\right] = \sigma^2 \tag{10.43}$$

$$H(X_g) = \frac{1}{2}\log 2\pi e\sigma^2 = \frac{1}{2}\log 2\pi e P_g \tag{10.44}$$

For the uniform case, the corresponding derivation gives

$$P_u = E\left[X_u^2\right] = \frac{a^2}{3} \tag{10.45}$$

$$H(X_u) = \log 2a = \frac{1}{2}\log 12P_u \tag{10.46}$$

In this region of the plot, for high SNR, the mutual information is dominated by the entropy of the input distribution. For these two distributions to have the same entropy, the relation on input power is

$$\gamma_s = \frac{P_u}{P_g} = \frac{\pi e}{6} \approx 1.62 = 1.53 \text{ dB} \tag{10.47}$$

The shaping gain is the gain made from using a Gaussian distribution on the transmitted signal compared to a uniform distribution. In the next example, it is shown that a fair amount of the gain can be reached just by considering a distribution that

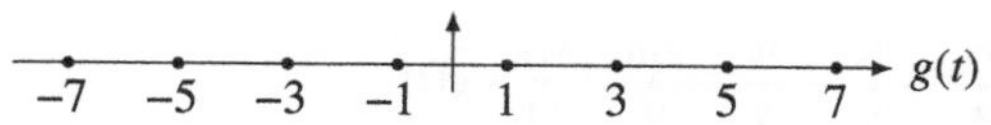

Figure 10.9 Signal alternatives in an 8-PAM constellation.

favors the low-energy signals before the high energy. The mapping from a uniform to nonuniform distribution indicates that the shaping process can be seen as the dual of source coding, in the sense that perfect source coding gives a uniform distribution of the code symbols. One easy way to get unequal vectors from equally distributed bits is to consider unequal lengths of the input vectors, and this mapping can be performed in a binary tree.

Example 10.2 First, the unshaped system is defined as an 8-PAM system. The signal alternatives can be viewed as in Figure 10.9. If they are equally likely the energy derived as the second moment of the signal amplitudes is $E[X^2] = 21$ and for each signal 3 bits are transmitted.

To find a constellation that gives a shaping gain, the signal energy should be lowered while the average number of transmitted bits and inter signal distance should be unchanged. Instead, the distribution of the signal alternatives should be chosen nonuniform. If the input sequence is considered as i.i.d. equiprobable bits, one way to alter the distribution is to have unequal length of the vectors mapping to the signal alternative. Here, these vectors are determined from the paths in a binary tree where there are no unused leaves. By choosing some vectors shorter than 3 and others longer, and by mapping high probable vectors to low-energy signals, the total energy can be lowered. The tree in Figure 10.10 shows the mapping between signal alternatives s_i and the input vectors decided by the tree paths. Since the binary information is assumed to be equiprobable, the probabilities for the nodes at each level are shown under the tree. The average length of the information vectors can then be determined by the path length lemma as

$$E[L] = 1 + 2\tfrac{1}{2} + 2\tfrac{1}{4} + 2\tfrac{1}{8} + 4\tfrac{1}{16} = 3 \tag{10.48}$$

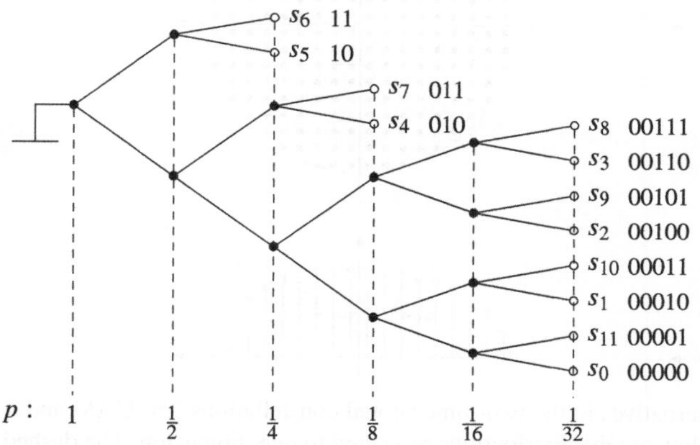

Figure 10.10 A binary tree for determining a shaping constellation.

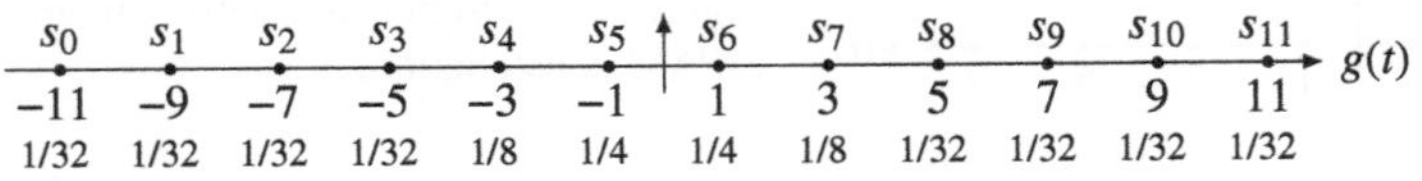

Figure 10.11 Signal alternatives and the probabilities in the shaped constellation.

so the average number of bits per signal is unchanged. In the tree, there are 12 leaves corresponding to signal alternatives. Hence, the price paid is an expansion of the signal constellation, but the idea is to use the added high-energy alternative with a low probability so on average there is a gain. The amplitudes and the corresponding probabilities of the signal alternatives are shown in Figure 10.11. The energy derived as the second moment is then

$$E[X_s^2] = 2\tfrac{1}{4} + 2\tfrac{1}{8}3^2 + 2\tfrac{1}{32}(5^2 + 7^2 + 9^2 + 11^2) = 20 \tag{10.49}$$

and the shaping gain is

$$\gamma_s = 10\log_{10}\frac{21}{20} = 0.21\text{dB} \tag{10.50}$$

In Problem 10.6, it is shown that the same construction when letting the tree grow even further can give an asymptotic shaping gain of $\gamma_s^{(\infty)} = 0.9177$dB.

In the example, it was seen that the average energy can be decreased by shaping the probability distribution over the signal constellation. The shaping procedure can also be seen in another way. A vector of two symbols, each modulated by a 16-PAM signal constellation, results in a 256-QAM signal constellation (see upper left constellation of Figure 10.12). Since the QAM signal space has a square form, the energy in the corner signal alternatives is rather high. If instead, the 256 signal alternatives are

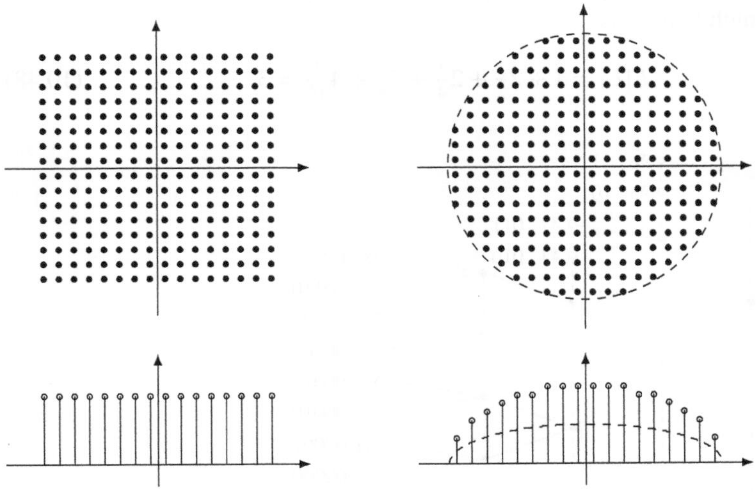

Figure 10.12 The signal alternatives in the two-dimensional constellations 256-QAM and a spherical-shaped version. Below are the distributions projected to one dimension. The dashed distribution is the projection of the continuous spherical constellation.

chosen within a circle, the upper right constellation in the figure, the total energy can be decreased. Assuming the distance between two signal points in the figure is 2 and that they are used with equal probability, the average energy derived as the second moment for the squared constellation is

$$E_{\text{QAM}} = 170 \tag{10.51}$$

Similarly, when the signal alternatives in the squared constellation are equally probable, the energy is

$$E_{\text{Sphere}} = 162.75 \tag{10.52}$$

The resulting shaping gain is

$$\gamma_s = 10\log_{10}\frac{E_{\text{QAM}}}{E_{\text{Sphere}}} = 0.189\text{dB} \tag{10.53}$$

In Figure 10.12, it is assumed that the signal alternatives are equally likely in both the squared and the spherical case. The distributions below the signal constellations are the probability functions projected in one dimension. Clearly for the square case, there are 16 equally likely alternatives. In the spherical case, there are 18 alternatives where the low-energy alternatives have highest probability. Hence, by choosing a spherical constellation in two dimensions, the distribution is shaped when projected into one dimension.

To find the maximum shaping, a cubic constellation in N dimensions, representing N PAM constellations, is compared with a spherical constellation in N dimensions. When the number of signal alternatives grows, the discrete constellations can be replaced with continuous distributions without any essential loss of accuracy. Then the second moment of a uniform distribution over an N-dimensional cube should be compared with the second moment of a uniform distribution over an N-dimensional sphere. To compare the two distributions, they should have the same volume and therefore they are normalized to unit volume.

Starting with the cubic constellation, the volume of an N dimensional cube with side A is

$$V_{\square} = \int_{\square} d\boldsymbol{x} = \int_{-A/2}^{A/2}\cdots\int dx_1 \ldots dx_N = A^N \tag{10.54}$$

where $\boldsymbol{x} = (x_1, \ldots, x_N)$ is an N-dimensional vector. Normalizing to a unit volume cube gives that $A = 1$. Since the N-cube is the boundary for a uniform distribution, the probability function is $f_{\square}(x) = 1/V_{\square} = 1$. Hence, the second moment, or the energy, for the cubic constellation in N dimensions can be derived as

$$\begin{aligned} E_{\square}^{(N)} &= \int_{\square} |\boldsymbol{x}|^2 d\boldsymbol{x} = \int_{-1/2}^{1/2}\cdots\int \left(x_1^2 + \cdots + x_N^2\right) dx_1 \ldots dx_N \\ &= N\int_{-1/2}^{1/2} x^2 dx = N\left[\frac{x^3}{3}\right]_{-1/2}^{1/2} = N\frac{1}{12} \end{aligned} \tag{10.55}$$

To do similar derivations for the spheric constellation in N dimensions, a useful integral relation from [52], formula 4.642, is noted

$$\int_{|x|^2\leq R^2} f(|x|)dx = \frac{2\pi^{N/2}}{\Gamma(\frac{N}{2})}\int_0^R x^{N-1}f(x)dx \tag{10.56}$$

where $\Gamma(n)$ is the gamma function (see Section A.2). By letting $f(x)=1$, the volume of an N-dimensional sphere is

$$\begin{aligned} V_{\bigcirc} &= \int_{\bigcirc} dx = \int_{|x|^2\leq R^2} dx = \frac{2\pi^{N/2}}{\Gamma(\frac{N}{2})}\int_0^R x^{N-1}dx \\ &= \frac{2\pi^{N/2}}{\Gamma(\frac{N}{2})}\left[\frac{x^N}{N}\right]_0^R = \frac{2\pi^{N/2}}{\Gamma(\frac{N}{2})}\frac{R^N}{N} \end{aligned} \tag{10.57}$$

Setting $V_{\bigcirc}=1$ yields the radius

$$R = \frac{1}{\sqrt{\pi}}\left(\tfrac{N}{2}\Gamma(\tfrac{N}{2})\right)^{1/N} = \frac{1}{\sqrt{\pi}}\left(\Gamma(\tfrac{N}{2}+1)\right)^{1/N} \tag{10.58}$$

and then the normalized energy can be derived as

$$\begin{aligned} E_{\bigcirc}^{(N)} &= \int_{\bigcirc} |x|^2 dx = \frac{2\pi^{N/2}}{\Gamma(\frac{N}{2})}\int_0^R x^{N-1}x^2 dx \\ &= \frac{2\pi^{N/2}}{\Gamma(\frac{N}{2})}\frac{R^{N+2}}{N+2} = \underbrace{\frac{2\pi^{N/2}R^N}{\Gamma(\frac{N}{2})N}}_{V_{\bigcirc}=1}\frac{N}{N+2}R^2 \\ &= \frac{N}{(N+2)\pi}\left(\Gamma(\tfrac{N}{2}+1)\right)^{2/N} \end{aligned} \tag{10.59}$$

The shaping gain for the N-dimensional case when comparing the cubic and the spherical constellations is

$$\gamma_s^{(N)} = \frac{E_{\square}^{(N)}}{E_{\bigcirc}^{(N)}} = \frac{\pi(N+2)}{12\Gamma(\frac{N}{2}+1)^{2/N}} \tag{10.60}$$

The gamma function generalizes the factorial function to positive real values with a smooth curve where $n! = \Gamma(n+1)$, for a positive integer n. Therefore, it is reasonable to use Stirling's approximation to get

$$\Gamma(n+1) = n! \approx \sqrt{2\pi n}\left(\tfrac{n}{e}\right)^n \tag{10.61}$$

Hence, for large N,

$$\begin{aligned} \gamma_s &\approx \frac{\pi(N+2)}{12\left(\left(2\pi\frac{N}{2}\right)^{1/2}\left(\frac{N}{2e}\right)^{N/2}\right)^{2/N}} \\ &= \frac{\pi e}{6}\frac{N+2}{N}\left(\frac{1}{\pi N}\right)^{1/N} \to \frac{\pi e}{6}, \quad N\to\infty \end{aligned} \tag{10.62}$$

which is the same ultimate shaping gain as when comparing uniform and Gaussian distributions for the input symbols. Actually, as seen in Problem 10.7, the projection from a uniform distribution over a multidimensional sphere to one dimension will be a Gaussian distribution when the dimensionality grows to infinity. Therefore, comparing the shaping gain between multidimensional cubic and spherical uniform distributions is the same as comparing the one-dimensional uniform and Gaussian distributions.

10.4 SNR GAP

When describing the capacity formula for discrete input constellations like M-PAM, it is also natural to consider the *SNR gap*. This is a measure of how far from the capacity limit a system is working for a specific achieved probability of error. Then the SNR gap describes the possible gain in SNR by approaching the capacity.

Previously, the signal constellation for 2-, 4-, and 8-PAM has been considered (see Figures 10.1 and 10.9). In general, for an M-PAM constellation the signal amplitudes are determined by

$$A_i = M - 1 - 2i, \quad i = 0, 1, \dots M - 1 \tag{10.63}$$

Then the valid amplitudes will be as described in Figure 10.13. The pulse shape is determined by the function $g(t) = \sqrt{E_g}\phi(t)$, where $\phi(t)$ has unit energy.

The received signal, distorted by AWGN, is given as $y = A_i + z$, where $z \sim \mathrm{N}(0, \sqrt{N_0/2})$. An ML decoding rule chooses the signal amplitude closest to the received signal in terms of Euclidian distance in Figure 10.13. There will be a decoding error in the case when the received signal is not closest to the transmitted signal alternative. For the $M - 2$ inner signal alternatives, $i \in \{1, \dots, M - 2\}$, this will happen when the noise component z is either larger than $\sqrt{E_g}$ or smaller than $-\sqrt{E_g}$. In both cases, the probability is $P\big(z > \sqrt{E_g}\big)$. For the outer signal alternatives, $i \in \{0, M - 1\}$, it will only be error in one of the cases. That means the error probability conditioned on the signal alternative is

$$P_{e|i} = \begin{cases} 2P\big(z > \sqrt{E_g}\big), & i = 1, \dots, M - 2 \\ P\big(z > \sqrt{E_g}\big), & i = 0, M - 1 \end{cases} \tag{10.64}$$

With equally likely signal alternatives, the average error probability is

$$\begin{aligned} P_e &= \sum_i \frac{1}{M} P_{e|i} = \frac{1}{M}\Big((M-2)2P\big(z > \sqrt{E_g}\big) + 2P\big(z > \sqrt{E_g}\big)\Big) \\ &= 2\frac{M-1}{M} P\big(z > \sqrt{E_g}\big) = 2\Big(1 - \frac{1}{M}\Big) P\big(z > \sqrt{E_g}\big) \end{aligned} \tag{10.65}$$

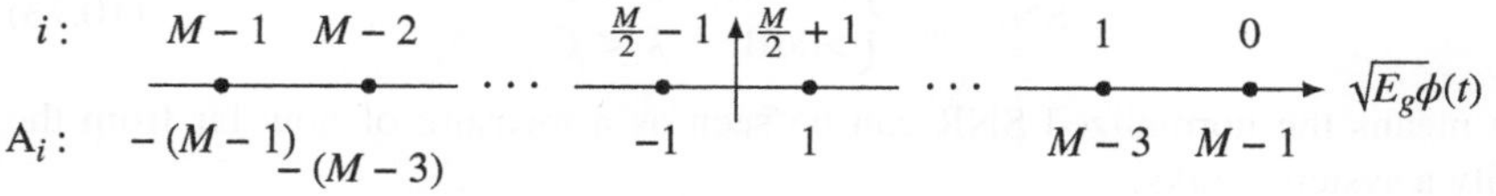

Figure 10.13 Signal alternatives in an M-PAM constellation.

The above probability is given in terms of the energy in the pulse shape E_g. the energy in signal alternative i is $A_i^2 E_g$, and, hence, the average signal energy is given by

$$E_s = \frac{1}{M}\sum_{i=0}^{M-1} A_i^2 E_g = \frac{E_g}{M}\sum_{i=0}^{M-1}(M-1-2i)^2 = \frac{E_g}{3}(M^2-1) \qquad (10.66)$$

or, equivalently,

$$E_g = \frac{3E_s}{M^2-1} \qquad (10.67)$$

Then, together with the noise variance of $N_0/2$, the signal error probability can be expressed as

$$\begin{aligned} P_e &= 2\Big(1-\frac{1}{M}\Big)P\Big(z > \sqrt{\tfrac{3E_s}{M^2-1}}\Big) \\ &= 2\Big(1-\frac{1}{M}\Big)Q\Big(\sqrt{\tfrac{3E_s}{(M^2-1)N_0/2}}\Big) \\ &= 2\Big(1-\frac{1}{M}\Big)Q\Big(\sqrt{\tfrac{3}{M^2-1}2\tfrac{E_s}{N_0}}\Big) \\ &= 2\Big(1-\frac{1}{M}\Big)Q\Big(\sqrt{\tfrac{3}{M^2-1}\tfrac{P}{WN_0}}\Big) \end{aligned} \qquad (10.68)$$

When transmitting binary vectors, the number of signal alternatives should be a power of two, $M = 2^k$, where k is the number of transmitted bits per channel use. Comparing with the capacity, this value should be lower, in bits per transmission,

$$k \le C = \frac{1}{2}\log(1+\text{SNR}) \qquad (10.69)$$

By rearranging the relation between the capacity and the transmitted bits, it is seen that

$$\frac{\text{SNR}}{2^{2k}-1} \ge 1 \qquad (10.70)$$

Therefore, it is reasonable to define a normalized SNR as

$$\text{SNR}_{\text{norm}} = \frac{\text{SNR}}{2^{2k}-1} \qquad (10.71)$$

where the SNR is

$$\text{SNR} = \frac{P}{WN_0} = 2\frac{E_s}{N_0} = 2k\frac{E_b}{N_0} \qquad (10.72)$$

As $k = C$ the normalized SNR is one since $C = \frac{1}{2}\log(1+\text{SNR})$ gives $\frac{\text{SNR}}{2^{2C}-1} = 1$. Thus,

$$\text{SNR}_{\text{norm}} \begin{cases} = 0\text{dB}, & k = C \\ > 0\text{dB}, & k < C \end{cases} \qquad (10.73)$$

which means the normalized SNR can be seen as a measure of how far from the capacity a system works.

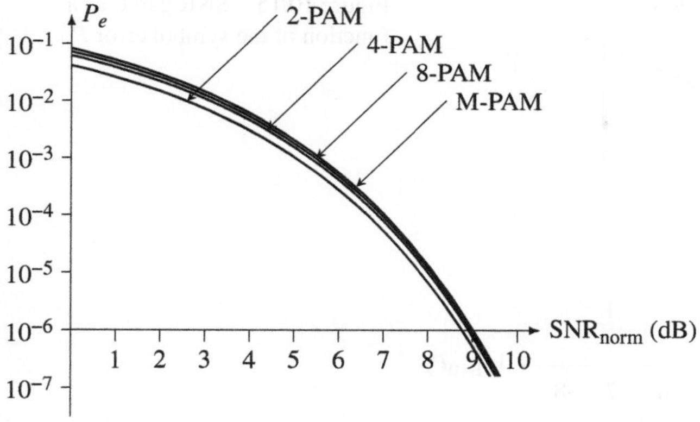

Figure 10.14 Symbol error probability for M-PAM signals as a function of the normalized SNR.

Since $M = 2^k$ the normalized SNR can be written as $\mathrm{SNR}_{\mathrm{norm}} = \frac{\mathrm{SNR}}{M^2-1}$, and the error probability for the M-PAM constellation becomes

$$P_e = 2\Big(1 - \frac{1}{M}\Big)Q\Big(\sqrt{3 \cdot \mathrm{SNR}_{\mathrm{norm}}}\Big) \tag{10.74}$$

For large M, it is simplified to

$$P_e = 2Q\Big(\sqrt{3 \cdot \mathrm{SNR}_{\mathrm{norm}}}\Big) \tag{10.75}$$

In Figure 10.14, the error probability is plotted as a function of the normalized SNR for 2-PAM, 4-PAM, 8-PAM and M-PAM, where M is large. At an error probability of 10^{-6}, the normalized SNR is close to 9 dB for large M. For a 2-PAM system, it is for the same error probability 8.8 dB. The conclusion from this is that a PAM system working at an error probability of 10^{-6} has a gap to the capacity limit of 9 dB.

Quite often, the SNR gap is used when estimating the bit rate achieved by a PAM (or QAM) system. Then it is viewed from another angle. Starting with (10.75) the symbol error probability for large M, the normalized SNR can be written as

$$\mathrm{SNR}_{\mathrm{norm}} = \frac{1}{3}\Big(Q^{-1}(P_e/2)\Big)^2 \tag{10.76}$$

Since the normalized SNR is $\mathrm{SNR}_{\mathrm{norm}} = \frac{\mathrm{SNR}}{2^{2k}-1}$, this gives

$$2^{2k} = 1 + \frac{\mathrm{SNR}}{\frac{1}{3}\big(Q^{-1}(P_e/2)\big)^2} \tag{10.77}$$

or, equivalently, the number of bits per transmission

$$k = \frac{1}{2}\log\bigg(1 + \frac{\mathrm{SNR}}{\frac{1}{3}\big(Q^{-1}(P_e/2)\big)^2}\bigg) = \frac{1}{2}\log\Big(1 + \frac{\mathrm{SNR}}{\Gamma}\Big) \tag{10.78}$$

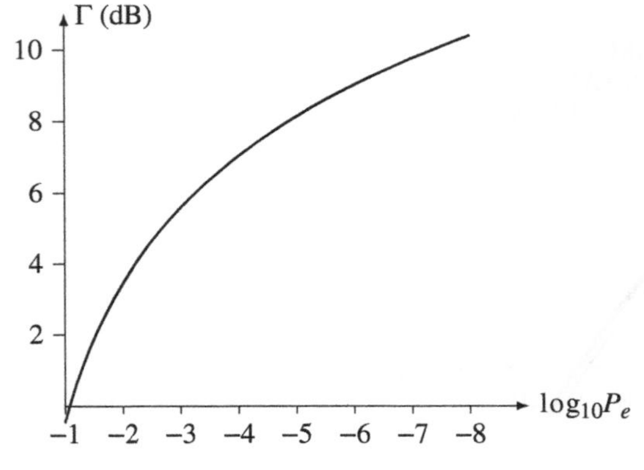

Figure 10.15 SNR gap Γ as a function of the symbol error P_e.

where $\Gamma = \frac{1}{3}\left(Q^{-1}(P_e/2)\right)^2$ is the same SNR gap for PAM (or QAM) constellations as derived earlier from Figure 10.14, in terms of normalized SNR. Going back to Figure 10.6, the circles on the curves for the mutual information for the M-PAM systems correspond to the SNR where $P_e = 10^{-6}$. For high SNR, the horizontal difference in SNR between the capacity and the circular mark is the SNR gap Γ. If the required error probability is decreased, the uncoded M-PAM system can tolerate a lower SNR and the gap is decreased. Similarly, if the required error probability is increased, the SNR for the M-PAM system is moved to the right and the gap is increased.

It is here worth noticing that the capacity as well as the mutual information plots in the figure are bounds for when it is possible to achieve arbitrarily low error probability, while the circular point denote the uncoded system when the error is 10^{-6}. To maintain the same error probability for a lower SNR some coding and/or shaping is required. The capacity limit is in this aspect the limit of the lowest possible SNR for which it is possible to transmit this number of bits.

In Figure 10.15, the SNR gap Γ is plotted as a function of the symbol error rate P_e. For $P_e = 10^{-6}$ then it becomes $\Gamma \approx 9$ dB. For the bit rate in bits per seconds, the Nyquist sampling rate of $2W$ can be assumed to get

$$R_b = W \log\left(1 + \frac{\text{SNR}}{\Gamma}\right) = W \log\left(1 + \frac{P}{\Gamma W N_0}\right) \tag{10.79}$$

Example 10.3 Consider a communication system using the bandwidth 20 MHz. Assume that the received signal level is −70 dBm over the complete band, which is a quite good signal strength in, e.g., long-term evolution (LTE). That gives the signal power $P = 10^{-70/10}$ mW. As soon as an electrical current flows through a conductor, the thermal noise is added. So just by receiving the signal in the antenna, a noise level of −174 dBm/Hz is added to the signal. This means that $N_0 = 10^{-174/10}$ mW/Hz. The SNR over the bandwidth is then

$$\text{SNR} = \frac{P}{N_0 W} = \frac{10^{-7}}{10^{-17.4} \cdot 20 \times 10^6} = 1.26 \times 10^3 = 31 \text{ dB} \tag{10.80}$$

The capacity for this system is then

$$C = W\log(1 + \text{SNR}) = 206 \text{ Mb/s} \tag{10.81}$$

Assuming a communication system based om M-PAM or M-QAM, working at an error rate of $P_e = 10^{-6}$ implies an SNR gap of $\Gamma = 9$ dB $= 7.94$. Thus, an estimated bit rate for the system is

$$R_b = W\log\left(1 + \frac{\text{SNR}}{\Gamma}\right) = 146 \text{ Mb/s} \tag{10.82}$$

This is the estimated bit rate for an uncoded system. If an error-correcting code is added to the system, the effect on the bit rate can be estimated by adding the corresponding coding gain. For many codes, this is somewhere between 3 and 4 dB. Assuming a relatively powerful code, the coding gain is set to $\gamma_c = 4$ dB. Expressed in dB, the effective SNR is

$$\text{SNR}_{\text{eff}} = \text{SNR} - \Gamma + \gamma_c = 31 - 9 + 4 = 26 \text{ dB} = 397 \tag{10.83}$$

The resulting estimated bit rate is

$$R_b = W\log(1 + \text{SNR}_{\text{eff}}) = 173 \text{ Mb/s} \tag{10.84}$$

PROBLEMS

10.1 In a communication system, a binary signaling is used, and the transmitted variable X has two equally likely amplitudes $+1$ and -1. During transmission, a uniform noise is added to the signal, and the received variable is $Y = X + Z$, where $Z \sim U(\alpha)$, $E[Z] = 0$. Derive the maximum transmitted number of bits per channel use, when

(a) $\alpha < 2$

(b) $\alpha \geq 2$

10.2 In the channel model described in Problem 10.1, consider the case when $\alpha = 4$. A hard decoding of the channel can be done by assigning

$$\tilde{Y} = \begin{cases} 1, & Y \geq 1 \\ \Delta, & -1 < Y \leq 1 \\ -1, & Y < -1 \end{cases}$$

Derive the capacity for the channel from X to $\tilde{Y}$ and compare with the result in Problem 10.1.

10.3 An additive channel, $Y = X + Z$, has the input alphabet $X \in \{-2, -1, 0, 1, 2\}$ and Z is uniformly distributed $Z \sim U(-1, 1)$. Derive the capacity.

10.4 A communication scheme uses 4-PAM system, meaning there are four different signal alternatives differentiated by their amplitudes (see Figure 10.16). During the transmission, there is a noise, Z, added to the signal so the received signal is $Y = X + Z$. The noise has the distribution as viewed in Figure 10.17.

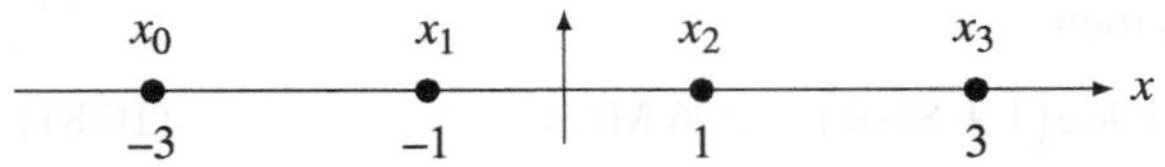

Figure 10.16 4-PAM signaling.

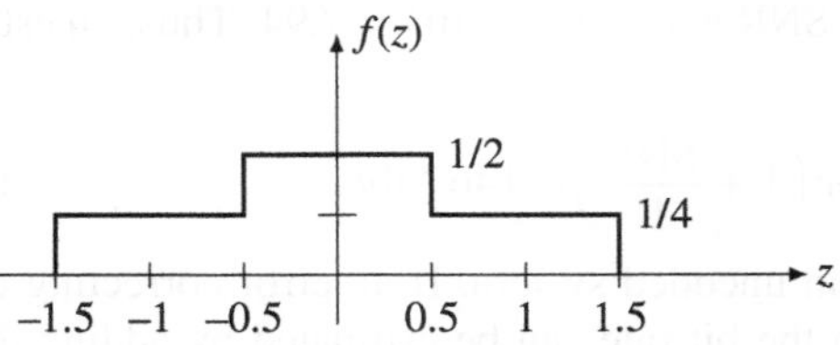

Figure 10.17 Density function of the additive noise.

(a) Assuming the signal alternatives are equally likely, how much information can be transmitted in each transmission (channel use)?

(b) What is the capacity for the transmission, i.e. what is the maximum mutual information using the given signal alternatives and the noise. For what distribution on X can it be obtained? How is the average power of the transmitted signal affected by the optimization?

10.5 Consider a 4-PAM communication system with the signal alternatives

$$s_x(t) = A_x\sqrt{E_g}\phi(t), \quad x = 0, 1, 2, 3$$

where $A_x \in \{-3, -1, 1, 3\}$ are the amplitudes and $\phi(t)$ a normalized basis function. During the transmission, the noise Z is added with the distribution $Z \sim N(0, \sqrt{N_0/2})$. At the receiver, each signal is decoded back to $\{0, 1, 2, 3\}$ according to the decision regions in Figure 10.18.

Denote the probability for the receiver to make an erroneous decision by

$$\varepsilon = P\big(Z > \sqrt{E_g}\big)$$

It can be assumed that the probabilities of making errors to nonneighboring decision regions are negligible, i.e.

$$P\big(Z > 3\sqrt{E_g}\big) \approx 0$$

(a) Construct a corresponding discrete memoryless channels (DMC)?

(b) Assume that the symbols are transmitted with equal probability. Express the maximum information transmission per channel use for the DMC, in terms of ε and the binary entropy function. Sketch a plot for different ε.

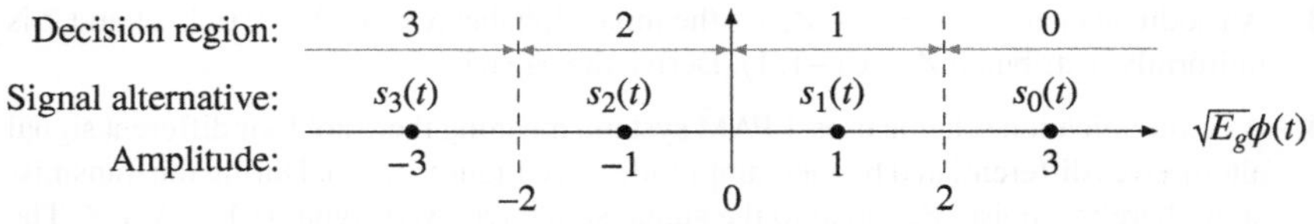

Figure 10.18 4-PAM modulation and decision regions.

10.6 In Example 10.2, a shaping algorithm based on a binary tree construction is given. In this problem, the same construction is used and the number of signal alternative expanded.

(a) Below is a tree given with two additional nodes compared with the example. What is the shaping gain for this construction?

(b) Letting the shaping constellation and tree in Example 10.2 have two levels and in subproblem (a) have three levels. Consider the same construction with k levels and show that $L = 3$ for all $k \geq 2$.

(c) For the constellation in subproblem (b) show that as $k \to \infty$ the second moment is $E[X_s^2] = 17$, and thus the asymptotic shaping gain is $\gamma_s^{(\infty)} = 0.9177\text{dB}$.
Note: It might be useful to consider the following standard sums for $|\alpha| < 1$,

$$\sum_i^{\infty} \alpha^i = \frac{\alpha}{1-\alpha} \quad \sum_i^{\infty} i\alpha^i = \frac{\alpha}{(1-\alpha)^2} \quad \sum_i^{\infty} i^2\alpha^i = \frac{\alpha+\alpha^2}{(1-\alpha)^3} \quad \sum_i^{\infty} i^3\alpha^i = \frac{\alpha+4\alpha^2+\alpha^3}{(1-\alpha)^4}$$

10.7 The maximum shaping gain, γ_s, can be derived in two ways. First, it is the relation in power between a uniform distribution and a Gaussian distribution with equal entropies. Second, it is the relation between second moments of an N-dimensional square uniform distribution and an N-dimensional spheric uniform distribution, as $N \to \infty$. This will give the maximum shaping gain since the most efficient way to pack the points is as a sphere. In this problem, it will be shown that the results are equivalent, since the spherical distribution projected to one dimension becomes Gaussian as $N \to \infty$.

(a) What is the radius in an N-dimensional sphere if the volume is one, i.e. if it constitutes a uniform probability distribution?

(b) If $X = (X_1, \dots, X_N)$ is distributed according to an N-dimensional spherical uniform distribution, show that its projection in one dimension is

$$f_X(x) = \int_{|\tilde{x}|^2 \leq R^2 - x^2} 1 d\tilde{x} = \frac{\pi^{\frac{N-1}{2}}}{\Gamma\left(\frac{N}{2}+\frac{1}{2}\right)} \left(\frac{\Gamma\left(\frac{N}{2}+1\right)^{2/N}}{\pi} - x^2 \right)^{\frac{N-1}{2}}$$

where $\tilde{x}$ is an $N-1$ dimensional vector.

(c) Using the first order Stirling's approximation

$$\Gamma(x) \approx \left(\frac{x-1}{e}\right)^{x-1}$$

show that the result in (b) can be written as

$$f_X(x) \approx \left(1 + \frac{\frac{1}{2} - \pi e x^2}{\frac{N-1}{2}}\right)^{\frac{N-1}{2}}$$

for large N.

(d) Let the dimensionality N grow to infinity and use $\lim_{N\to\infty}(1+\frac{x}{N})^N = e^x$ to show that $X \sim N\left(0, \sqrt{\frac{1}{2\pi e}}\right)$, i.e. that

$$\lim_{N\to\infty} f_X(x) = \frac{1}{\sqrt{2\pi\sigma^2}} e^{-x^2/2\sigma^2}$$

where $\sigma^2 = \frac{1}{2\pi e}$.

10.8 Assume a transmission system uses a bandwidth of $W = 10$ kHz and the allowed signaling power is $P = 1$ mW. The channel noise can be assumed to be Gaussian with $N_0 = 1$ nW/Hz.

(a) Derive the channel capacity for the system.

(b) Give an estimate of the achievable bit rate when PAM modulation is used, if the targeted error probability is $P_e = 10^{-6}$.

10.9 Sketch a plot for the ratio between the estimated achievable bit rate and the capacity for a channel with Gaussian noise. The SNR $= \frac{P}{WN_0}$ should be in the range from -10 to 30 dB. Make plots for different allowed bit error probabilities, e.g., $P_e \in \{10^{-3}, 10^{-6}, 10^{-9}\}$.

10.10 A signal is based on an orthogonal frequency division multiplexing (OFDM) modulation with 16 subchannels of width $\Delta f = 10$ kHz. The signal power level in the whole spectra is -70 dBm/Hz. On the transmission channel, the noise level is constant at -140 dBm/Hz, but the signal attenuation is increasing with the frequency as $|H_i|^2 = 5i + 10$ dB, $i = 0, \dots, 15$.

(a) Derive the capacity for the channel.

(b) If the required error rate on the channel is 10^{-6}, and it is expected that the error-correcting code gives a coding gain of 3 dB, what is the estimated obtained bit rate for the system?

10.11 Consider a frequency divided system like orthogonal frequency division multiplexing (OFDM), where the total bandwidth of $W = 10$ MHz, is split in 10 equally wide independent subbands. In each of the subbands, an M-PAM modulation should be used and the total allowed power is $P = 1$ W. On the subbands, the noise-to-attenuation ratios are given by the vector

$$\frac{N_{o,i}}{|G_i|^2} = \begin{pmatrix} -53 & -43 & -48 & -49 & -42 & -54 & -45 & -45 & -52 & -49 \end{pmatrix} \text{ [dBm/Hz]}$$

where $N_{0,i}$ is the noise on subband i and G_i the signal attenuation on subband i. The system is supposed to work at an average error probability of $P_e = 10^{-6}$. The aim of this problem is to maximize the total information bit rate R_b for the system.

(a) Show that the total information bit rate for the system can be maximized using the water-filling procedure.

(b) Derive the maximum information bit rate for the system.

(c) Assume that you add an error-correcting code in each subband, with a coding gain of $\gamma_c = 3$ dB. How does that influence the maximum information bit rate?

CHAPTER 11

INFORMATION THEORY AND DISTORTION

IN THE PREVIOUS chapters, the aim was to achieve perfect reconstruction for source coding and arbitrary small error probability in error coding. This is the basis for the source coding theorem and the channel coding theorem. However, in practical system design, this is not always the case.

For example in image compression or voice coding, it is affordable to have a certain amount of losses in the reconstruction, as long as the perceived quality is not affected. This is the idea behind lossy coding instead of lossless coding, like Huffman coding or the Lempel–Ziv (LZ) algorithms. The gain with allowing some distortion to the original image is that the compression ratio can be made much better. In image coding or video coding, algorithms like JPEG or MPEG are typical examples.

In his 1948 paper [1], Shannon started the study on how to incorporate an allowed distortion in the theory. Ten years later, in 1957, he published the full theory including distortion [53]. In this paper, the rate-distortion function is defined and shown to bound the compression capability in the same manner as the entropy does in the lossless case. In both [68, 78] the rate-distortion theory is described in more details.

This chapter will give the basic theory for rate distortion, and see how this influence the bounds on source coding and channel coding. For source coding, distortion means lossy compression, which is used in, e.g., audio and video coding. These algorithms are often based on quantization, which is described later in the chapter. Finally, transform coding is described, which is the basis of, e.g., JPEG and MPEG compression.

11.1 RATE-DISTORTION FUNCTION

To start the study of rate distortion, it must first be determined what is meant by distortion of a source. In Figure 11.1, a model for a source coding is depicted. The source symbol is a vector of length n, $\boldsymbol{X} = X_1, \dots X_n$. This is encoded to a length ℓ vector $\boldsymbol{Y} = Y_1 \dots Y_\ell$, which is then decoded back (reconstructed) to a length n vector $\hat{\boldsymbol{X}} = \hat{X}_1, \dots \hat{X}_n$. The codeword length ℓ is regarded as a random variable, and its expected

Information and Communication Theory, First Edition. Stefan Höst.

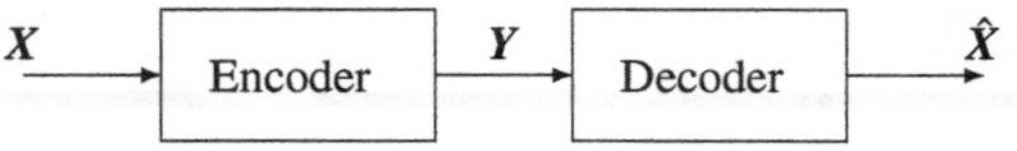

Figure 11.1 A communication model for introducing distortion.

value is denoted by $L = E[\ell]$. This is the same model as used in the lossless case in Chapter 4. The difference is that the mapping from $\boldsymbol{X}$ to $\hat{\boldsymbol{X}}$ includes an allowance of a mismatch, i.e. in general they will not be equal. The rate of the code is defined as

$$R = \frac{L}{n} \tag{11.1}$$

This is the transmission rate and should not be confused with the compression ratio used earlier, which is its inverse. For simplicity, assume that $\boldsymbol{Y}$ is a binary vector.

To measure the introduced mismatch between the source symbol and the reconstructed symbol, a *distortion measure* is required. It is here assumed that the distortion measure is additive and that the average distortion per symbol can be written as

$$d(\boldsymbol{x}, \hat{\boldsymbol{x}}) = \sum_{i=1}^{n} d(x_i, \hat{x}_i) \tag{11.2}$$

where $d(x, \hat{x})$ is the single-letter distortion. Without loss of generality, it can be assumed that the minimum distortion is zero, $\min_{\hat{x}} d(x, \hat{x}) = 0$, for all x. There are several such measures, but the two most well known are the Hamming distortion and the squared distance. The first one is typically used for discrete sources, especially for the binary case, whereas the second is mostly used for continuous sources.

Definition 11.1 The *Hamming distortion* between two discrete letters x and $\hat{x}$ is

$$d(x, \hat{x}) = \begin{cases} 0, & x = \hat{x} \\ 1, & x \neq \hat{x} \end{cases} \tag{11.3}$$

□

For the binary case, the Hamming distortion can be written as

$$d(x, \hat{x}) = x \oplus \hat{x} \tag{11.4}$$

where $\oplus$ denotes addition modulo 2.

Definition 11.2 The *squared distance distortion* between two variables x and $\hat{x}$ is

$$d(x, \hat{x}) = (x - \hat{x})^2 \tag{11.5}$$

□

In principle, all vector norms can be used as distortion measures, but, for example, the maximum as $d(\boldsymbol{x}, \hat{\boldsymbol{x}}) = \max_i |x_i - \hat{x}_i|$ does not work with the assumption of additive distortion measures. In the following, the derivations will be performed for discrete sources, but in most cases it is straightforward to generalize for continuous sources.

In the model for the lossy source coding scheme, the distortion is introduced in the encoder/decoder mapping. A mathematical counterpart of the decoder is the probability for the reconstructed symbol $\hat{X}$ conditioned on the source symbol X,

$p(\hat{x}|x)$. Then the distortion is modeled as a probabilistic mapping between the inputs and outputs. From the assumption of additive distortion, the average distortion over a vector of length n is

$$E\big[d(\boldsymbol{X},\hat{\boldsymbol{X}})\big] = E\Big[\frac{1}{n}\sum_{i=1}^{n} d(X_i,\hat{X}_i)\Big] = \frac{1}{n}\sum_{i=1}^{n} E\big[d(X_i,\hat{X}_i)\big] = E\big[d(X,\hat{X})\big] \tag{11.6}$$

By specifying a maximum distortion per symbol as δ, the averaged distortion should be bounded by $E\big[d(\boldsymbol{X},\hat{\boldsymbol{X}})\big] \leq \delta$. The expected distortion is averaged over the joint probability of the input sequence and the output sequence, $p(\boldsymbol{x},\hat{\boldsymbol{x}}) = p(\boldsymbol{x})p(\hat{\boldsymbol{x}}|\boldsymbol{x})$. Among those the input distortion is fixed by the source, meaning that the requirement of a maximum symbol distortion gives a set of conditional distributions as

$$\big\{p(\hat{\boldsymbol{x}}|\boldsymbol{x}) : E[d(\boldsymbol{X},\hat{\boldsymbol{X}})] \leq \delta\big\} \tag{11.7}$$

According to (11.6), this can for an additive distortion measure be written as

$$\big\{p(\hat{x}|x) : E[d(X,\hat{X})] \leq \delta\big\} \tag{11.8}$$

From the assumption that $\boldsymbol{Y}$ is a binary vector with average length L, the number of codewords is $2^L = 2^{nR}$. Each code vector is decoded to an estimated reconstruction vector $\hat{\boldsymbol{X}}$, and there are equally many possible reconstructed vectors. Thus, the mutual information between the input and the output can be bounded as

$$I(\boldsymbol{X};\hat{\boldsymbol{X}}) = H(\hat{\boldsymbol{X}}) - H(\hat{\boldsymbol{X}}|\boldsymbol{X}) \leq H(\hat{\boldsymbol{X}}) \leq \log 2^{nR} = nR \tag{11.9}$$

Equivalently, the rate can be bounded by the mutual information as

$$R \geq \frac{1}{n} I(\boldsymbol{X};\hat{\boldsymbol{X}}) \tag{11.10}$$

That is, to get a measure of the lowest possible rate, the mutual information should be minimized with respect to a certain maximum distortion. Since the maximum distortion level corresponds to a set of conditional distributions, the following definition is reasonable.

Definition 11.3 The *rate-distortion function* for a source with output vector $\boldsymbol{X}$ and a distortion measure $d(\boldsymbol{x},\hat{\boldsymbol{x}})$ is

$$R(\delta) = \min_{p(\hat{\boldsymbol{x}}|\boldsymbol{x}):E[d(\boldsymbol{X},\hat{\boldsymbol{X}})]\leq\delta} \frac{1}{n} I(\boldsymbol{X};\hat{\boldsymbol{X}}) \tag{11.11}$$

□

For an identical and independently distributed (i.i.d.) source, $I(\boldsymbol{X};\hat{\boldsymbol{X}}) = nI(X;\hat{X})$, together with (11.8) gives the following theorem.

Theorem 11.1 The rate-distortion function for an i.i.d. source with output variable X and the distortion measure $d(x,\hat{x})$ is

$$R(\delta) = \min_{p(\hat{x}|x):E[d(X,\hat{X})]\leq\delta} I(X;\hat{X}) \tag{11.12}$$

□

Before showing that $R(\delta)$ is the minimum average number of bits needed to represent a source symbol when the acceptable distortion is δ, a closer look at the actual derivation of the rate-distortion function and some of its properties is in place. If $\delta_1 \leq \delta_2$, the set of distributions $\{p(\hat{x}|x) : E[d(X,\hat{X})] \leq \delta_1\}$ is a subset of $\{p(\hat{x}|x) : E[d(X,\hat{X})] \leq \delta_2\}$, and

$$R(\delta_1) \geq R(\delta_2) \tag{11.13}$$

Hence, the rate-distortion function is a decreasing function in δ. To see how the rate-distortion function can behave, the next example derives it for a binary source.

Example 11.1 Consider a binary i.i.d. source with output symbol $X \in \{0,1\}$ and $p(X=0) = p$, where $p \leq 1/2$. The aim of this example is to derive the rate-distortion function for a binary source and Hamming distortion of maximum $\delta \leq 1/2$. To derive the rate-distortion function, it is possible to apply standard optimization technology, but already in this simple case it becomes relatively complex. Instead, first note that $E[d(x,\hat{x})] = P(X \neq \hat{X}) = P(X \oplus \hat{X} = 1) \leq \delta$. Then a lower bound on the mutual information can be derived as

$$\begin{aligned} I(X;\hat{X}) &= H(X) - H(X|\hat{X}) \\ &= h(p) - H(X \oplus \hat{X}|\hat{X}) \\ &\geq h(p) - H(X \oplus \hat{X}) \geq h(p) - h(\delta) \end{aligned} \tag{11.14}$$

For this lower bound to equal the rate-distortion function, it is needed that $H(X|\hat{X}) = h(\delta)$, which gives the distribution

$P(X|\hat{X})$

$\hat{X}$	$X=0$	$X=1$
0	$1-\delta$	δ
1	δ	$1-\delta$

To get the distribution on $\hat{X}$ assign $P(\hat{X}=0) = q$,

$$\begin{aligned} p &= P(X=0) \\ &= P(X=0|\hat{X}=0)P(\hat{X}=0) + P(X=0|\hat{X}=1)P(\hat{X}=1) \\ &= (1-\delta)q + \delta(1-q) = (1-2\delta)q + \delta \end{aligned} \tag{11.15}$$

or, equivalently,

$$q = \frac{p-\delta}{1-2\delta} \quad \text{and} \quad 1-q = \frac{1-p-\delta}{1-2\delta} \tag{11.16}$$

For the case when $0 \leq \delta \leq p \leq 1/2$, the probability of $\hat{X}$ in (11.16) is bounded by $0 \leq q \leq p$.Thus, q and $1-q$ forms a distribution, and according to (11.14) the rate-distortion function is $R(\delta) = h(p) - h(\delta)$, $0 \leq \delta \leq p$.

For the case when $p < \delta \leq 1/2$, let $P(\hat{X}=1|X) = 1$; q and $1-q$ do not form a distribution since $p - q < 0$. Instead, always set the reconstructed symbol to $\hat{X} = 1$

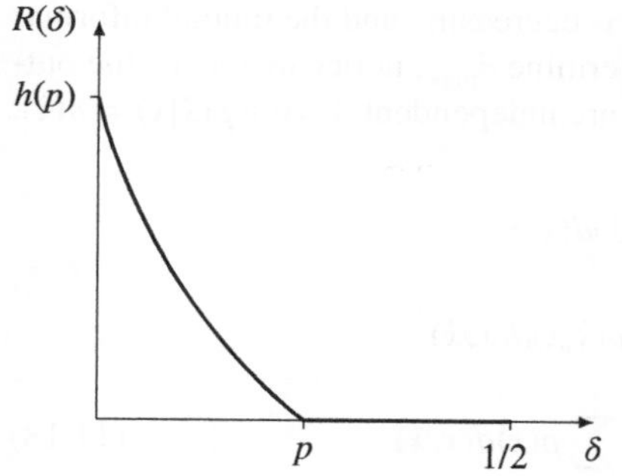

Figure 11.2 Rate-distortion function for a binary i.i.d. source.

to get $E[d(X,\hat{X})] = p \le \delta$ and the distortion requirement is fulfilled. Since $\hat{X} = 1$ is independent of X, the mutual information is $I(X;\hat{X}) = 0$, which gives $R(\delta) = 0$. Summarizing, for a binary i.i.d. source with $P(X = 0) = p$, the rate-distortion function is

$$R(\delta) = \begin{cases} h(p) - h(\delta), & 0 \le \delta \le p \le 1/2 \\ 0, & p < \delta \le 1/2 \end{cases} \tag{11.17}$$

In Figure 11.2, this function is shown as a plot.

It is interesting to notice in Figure 11.2 that for no distortion, i.e. $\delta = 0$, the rate-distortion function equals the entropy for the source. Since the rate-distortion function was defined as a lower bound on the transmission rate, and that the symbols are binary, this is the amount of information in one source symbol. Thus, it falls back to the lossless case and the source coding theorem as seen before.

In the previous example, the relation between $p(x)$, $p(x|\hat{x})$, and $p(\hat{x})$ is often described by using a *backward test channel* from $\hat{X}$ to X, as shown in Figure 11.3. It should be noted that this channel does not have anything to do with transmission; it should be seen as a mathematical model showing the relations. It has its purpose in giving an overview of the distributions involved in the problem.

It turns out that the rete-distortion function plotted in Figure 11.2 has a typical behavior. It starts at some value for $\delta = 0$ and decreases as a convex function down until $\delta = \delta_{\max}$ where $R(\delta_{\max}) = 0$. As was seen in (11.13), the rate-distortion function $R(\delta)$ is a decreasing function, but not necessarily strictly decreasing. At some value $\delta_{\max}$, the allowed distortion is so large that the reconstructed value $\hat{X}$ can take a predetermined value. Then it is not needed to transmit any codeword, and the rate becomes

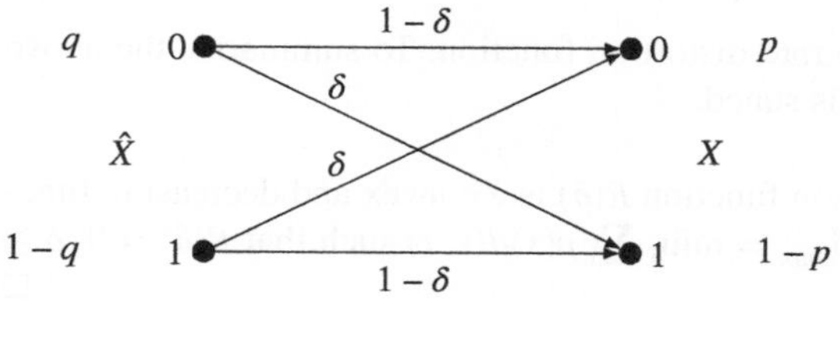

Figure 11.3 Test channel for describing the distributions in Example 11.1.

$R(\delta_{max}) = 0$. Since the rate-distortion function is decreasing and the mutual information is nonnegative, $R(\delta) = 0$, $\delta \geq \delta_{max}$. To determine δ_{max}, notice that since the output is predetermined, the input X and output $\hat{X}$ are independent, giving $p(\hat{x}|x) = p(\hat{x})$, and

$$\begin{aligned} E\big[d(X,\hat{X})\big] &= \sum_{x,\hat{x}} p(x,\hat{x})d(x,\hat{x}) \\ &= \sum_{x,\hat{x}} p(x)p(\hat{x}|x)d(x,\hat{x}) \\ &= \sum_{\hat{x}} p(\hat{x}) \sum_{x} p(x)d(x,\hat{x}) \end{aligned} \tag{11.18}$$

To get the minimum, find an $\hat{x}$ that minimizes $\sum_x p(x)d(x,\hat{x})$ and set $p(\hat{x}) = 1$ for this value, yielding

$$\delta_{max} = \min_{\hat{x}} \sum_{x} p(x)d(x,\hat{x}) \tag{11.19}$$

To show that the rate-distortion function is convex, let $p_1(\hat{x}|x)$ and $p_2(\hat{x}|x)$ denote the distributions achieving $R(\delta_1)$ and $R(\delta_2)$, i.e.

$$R(\delta_1) = I_{p_1}(X;\hat{X}) \quad \text{where} \quad E_{p_1}\big[d(X,\hat{X})\big] \leq \delta_1 \tag{11.20}$$

$$R(\delta_2) = I_{p_2}(X;\hat{X}) \quad \text{where} \quad E_{p_2}\big[d(X,\hat{X})\big] \leq \delta_2 \tag{11.21}$$

Consider the probability $p(\hat{x}|x) = \alpha_1 p_1(\hat{x}|x) + \alpha_2 p_2(\hat{x}|x)$, where $\alpha_1 \geq 0$, $\alpha_2 \geq 0$ and $\alpha_1 + \alpha_2 = 1$. Then

$$\begin{aligned} E_p\big[d(X,\hat{X})\big] &= \sum_{x,\hat{x}} p(x)p(\hat{x}|x)d(x,\hat{x}) \\ &= \sum_{x,\hat{x}} p(x)\big(\alpha_1 p_1(\hat{x}|x) + \alpha_2 p_2(\hat{x}|x)\big)d(x,\hat{x}) \\ &= \alpha_1 \sum_{x,\hat{x}} p(x)p_1(\hat{x}|x)d(x,\hat{x}) + \alpha_2 \sum_{x,\hat{x}} p(x)p_2(\hat{x}|x)d(x,\hat{x}) \\ &= \alpha_1 E_{p_1}\big[d(X,\hat{X})\big] + \alpha_1 E_{p_1}\big[d(X,\hat{X})\big] \\ &\leq \alpha_1\delta_1 + \alpha_2\delta_2 \end{aligned} \tag{11.22}$$

With $\delta = \alpha_1\delta_1 + \alpha_2\delta_2$, this shows $p(\hat{x}|x)$ is one of the distribution in the minimization to reach $R(\delta)$. From the convexity of the mutual information

$$\begin{aligned} R(\delta) &\leq I_p(X;\hat{X}) \leq \alpha_1 I_{p_1}(X;\hat{X}) + \alpha_2 I_{p_2}(X;\hat{X}) \\ &= \alpha_1 R(\delta_1) + \alpha_2 R(\delta_2) \end{aligned} \tag{11.23}$$

which shows the convexity of the rate-distortion function. To summarize the above reasoning, the following theorem is stated.

Theorem 11.2 The rate-distortion function $R(\delta)$ is a convex and decreasing function. Furthermore, there exists a $\delta_{max} = \min_{\hat{x}} \sum_x p(x)d(x,\hat{x})$ such that $R(\delta) = 0$, $\delta \geq \delta_{max}$. □

So far the rate-distortion function has been considered for discrete random variables, but the same definition makes sense for continuous variables. The same theory as above will hold for this case. One important case is naturally the Gaussian distribution, that is treated in the next example.

Example 11.2 Consider an i.i.d. source where the output is a Gaussian variable $X \sim N(0, \sigma_X)$. The reconstructed variable is $\hat{X}$, where it is assumed that $E[\hat{X}] = 0$ is inherited from X. It is also assumed that the squared distance distortion measure $d(x, \hat{x}) = (x - \hat{x})^2$ is used. Similar to the previous example with the binary source, instead of going directly to standard optimization methods, derive a lower bound for the mutual information and find a distribution to fulfill it. Starting with the mutual information

$$\begin{aligned} I(X;\hat{X}) &= H(X) - H(X|\hat{X}) \\ &= \frac{1}{2}\log\left(2\pi e\sigma_X^2\right) - H(X-\hat{X}|\hat{X}) \\ &\geq \frac{1}{2}\log\left(2\pi e\sigma_X^2\right) - H(X-\hat{X}) \end{aligned} \tag{11.24}$$

From $E[\hat{X}] = 0$, it follows that $E[X - \hat{X}] = 0$ and $V[X - \hat{X}] = E[(X - \hat{X})^2] = E[d(X, \hat{X})] \leq \delta$. Define a random variable $Z \sim N(0, \sigma_Z)$, where $\sigma_Z^2 = V[X - \hat{X}]$. The rate-distortion function $R(\delta)$ is found by minimizing $I(X; \hat{X})$ over the distributions $f(\hat{x}|x) : \sigma_Z^2 \leq \delta$. Since the Gaussian distribution maximizes the differential entropy $H(X - \hat{X}) \leq \frac{1}{2}\log(2\pi e\sigma_Z^2) \leq \frac{1}{2}\log(2\pi e\delta)$. Hence, the bound on the mutual information becomes

$$I(X;\hat{X}) \geq \frac{1}{2}\log\left(2\pi e\sigma_X^2\right) - \frac{1}{2}\log(2\pi e\delta) = \frac{1}{2}\log\left(\frac{\sigma_X^2}{\delta}\right) \tag{11.25}$$

To see that this bound is actually tight, and equals $R(\delta)$, notice that $X = \hat{X} + Z$ and choose $\hat{X} \sim N\left(0, \sqrt{\sigma_X^2 - \delta}\right)$ and $Z \sim N(0, \sqrt{\delta})$. Then $X \sim N(0, \sigma_X)$ and the average distortion $E[d(X, \hat{X})] = V[Z] = \delta$, meaning the minimization criterion is fulfilled. Hence, for $0 \leq \delta \leq \sigma_X^2$ the rate-distortion function is $R(\delta) = \frac{1}{2}\log\left(\frac{\sigma_X^2}{\delta}\right)$. For $\delta \geq \sigma_X^2$ choose $\hat{X} = 0$ independently of X, implying $I(X; \hat{X}) = 0$. The minimization criterion is fulfilled since $E[(X - \hat{X})^2] = E[X^2] = \sigma_X^2 \leq \delta$. Summarizing, the rate distortion function for an i.i.d. Gaussian source is

$$R(\delta) = \begin{cases} \frac{1}{2}\log\left(\frac{\sigma_X^2}{\delta}\right), & 0 \leq \delta \leq \sigma_X^2 \\ 0, & \delta \geq \sigma_X^2 \end{cases} \tag{11.26}$$

The function is plotted in Figure 11.4.

The importance of the rate-distortion function was partly seen in (11.10) where the rate is lower bounded by the mutual information. Together with the definition of the

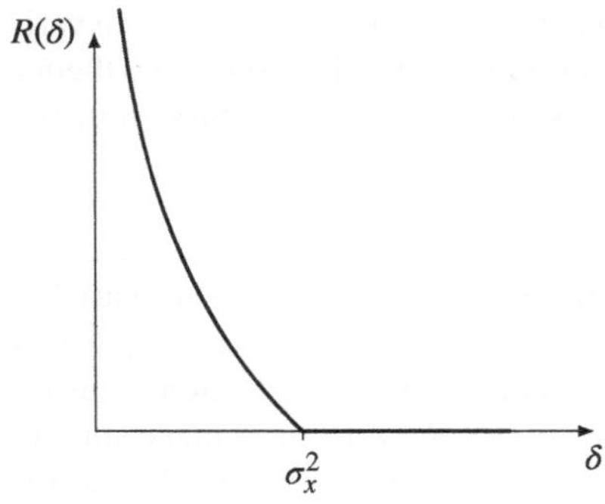

Figure 11.4 Rate-distortion function for a Gaussian source.

rate-distortion function, it means for an i.i.d. source that $R(\delta) \leq I(X;\hat{X}) \leq R$. Hence, for any source code with average distortion $E\big[d(X,\hat{X})\big] \leq \delta$ the rate is bounded by $R \geq R(\delta)$. The next theorem, called the *rate-distortion theorem*, is the direct counterpart of the source coding theorem, stating also the existence of such code.

Theorem 11.3 Let $\boldsymbol{X} = X_1X_2\dots X_n$ be generated by an i.i.d. source, $\hat{\boldsymbol{X}} = \hat{X}_1\hat{X}_2\dots\hat{X}_n$ the reconstructed sequence after source coding, and δ the allowed distortion when the additive distortion measure $d(x,\hat{x})$ is used. In the limit as $n \to \infty$, there exists a source code with rate R if and only if

$$R \geq R(\delta) = \min_{p(\hat{x}|x):E[d(X,\hat{X})]\leq\delta} I(\boldsymbol{X};\hat{\boldsymbol{X}}) \tag{11.27}$$

□

Proof: The first part of the theorem, that the rate of a given code satisfying the distortion requirement is bounded by the rate-distortion function, is already shown above. The existence part, that for a given rate satisfying the bound there exists a code, is a bit more tedious. The idea is to extend the concept of jointly typical sequences and construct an encoding/decoding pair satisfying the bound as the length of the source vector grows to infinity. As a start, a set of *distortion typical* sequences are defined. ■

Definition 11.4 The set of all *distortion typical* sequences $A_{\varepsilon,\delta}(X,\hat{X})$ is the set of all pairs of n-dimensional vectors of i.i.d. variables

$$\boldsymbol{x} = (x_1, x_2, \dots, x_n) \quad \text{and} \quad \hat{\boldsymbol{x}} = (\hat{x}_1, \hat{x}_2, \dots, \hat{x}_n) \tag{11.28}$$

such that they are jointly typical $(\boldsymbol{x},\hat{\boldsymbol{x}}) \in A_\varepsilon(X,\hat{X})$ (see Definition 6.5), and

$$\Big|\frac{1}{n}d(\boldsymbol{x},\hat{\boldsymbol{x}}) - E\big[d(X,\hat{X})\big]\Big| \leq \varepsilon \tag{11.29}$$

where $E\big[d(X,\hat{X})\big] \leq \delta$. □

From the weak law of large numbers, it follows directly that

$$\frac{1}{n}d(\boldsymbol{x},\hat{\boldsymbol{x}}) = \frac{1}{n}\sum_{i=1}^{n} d(x_i,\hat{x}_i) \xrightarrow{p} E\big[d(X,\hat{X})\big], \quad n \to \infty \tag{11.30}$$

Following Theorem 6.7, it can be seen that there exists a set of integers n_i, $i = 1,2,3,4$, such that

$$P_1 = P\Big(\Big|-\tfrac{1}{n}\log p(\boldsymbol{x}) - H(X)\Big| > \varepsilon\Big) < \frac{\varepsilon}{4}, \qquad n > n_1 \tag{11.31}$$

$$P_2 = P\Big(\Big|-\tfrac{1}{n}\log p(\hat{\boldsymbol{y}}) - H(\hat{X})\Big| > \varepsilon\Big) < \frac{\varepsilon}{4}, \qquad n > n_2 \tag{11.32}$$

$$P_3 = P\Big(\Big|-\tfrac{1}{n}\log p(\boldsymbol{x},\hat{\boldsymbol{x}}) - H(X,\hat{X})\Big| > \varepsilon\Big) < \frac{\varepsilon}{4}, \qquad n > n_3 \tag{11.33}$$

$$P_4 = P\Big(\Big|\tfrac{1}{n}d(\boldsymbol{x},\hat{\boldsymbol{x}}) - E\big[d(X,\hat{X})\big]\Big| > \varepsilon\Big) < \frac{\varepsilon}{4}, \qquad n > n_4 \tag{11.34}$$

where $E\big[d(X,\hat{X})\big] \le \delta$. Then, for $n > \max\{n_1,n_2,n_3,n_4\}$, by the use of union bound,

$$P\Big((\boldsymbol{x},\hat{\boldsymbol{x}}) \notin A_{\varepsilon,\delta}(X,\hat{X})\Big) < \varepsilon \tag{11.35}$$

and, hence, for arbitrary $\varepsilon > 0$ and sufficiently large n

$$P\Big((\boldsymbol{x},\hat{\boldsymbol{x}}) \in A_{\varepsilon,\delta}(X,\hat{X})\Big) \ge 1-\varepsilon \tag{11.36}$$

From the alternative definition of typical sequences, Definition 6.6, the conditional probability $P(\hat{\boldsymbol{x}}|\boldsymbol{x})$ is bounded as

$$\begin{aligned} p(\hat{\boldsymbol{x}}|\boldsymbol{x}) &= \frac{p(\boldsymbol{x},\hat{\boldsymbol{x}})}{p(\boldsymbol{x})} = p(\hat{\boldsymbol{x}})\frac{p(\boldsymbol{x},\hat{\boldsymbol{x}})}{p(\boldsymbol{x})p(\hat{\boldsymbol{x}})} \\ &\le p(\hat{\boldsymbol{x}})\frac{2^{-n(H(X,\hat{X})-\varepsilon)}}{2^{-n(H(X)+\varepsilon)}2^{-n(H(\hat{X})+\varepsilon)}} \\ &= p(\hat{\boldsymbol{x}})2^{n(H(X)+H(\hat{X})-H(X,\hat{X})+3\varepsilon)} = p(\hat{\boldsymbol{x}})2^{n(I(X;\hat{X})+3\varepsilon)} \end{aligned} \tag{11.37}$$

or, in other words,

$$p(\hat{\boldsymbol{x}}) \ge p(\hat{\boldsymbol{x}}|\boldsymbol{x})2^{-n(I(X;\hat{X})+3\varepsilon)} \tag{11.38}$$

To continue the encoding and decoding procedure should be specified. First, let $p^*(\hat{x}|x)$ be a distribution that gives the rate-distortion function for distortion δ, i.e.

$$p^*(\hat{x}|x) = \arg \min_{p(\hat{x}|x):E[d(X,\hat{X})]\le\delta} I(X;\hat{X}) \tag{11.39}$$

From the source statistics $p(x)$, let

$$p^*(\hat{x}) = \sum_x p^*(\hat{x}|x)p(x) \tag{11.40}$$

Going back to Figure 11.1, a binary source vector $\boldsymbol{x}$ of length n is encoded to a codeword $\boldsymbol{y}$. The decoder maps the codeword to a reconstructed binary vector $\hat{\boldsymbol{x}}$ of length n. When the rate is R, there are 2^{nR} codewords $\boldsymbol{y}$, and equally many reconstructed vectors $\hat{\boldsymbol{x}}$. To define a decoding rule, generate 2^{nR} reconstruction vectors using the distribution

$$p^*(\hat{\boldsymbol{x}}) = \prod_{i=1}^{n} p^*(\hat{x}_i) \tag{11.41}$$

and pair these with the codewords. Denote the decoding function $\hat{\boldsymbol{x}} = g(\boldsymbol{y})$. The encoding rule can be based on typical sequences. Given a vector $\boldsymbol{x}$, find a codeword $\boldsymbol{y}$ such that $(\boldsymbol{x}, g(\boldsymbol{y})) \in A_{\varepsilon,\delta}(X, \hat{X})$. If there are more than one possible codeword, choose one of them at random, and if there is no codeword forming a typical pair with $\boldsymbol{x}$ choose $\boldsymbol{y} = \boldsymbol{0}$. To see what this means for the average distortion first define the event that $\boldsymbol{x}$ and $\hat{\boldsymbol{x}} = g(\boldsymbol{y})$ are distortion typical sequences,

$$E_{\boldsymbol{y}|\boldsymbol{x}}\big\{(\boldsymbol{x},\hat{\boldsymbol{x}}) \in A_{\varepsilon,\delta}(X,\hat{X})\big|\hat{\boldsymbol{x}} = g(\boldsymbol{y})\big\} \tag{11.42}$$

Then the event that $\boldsymbol{x}$ does not have any matching codeword becomes

$$E_{e|\boldsymbol{x}} = \bigcap_{\boldsymbol{y}} E^c_{\boldsymbol{y}|\boldsymbol{x}} \tag{11.43}$$

Since the reconstructed vectors are generated i.i.d., the corresponding codewords are independent and

$$\begin{aligned}
P(E_{e|\boldsymbol{x}}) &= P\Big(\bigcap_{\boldsymbol{y}} E^c_{\boldsymbol{y}|\boldsymbol{x}}\Big)\\
&= \prod_{\boldsymbol{y}} P(E^c_{\boldsymbol{y}|\boldsymbol{x}})\\
&= \prod_{\boldsymbol{y}} (1 - P(E_{\boldsymbol{y}|\boldsymbol{x}}))\\
&= \prod_{\boldsymbol{y}} \Big(1 - \sum_{\hat{\boldsymbol{x}}:(\boldsymbol{x},\hat{\boldsymbol{x}})\in A_{\varepsilon,\delta}} p(\hat{\boldsymbol{x}})\Big)\\
&\leq \prod_{\boldsymbol{y}} \Big(1 - \sum_{\hat{\boldsymbol{x}}:(\boldsymbol{x},\hat{\boldsymbol{x}})\in A_{\varepsilon,\delta}} p(\hat{\boldsymbol{x}}|\boldsymbol{x}) 2^{-n(I(X;\hat{X})+3\varepsilon)}\Big)\\
&= \Big(1 - \sum_{\hat{\boldsymbol{x}}:(\boldsymbol{x},\hat{\boldsymbol{x}})\in A_{\varepsilon,\delta}} p(\hat{\boldsymbol{x}}|\boldsymbol{x}) 2^{-n(I(X;\hat{X})+3\varepsilon)}\Big)^{2^{nR}}\\
&= \Big(1 - 2^{-n(I(X;\hat{X})+3\varepsilon)} \sum_{\hat{\boldsymbol{x}}:(\boldsymbol{x},\hat{\boldsymbol{x}})\in A_{\varepsilon,\delta}} p(\hat{\boldsymbol{x}}|\boldsymbol{x})\Big)^{2^{nR}}
\end{aligned} \tag{11.44}$$

For $1 - \alpha x > 0$, the IT-inequality gives $\ln(1 - \alpha x) \leq -\alpha x$. Thus, $(1 - \alpha x)^M = e^{M \ln(1-\alpha x)} \leq e^{-M\alpha x}$. Furthermore, for $0 \leq x \leq 1$ it can be found that

$$e^{-M\alpha x} \leq 1 - x + e^{M\alpha} \tag{11.45}$$

To see this, first notice that the bound is clearly fulfilled for the end points $x = 0$ and $x = 1$. At the considered interval, the left-hand side, $e^{-M\alpha x}$ is convex, whereas the right-hand side is linearly decreasing with x, and, hence, the bound must be fulfilled in between the end points as well. So, for $0 \le x \le 1$, $0 \le \alpha \le 1$ and $M \ge 0$

$$(1-\alpha x)^M \le 1 - x + e^{M\alpha} \tag{11.46}$$

Applying to (11.44) and identifying $M = 2^{nR}$, $x = \sum p(\hat{\boldsymbol{x}}|\boldsymbol{x})$ and $\alpha = 2^{-n(I(X;\hat{X})+3\varepsilon}$ gives

$$P(E_{e|\boldsymbol{x}}) \le 1 - \sum_{\hat{\boldsymbol{x}}:(\boldsymbol{x},\hat{\boldsymbol{x}})\in A_{\varepsilon,\delta}} p(\hat{\boldsymbol{x}}|\boldsymbol{x}) + e^{2^{-n(I(X;\hat{X})+3\varepsilon)}2^{nR}} \tag{11.47}$$

Averaging over all $\boldsymbol{x}$ gives the total probability of no match as

$$\begin{aligned} P(E_e) &= \sum_{\boldsymbol{x}} p(\boldsymbol{x})P(E_{e|\boldsymbol{x}}) \\ &\le 1 - \sum_{(\boldsymbol{x},\hat{\boldsymbol{x}})\in A_{\varepsilon,\delta}} p(\boldsymbol{x})p(\hat{\boldsymbol{x}}|\boldsymbol{x}) + e^{2^{n(R-I(X;\hat{X})-3\varepsilon)}} \\ &= P\big((\boldsymbol{x},\hat{\boldsymbol{x}}) \notin A_{\varepsilon,\delta}\big) + e^{2^{n(R-R(\delta)-3\varepsilon)}} \end{aligned} \tag{11.48}$$

where it is used that $p(\hat{x}|x) = p^*(\hat{x}|x)$ to get $I(X;\hat{X}) = R(\delta)$ in the last equality. From the definition of $A_{\varepsilon,\delta}(X,\hat{X})$, the term $P\big((\boldsymbol{x},\hat{\boldsymbol{x}}) \notin A_{\varepsilon,\delta}\big) \le \varepsilon$, where ε can be chosen arbitrarily small. For $R > R(\delta)$ and small enough ε, the exponent in the second term $R - R(\delta) - 3\varepsilon < 1$, and the term will decrease toward zero as n grows. Thus, with $R > R(\delta)$ it is possible to find a code where $P(E_e) \to 0$ as $n \to \infty$.

To derive the average distortion, consider first the vector pairs $(\boldsymbol{x},\hat{\boldsymbol{x}}) \in A_{\varepsilon,\delta}(X,\hat{X})$. Then the distortion is bounded by

$$\frac{1}{n}d(\boldsymbol{x},\hat{\boldsymbol{x}}) \le E_{p^*}\big[d(X,\hat{X})\big] + \varepsilon \le \delta + \varepsilon \tag{11.49}$$

For the vector pairs not included in the set of distortion typical sequences, the distortion is bounded by $\frac{1}{n}d(\boldsymbol{x},\hat{\boldsymbol{x}}) \le \hat{\delta}$, where $\hat{\delta} = \max_{(x,\hat{x})} d(x,\hat{x})$ is assumed to be finite. Then the average distortion is

$$\begin{aligned} \frac{1}{n}E\big[d(X,\hat{X})\big] &\le (\delta+\varepsilon)P\big(E_e^c\big) + \hat{\delta}P(E_e) \\ &\le \delta + \varepsilon + \hat{\delta}P(E_e) = \delta + \overline{\varepsilon} \end{aligned} \tag{11.50}$$

where $\overline{\varepsilon} = \varepsilon + \hat{\delta}P(E_e)$ can be chosen arbitrarily small, for large enough n. This completes the proof.

From the above rate-distortion theorem, the rate-distortion function plays the same role for lossy source coding as the entropy does for lossless source coding. It is the limit for when it is possible to find a code. It does not, however, say much on how to construct the code since the construction in the proof is not practically implementable. Especially in the area of image, video, and voice coding there are active research ongoing. Another, closely related, topic is quantization, which in its nature is both lossy and a compression. In Section 11.3, quantization is treated more in detail, and in Section 11.4 transform coding is described, including overviews of JPEG and MPEC coding.

As in the case of the source coding theorem, the rate-distortion theorem can be generalized to hold for stationary ergodic sources. The theory for this is out of the scope for this text.

11.2 LIMIT FOR FIX P_b

In the previous section, it was shown that the rate-distortion function has the same interpretation for lossy source coding as the entropy has for lossless source coding. In this section, it will be seen that it can also be applied to the case of channel coding when a certain bit error rate can be acceptable. For this purpose, the system model has to be expanded a bit to include the channel. In Figure 11.5, the source vector $\boldsymbol{X} = X_1 \dots X_k$ is of length k and the code vector $\boldsymbol{Y} = Y_1 \dots Y_n$ of length n, which gives the encoding rate $R = \frac{k}{n}$. After transmission over the channel, the received vector is $\hat{\boldsymbol{Y}} = \hat{Y}_1 \dots \hat{Y}_n$. Then the decoding outputs the estimated vector as $\hat{\boldsymbol{X}} = \hat{X}_1 \dots \hat{X}_k$.

In Section 6.5, it was shown that reliable communication is possible if and only if the rate is bounded by the capacity,

$$R < C = \max_{p(y)} I(Y; \hat{Y}) \tag{11.51}$$

The term *reliable communication* refers to the case when the error probability after decoding can be made arbitrarily low. As with the case of lossless compared to lossy compression, this puts some hard restrictions on the system. In a real system design, a certain level of error probability can often be accepted. It is possible to treat this error level as an acceptable level of distortion at the decoder output. The next theorem shows the relation between the channel capacity and the rate-distortion function.

Theorem 11.4 Given a source with probability distribution $p(\boldsymbol{x})$, that is encoded with a rate R channel code before transmitted over a channel. If the acceptable distortion is δ for a distortion measure $d(x, \hat{x})$, such system can be designed if and only if

$$R \leq \frac{C}{R(\delta)} \tag{11.52}$$

where C is the channel capacity for the channel and $R(\delta)$ the rate-distortion function. □

In this text, the proof of the theorem is omitted. Instead refer to [54].

The above theorem gives a relation between the channel capacity and the rate-distortion function. In the next, the influence of the acceptable distortion on the fundamental limit in Section 9.3 is treated. The limit $E_b/N_0 \geq -1.59$ dB was derived

Figure 11.5 Block scheme with source and channel coding.

by considering a binary equiprobable source where the bits are encoded by a rate R channel code before the bits are transmitted over channel with signal-to-noise ratio E_b/N_0. For reliable communication, the limit on the coding rate can be written as

$$R < C = \frac{1}{2}\log\left(1 + 2R\frac{E_b}{N_0}\right) \tag{11.53}$$

Assuming a binary source with equally distributed bits, and an acceptable bit error probability P_b after decoding, the rate-distortion function is given by

$$R(P_b) = 1 - h(P_b) \tag{11.54}$$

Thus, the code rate bound becomes

$$R < \frac{C}{R(P_b)} = \frac{\frac{1}{2}\log\left(1 + 2R\frac{E_b}{N_0}\right)}{1 - h(P_b)} \tag{11.55}$$

Equivalently, rewritten as a bound on the signal-to-noise ratio for communication with a maximum bit error probability,

$$\frac{E_b}{N_0} > \frac{2^{2R(1-h(P_b))} - 1}{2R} \tag{11.56}$$

In Figure 11.6, the bound is plotted as the minimum signal-to-noise ratio for the bit error probability. In the figure, there are four plots, one each for the coding rates $R = 1/4$, $R = 1/2$, $R = 3/4$, and the fourth, left most, curve is the case when the encoding rate tends to zero. This is the case when the fundamental limit is given, and the function becomes

$$\lim_{R\to 0} \frac{2^{2R(1-h(P_b))} - 1}{2R} = \ln(2)(1 - h(P_b)) \tag{11.57}$$

which describes the lowest achievable E_b/N_0 for an acceptable bit error probability of P_b at the receiver. As the bit error rate becomes smaller, the entropy function in the

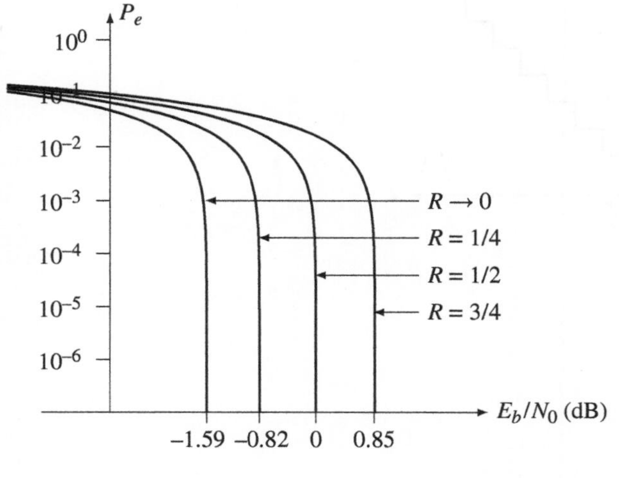

Figure 11.6 Plot of the achievable signal-to-noise ratio for certain bit error probability.

formula will close to zero and the curves fall down close to vertically below about $P_e = 10^{-3}$, where they equal the capacity limit for a given rate.

11.3 QUANTIZATION

In the next two sections, two examples of lossy compression are considered. First, it is quantization that represents a continuous variable by a discrete, and thus introducing disturbance, and second transform decoding. A well-known example of the latter is the image format JPEG that will be briefly described.

As said, quantization maps a continuous variable to a discrete version for the purpose of representing it with a finite length binary vector. An analog-to-digital converter consist of sampling and quantization, i.e. first mapping from continuous time to discrete time and then from continuous amplitude to discrete amplitude. This operation, as well as its inverse–digital-to-analog conversion , is a common component in circuits operating with signals from and to an outer unit, like a sensor of some kind.

In the sampling procedure, the optimal sampling frequency and the reconstruction formula are described by the sampling theorem used in Chapter 9. According to this, sampling and reconstruction do not introduce any distortion. However, to be able to represent the sample values in a computer using finite vectors they have to be quantized. This operation means representing a real value by a discrete variable, and it is inevitable that information is destroyed, and thus distortion introduced.

In this description, a uniform quantization, as shown in Figure 11.7, is used. The input to the quantizer is the continuous variable x. The mapping to the quantizer output x_Q is determined by a staircase function in the figure. In a uniform quantizer, the size of the steps is constant, say Δ.

In a uniform quantizer, the quantization intervals are equally sized, and the staircase function is centered around a linear function, the dashed line in Figure 11.7.

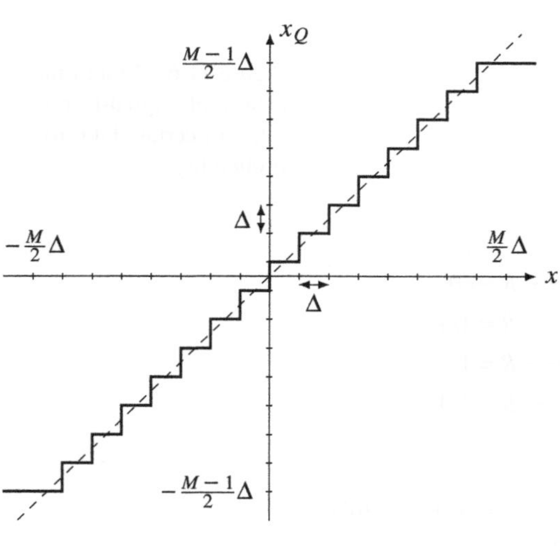

Figure 11.7 A uniform quantization function.

For a nonuniform quantizer, the steps are of unequal size, typically large steps for large values. Then the line center line is nonlinear, and typically have more of an S-shape. This can be used to form the quantizer for the statistics of the continuous source and to have different sizes of the quantization intervals. However, in most practical implementations a uniform quantizer is used. If the statistics differ much from the uniform distribution, the quantizer can either be followed by a source code like the Huffman code or preceded by a compensation filter.

In the figure, a quantizer with M output levels is shown. Assuming a maximum level of the output mapping of $D = \frac{M-1}{2}\Delta$, the granularity of the quantization becomes $\Delta = \frac{2D}{M-1}$. The mth output level then corresponds to the value

$$x_Q(m) = \left(m\Delta - \frac{M-1}{2}\Delta\right) = (2m - M + 1)\frac{\Delta}{2} \tag{11.58}$$

The mapping function shown in the figure is determined from finding an integer m such that

$$x_Q(m) - \frac{\Delta}{2} \le x < x_Q(m) - \frac{\Delta}{2} \tag{11.59}$$

then the output index is $y = m$. If the input value x can exceed the interval $x_Q(0) - \frac{\Delta}{2} \le x < x_Q(M-1) + \frac{\Delta}{2}$, the limits should be

$$y = \begin{cases} m, & x_Q(m) - \frac{\Delta}{2} \le x < x_Q(m) - \frac{\Delta}{2}, \quad 1 \le m \le M-2 \\ 0, & x < x_Q(0) + \frac{\Delta}{2} \\ M-1, & x \ge x_Q(M-1) - \frac{\Delta}{2} \end{cases} \tag{11.60}$$

From (11.58), this can equivalently be written as

$$y = \begin{cases} m, & (2m-M)\frac{\Delta}{2} \le x < (2m-M)\frac{\Delta}{2} + \Delta, \quad \text{for } 1 \le m \le M-2 \\ 0, & x < (2-M)\frac{\Delta}{2} \\ M-1, & x \ge (M-2)\frac{\Delta}{2} \end{cases} \tag{11.61}$$

The output values from the quantizer can be represented by a finite length binary vector. The price for representing a real value with finite levels is an error introduced in the signal. In Figure 11.8, this error, defined as the difference $x - x_Q$, is shown in the upper plot, and the corresponding distortion, $d(x, x_Q) = (x - x_Q)^2$ in the lower plot.

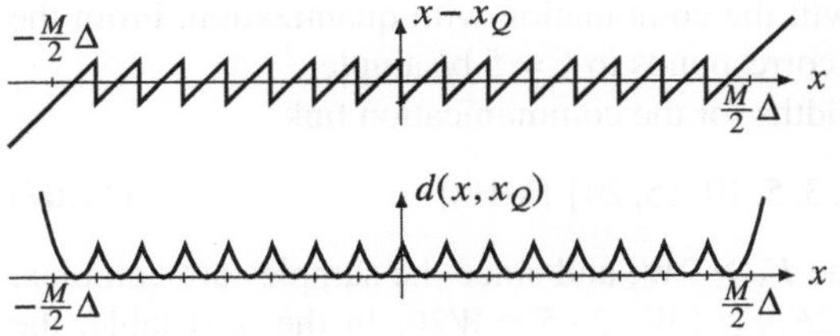

Figure 11.8 The quantization error, $x - x_Q$, and distortion, $d(x, x_Q) = (x - x_Q)^2$, for the uniform quantization function in Figure 11.7.

An estimate of the distortion introduced can be made by considering a uniformly distributed input signal, $X \sim \mathrm{U}(-M\frac{\Delta}{2}, M\frac{\Delta}{2})$. Then all quantization levels will have uniformly distributed input with $f(x) = \frac{1}{\Delta}$, and deriving the average distortion can be made with normalized $x_Q = 0$,

$$E\big[(X - X_Q(m))^2 | Y = m\big] = \int_{-\Delta/2}^{\Delta/2} x^2 \frac{1}{\Delta} dx = \frac{\Delta^2}{12} \tag{11.62}$$

From the uniform assumption $P(Y = m) = \frac{1}{M}$, and hence

$$E\big[(X - X_Q)^2\big] = \sum_{m=0}^{M-1} \frac{1}{M}\frac{\Delta^2}{12} = \frac{\Delta^2}{12} \tag{11.63}$$

When the quantized value is mapped to a vector of length k and $M = 2^k$ this is equivalent to, $E\big[(X - X_Q(m))^2\big] \approx 2^{2(k-1)}\frac{D^2}{12}$, where the approximation $M - 1 \approx M$ is used.

Viewing the distortion as a noise, it is convenient to consider the signal-to-quantization noise ratio (SQNR). Since the signal has zero mean, its variance is

$$E\big[X^2\big] = \int_{-M\frac{\Delta}{2}}^{M\frac{\Delta}{2}} x^2 \frac{1}{M\Delta} dx = \frac{(M\Delta)^2}{12} \tag{11.64}$$

Hence the signal-to-quantization noise is

$$\mathrm{SQNR} = \frac{E\big[X^2\big]}{E\big[(X - X_Q)^2\big]} = M^2 = 2^{2k} \tag{11.65}$$

Expressed in dB, this means

$$\mathrm{SQNR_{dB}} = 2k \cdot 10\log_{10} 2 \approx k \cdot 6 \text{ dB} \tag{11.66}$$

i.e., the SQNR increases with 6 dB with each bit in the quantization.

Example 11.3 In the 4G mobile standard LTE, the downstream signals are constructed with an OFDM (orthogonal frequency division multiplexing) modulation scheme. The modulation carries 2, 4, or 6 bits per ton and transmission. To get the maximum data rate of the system, a reasonable lower requirement on the signal-to-noise ratio is 30 dB. Then the quality of the total channel, both quantization and air channel, will not constrain the modulation due to the quantization. If the air channel is good enough for full speed, so will the combination with quantization. From the approximation of 6 dB per bit, this corresponds to $k = 5$ b/sample.

There are six possible bandwidths for the communication link,

$$W \in \{1.4, 3, 5, 10, 15, 20\} \text{ [MHz]} \tag{11.67}$$

Following the Nyquist sampling rate $F_S \geq 2W$, and since the samples are complex, the total required bit rate is $R_b = 2F_s k \geq 2W \cdot 2 \cdot 5 = W20$. In the next table, the resulting minimum bit rates for the LTE bands are shown. The calculations are based

TABLE 11.1 Bandwidth and data rates for the CPRI protocol.

W (MHz)	$R_{b,\min}$ (Mbps)	Common Public Radio Interface (Mbps)
1.4	28	614.4/8
3	60	614.4/3
5	100	614.4
10	200	1228.8
15	300	1228.8
20	400	2457.6

on uniformly distributed amplitude of the samples, which is not the case in reality. So, the result is a bit optimistic and a real signal would require some extra bits per real sample.

As a comparison, for each bit rate, the rates used by the fronthaul protocol Common Public Radio Interface (CPRI) is shown in Table 11.1. This is a standard developed for transporting samples within the base station, but also often considered for transporting LTE samples over fiber connections further distances. The relatively high bit rates come from the requirement of 15 b/real sample. In CPRI, the specified bit rates in Mbps are 614.4, 1228.8, 2457.6, 3072, 4915.2, 6144, and 9830.4. Then for the 1.4-MHz band, there can be eight signals in one 614.4 Mbps stream and for the 3-MHz band three signals in a 614.4-Mbps stream. For the others, it is one signal per stream.

In the case of uniform distribution, it is natural to set the reconstructed value to the center in the quantization interval. In the general case, for a given interval $\Delta_m \leq x < \Delta_{m+1}$ and a reconstruction value x_m in the interval m with the distribution $f(x|m)$, the average distortion is

$$d_m = E_{X|m}\big[(X - x_m)^2\big] = E_{X|m}\big[X^2\big] - 2x_m E_{X|m}\big[X\big] + x_m^2 \tag{11.68}$$

Thus, to find the reconstruction value that minimizes the distortion take the derivative with respect to x_m to get

$$\frac{\partial d_m}{\partial x_m} = -2E_{X|m}\big[X\big] + 2x_m = 0 \tag{11.69}$$

and hence the optimal reconstruction value is $x_m = E_{X|m}\big[X\big]$. For the uniform distribution, this is indeed the center in the interval as used above. For other distributions, the value can change. In the next example, the reconstruction value for a Gaussian source when using a 1-bit quantizer is derived.

Example 11.4 Assume a Gaussian source where $X \sim N(0, \sigma)$ and a 1-bit quantizer. The natural intervals are divided buy the value $x = 0$, i.e. for $x < 0$, $y = 0$, and for $x \geq 0$, $y = 1$. Since the two sides are symmetric, it is only needed to derive the optimal reconstruction level for the positive side. Therefore, the distribution is given by

$$f(x|y = 1) = \frac{2}{\sqrt{2\pi\sigma^2}} e^{-x^2/2\sigma^2} \tag{11.70}$$

and the reconstruction value is

$$x_1 = \int_0^\infty x \frac{2}{\sqrt{2\pi\sigma^2}} e^{-x^2/2\sigma^2} dx = \sqrt{\frac{2}{\pi}}\sigma \tag{11.71}$$

Consequently, the reconstruction value for the negative side

$$x_0 = -\sqrt{\frac{2}{\pi}}\sigma \tag{11.72}$$

With these levels, the average quantization distortion becomes $E\big[d(X, X_Q)\big] = \frac{\sigma^2}{\pi}(\pi - 2)$.

11.4 TRANSFORM CODING

There are two main families of source codes, lossless and lossy. The previously studied algorithms, Huffman and LZ, are both lossless source codes. Going back to Figure 4.1, it means that after the source vector $\boldsymbol{X}$ is encoded to $\boldsymbol{Y}$ and reconstructed to $\widehat{\boldsymbol{X}}$, there is no distortion, i.e. $\widehat{\boldsymbol{X}} = \boldsymbol{X}$. This is the typical behavior of a compression algorithm for, e.g., text or program codes, where there can be no errors introduced. On the other hand, when compressing, e.g., images, a certain amount of distortion can be allowed, as long as end user does not perceive a substantial quality deterioration. In this way, the achieved compression can be much higher than in the case of lossless coding [68, 69, 88].

To get a measure of the loss made in the coding, a distortion measure is introduced $d(\boldsymbol{X}, \widehat{\boldsymbol{X}})$. For the case of an image, the symbols are typically the pixels. By the additive assumption of the distortion measure, the average distortion for the complete image with N pixels is

$$\frac{1}{N}\sum_i d(x, \hat{x}) \tag{11.73}$$

It is not obvious how the distortion measure shall be chosen for image compression. Often the squared distance distortion is used, but this is also misleading. Different images can tolerate a different amount of distortion, and it also depends on where in the image it is located. To get a more reliable measure of how a certain coding and distortion affect the user experience, real user ratings has to be used. In those, people are asked to rate the distortion in the image or video that is compressed. These measures are very useful when developing the algorithms, but naturally not possible to use for all images.

If a vector of length N symbols are compressed, the optimal method for a given compression rate R is the one that gives the lowest distortion. Ironically, when taking the computational complexity into account, in many cases it is preferably to use a suboptimal method. The computational complexity of an optimal algorithm often grows exponentially with the length of the vector. On the other hand, the efficiency of the compression also grows with the length of the vector, but not exponentially. So to get a good compression, it requires a longer vector, which means a high complexity. Often

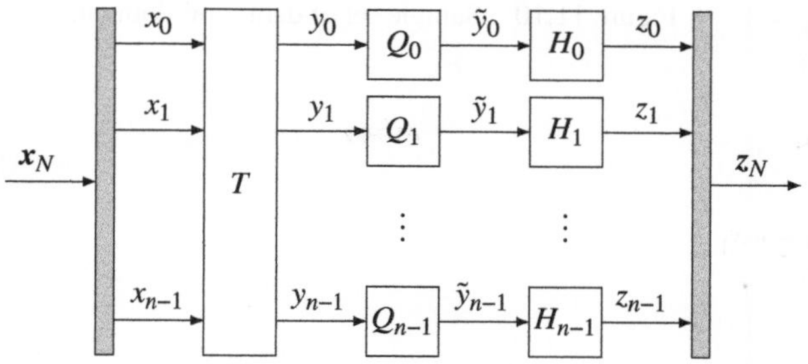

Figure 11.9 Block diagram for transform coding.

it is possible to construct a suboptimal compression method with a lower complexity growths and still almost as good results. Comparing this can actually mean that for a certain complexity, the best compression is achieved for a suboptimal method. In this section, one such method, called transform coding, will be described. It is the basis of the image format JPG, which will also briefly be described.

For images today, there are a couple of formats including compression. As described in Section 5.4, PNG uses a combination of LZ and Huffman. The other main format is JPG, which uses a transform coding. In Figure 11.9, the base model for the transform encoder is shown. Assuming the source vector $\boldsymbol{x}_N$ of length N is blocked into column vectors of length n, $\boldsymbol{x} = (x_0, \dots, x_{n-1})^T$. Each of these vectors is transformed using a linear transform represented by the $n \times n$ transform matrix T as

$$\boldsymbol{y} = T\boldsymbol{x} \tag{11.74}$$

The general idea behind the algorithm is that each symbol in the vector before the transform has similar statistical properties, whereas after the transform there are some symbols with high variance and some with low variance. If each symbol is quantized independently, the low variance symbols need few bits whereas the high variance symbols need more bits. However, the required number of bits grows essentially as the logarithm of the variance and in total the number of bits required for the transformed vector can be lower than the number of bits for the untransformed vector, without distortion growth. To get a good mapping between the quantized value and the binary vector, a Huffman code, or some other efficient lossless compression, is followed after the quantization. Finally, the compressed binary vector is again merged into a long sequence, z_N.

The reconstruction works in the opposite direction, where the first code sequence is split into the binary codewords by the Huffman decoders. These are then reconstructed according to the quantization levels and inversely transformed using

$$\tilde{\boldsymbol{x}} = T^{-1}\tilde{\boldsymbol{y}} \tag{11.75}$$

which forms the reconstructed sequence.

In compression of, e.g., images and video, it is often assumed that the vectors are real valued. Then it is also reasonable to assume real-valued transform matrices. If the inverse of the transform matrix equals the transpose, i.e. if $T^{-1} = T^T$, the transform is said to be *orthonormal.*[1]

[1] For the case of a complex transform matrix, it is instead the Hermitian transpose that is of interest.

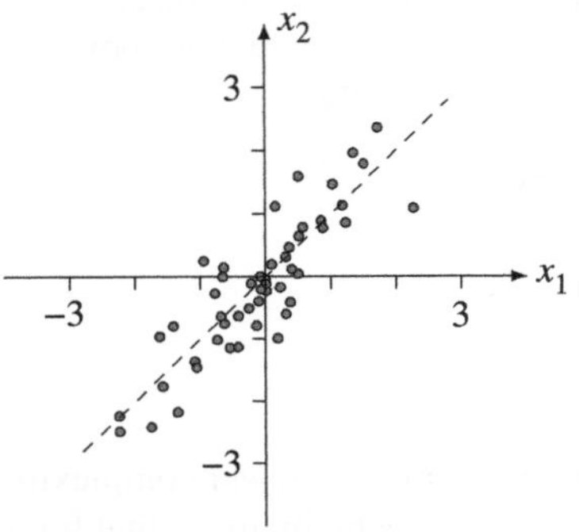

Figure 11.10 Sample set of data in $\mathcal{X}$-domain.

Example 11.5 Consider a two-dimensional random variable $X = (X_1 X_2)^T$, and 50 measured outcomes as shown in Figure 11.10. It is seen in the figure that the samples are located relatively concentrated along a linear function angled $\frac{\pi}{4}$ rad relative to the x_1 axis. That means the samples have about the same dynamics in both dimensions. Hence, to quantify the samples, the same number of bits should be used in both dimensions. Since the samples are concentrated along the dashed line in the figure, an alternative solution is to use a quantization along this line. In other words, the sample set can be rotated $\frac{\pi}{4}$ rad clockwise. To rotate the two-dimensional vector $\boldsymbol{x} = (x_1 x_2)^T$, an angle φ the transform matrix

$$T_\varphi = \begin{pmatrix} \cos(\varphi) & \sin(\varphi) \\ -\sin(\varphi) & \cos(\varphi) \end{pmatrix} \tag{11.76}$$

It is easily seen that $T_\varphi T_\varphi^T = I$, thus it is an orthonormal transform. The transform matrix for $\varphi = \frac{\pi}{4}$ is

$$T_{\frac{\pi}{4}} = \frac{1}{\sqrt{2}} \begin{pmatrix} 1 & -1 \\ 1 & 1 \end{pmatrix} \tag{11.77}$$

The transformed sample space $\boldsymbol{y} = T_{\frac{\pi}{4}} \boldsymbol{x}$ is shown in Figure 11.11. The dynamics is a bit higher in the y_1 dimension than in x_1 or x_2, but the dynamics in y_2 is much smaller. This can be utilized by using a smaller number of bits in the quantization for y_2.

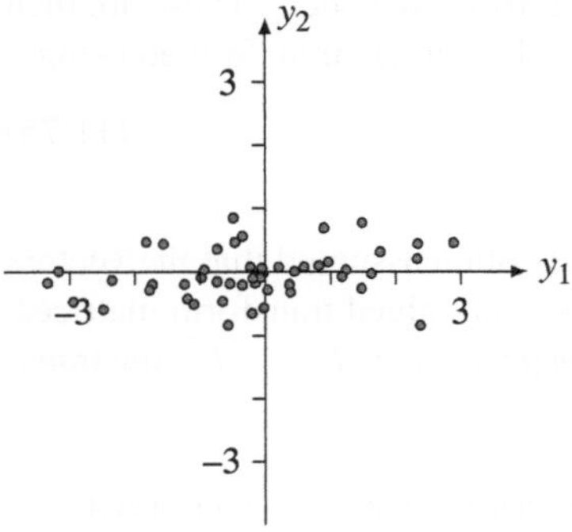

Figure 11.11 Sample set of data in $\mathcal{Y}$-domain.

Quantizing the original vectors $\boldsymbol{x}$ with 4 bits in each dimension gives the mean squared error distortion $d(\boldsymbol{x}, \tilde{\boldsymbol{x}}) = 0.027$. If instead the data are transformed to $\boldsymbol{y}$ and quantized with 4 bits in y_1 dimension and 3 bits in y_2 dimension gives the distortion after inverse transform $d(\boldsymbol{x}, T_{\frac{\pi}{4}}^T \tilde{\boldsymbol{y}}) = 0.029$. So for, approximately, the same distortion either 8 bits are used for quantization in $\mathcal{X}$ domain or 7 bits is used in the $\mathcal{Y}$ domain.

The previous example shows the idea behind transform coding. For a given n-dimensional sample set, there might be a transform that gives a sample set where the quantization can be performed more efficiently. It can be shown that for high-resolution quantization and optimal quantizer can be constructed by a uniform quantization followed by a Huffman code. By optimal quantizer is meant a quantizer that minimizes the distortion for a given rate.

Let $\boldsymbol{X}$ be an n-dimensional random column vector with mean and covariance matrix

$$E[\boldsymbol{X}] = \boldsymbol{0} \tag{11.78}$$

$$\Sigma_X = E[\boldsymbol{X}\boldsymbol{X}^T] = \left[E[X_i X_j]\right]_{i,j=0\ldots n-1} \tag{11.79}$$

where the variances $\sigma_{X_i}^2 = E[X_i^2]$ are the diagonal elements in Σ_X. Using the orthonormal transform matrix T, the transformed vector $\boldsymbol{Y} = T\boldsymbol{X}$ has zero mean, $E[\boldsymbol{Y}] = E[T\boldsymbol{X}] = TE[\boldsymbol{X}] = \boldsymbol{0}$. Thus, the dynamics of a vector is given by the variances, which gives the power of the vector,

$$P_X = \sum_i \sigma_{X_i}^2 = \sum_i E[X_i^2] = E[\boldsymbol{X}^T\boldsymbol{X}] \tag{11.80}$$

Similarly, the power of $\boldsymbol{Y}$ is

$$P_Y = E[(T\boldsymbol{X})^T T\boldsymbol{X}] = E[\boldsymbol{X}^T T^T T\boldsymbol{X}] = E[\boldsymbol{X}^T\boldsymbol{X}] = P_X \tag{11.81}$$

That means an orthogonal transform preserves the power of the vector. Finally, the covariance matrix of $\boldsymbol{Y}$ is given by

$$\Sigma_Y = E[\boldsymbol{Y}\boldsymbol{Y}^T] = E[T\boldsymbol{X}(T\boldsymbol{X})^T] = TE[\boldsymbol{X}\boldsymbol{X}^T]T^T = T\Sigma_X T^T \tag{11.82}$$

Since the power is preserved by an orthonormal transform, the quantization distortion then comes down to its distribution over the vectors. Dimensions with smaller variance can use fewer bits than dimensions with high variance. If the entries of $\boldsymbol{X}$ are equally distributed with a constant variance, and the transformed vector has variable variance over the vector, there should be possibilities for gains in the quantization.

To see this result in a bit more formal way, introduce the *transform coding gain* as the ratio between the average distortion for $\boldsymbol{X}$ and $\boldsymbol{Y}$,

$$G_{TC} = \frac{d(X)}{d(Y)} \tag{11.83}$$

for a certain average number of bits per symbol, k. Consider a Gaussian vector $\boldsymbol{X}$ with constant variance, i.e. $X_i \sim \mathrm{N}(0, \sigma_X)$. For a Gaussian variable, the distortion due to optimal uniform quantization at high resolution can be approximated by [55][2]

$$d(X) \approx c_G \sigma_X^2 2^{-2k} \tag{11.84}$$

where $c_G = \frac{\pi e}{6}$. The sum of the variances for $\boldsymbol{X}$ can be written as

$$E[\boldsymbol{X}^T\boldsymbol{X}] = \sum_i \sigma_X^2 = n\sigma_X^2 \tag{11.85}$$

Again assuming that T is an orthonormal transform matrix, $\boldsymbol{X}$ can be expressed as $\boldsymbol{X} = T^T\boldsymbol{Y}$. Thus,

$$\begin{aligned} n\sigma_X^2 &= E[\boldsymbol{X}^T\boldsymbol{X}] = E[(T^T\boldsymbol{Y})^T T^T\boldsymbol{Y}] \\ &= E[\boldsymbol{Y}^T T T^T \boldsymbol{Y}] = E[\boldsymbol{Y}^T\boldsymbol{Y}] = \sum_i \sigma_{Y_i}^2 \end{aligned} \tag{11.86}$$

which gives that the variance of X is the arithmetic mean of the variances for Y_i, i.e., $\sigma_X^2 = \frac{1}{n}\sum_i \sigma_{Y_i}^2$.

Since $\boldsymbol{X} \sim \mathrm{N}(\boldsymbol{0}, \Sigma_X)$, the entries of the transform vector $\boldsymbol{Y} = T\boldsymbol{X}$ are linear combinations of Gaussian variables, and hence, also Gaussian. Thus the transform vector is also Gaussian, $\boldsymbol{Y} \sim \mathrm{N}(\boldsymbol{0}, \Sigma_Y)$, with the distortion $d(Y_i) \approx c_G \sigma_{Y_i}^2 2^{-2k_i}$. If the average number of bits per dimension is $\frac{1}{n}\sum_i k_i = k$, the optimal bit allocation can be found through the Lagrange optimization function

$$J = \frac{1}{n}\sum_i c_G \sigma_{Y_i}^2 2^{-2k_i} + \lambda\Big(\frac{1}{n}\sum_i k_i - k\Big) \tag{11.87}$$

Differentiating and equal to zero yields

$$\frac{\partial}{\partial k_j} J = -\frac{2\ln 2}{n} c_G \sigma_{Y_j}^2 2^{-2k_j} + \frac{\lambda}{n} = 0 \tag{11.88}$$

or, equivalently,

$$d(Y_j) = \frac{\lambda}{2\ln 2} = d(Y) \tag{11.89}$$

Hence, the bit loading should be performed such that the distortion is equal in all dimensions. Rewriting (11.89) gives that $\lambda = d(Y) 2\ln 2$, and thus

$$-2\ln 2 c_G \sigma_{Y_j}^2 2^{-2k_j} = -d(Y) 2\ln 2 \tag{11.90}$$

which gives the optimal bit allocation in dimension j as

$$k_i = \frac{1}{2}\log \frac{c_G \sigma_{Y_j}^2}{d(Y)} \tag{11.91}$$

[2] In general, the minimum high-resolution distortion can be written as $d(X) \approx c\sigma_X^2 2^{-2k}$, where c is a constant dependent on the distribution of X and the quantization method. For uniform quantization of a Gaussian variable, the constant is $c_G = \frac{\pi e}{6}$.

Hence, the average number of bits per dimension is

$$k = \frac{1}{2n} \sum_i \log \frac{c_G \sigma_{Y_i}^2}{d(Y)} = \frac{1}{2} \log \frac{c_G \left(\prod_i \sigma_{Y_i}^2\right)^{1/n}}{d(Y)} \tag{11.92}$$

which leads to an expression for the distortion as

$$d(Y) = c_G \left(\prod_i \sigma_{Y_i}^2\right)^{1/n} 2^{-2k} = c_G \sigma_Y^2 2^{-2k} \tag{11.93}$$

where σ_Y^2 is the geometric mean for the variances of Y_i, i.e., $\sigma_Y^2 = \left(\prod_i \sigma_{Y_i}^2\right)^{1/n}$. Inserting in the transform coding gain (11.83),

$$G_{TC} = \frac{d(X)}{d(Y)} \approx \frac{c_G \sigma_X^2 2^{-2k}}{c_G \sigma_Y^2 2^{-2k}} = \frac{\sigma_X^2}{\sigma_Y^2} = \frac{\frac{1}{n}\sum_i \sigma_{Y_i}^2}{\left(\prod_i \sigma_{Y_i}^2\right)^{1/n}} \tag{11.94}$$

it is seen that the ratio between the arithmetic mean and the geometric mean for variances of the transformed vector is a relevant measure on the efficiency of the transform. Since the arithmetic mean is independent of the transform, the sum of the variances is constant. The geometric mean, however, depends on the transform. That is, if the variance of one dimension decreases, to preserve the sum the variance needs to increase in at least one other dimension. This means that the transform coding gain G_{TC} is maximized if the geometric mean is minimized over all orthogonal transforms.

For any positive valued sequence $a_1, \dots, a_n$, it can be shown from Jensen's inequality that the arithmetic mean exceeds the geometric mean (see Problem 11.8), i.e. that

$$\frac{1}{n} \sum_i a_i \geq \left(\prod_i a_i\right)^{1/n} \tag{11.95}$$

with equality if and only if $a_i = a_j, \forall i,j$. That is, if $\boldsymbol{X}$ is Gaussian with constant variance, optimal transform coding using an orthonormal transform can only decrease the total distortion. An important requirement for result to hold is that the variances in the original vector are equal.

The covariance matrix is by definition symmetric and therefore also normal, i.e. $\Sigma\Sigma^T = \Sigma^T\Sigma$, and as such it has orthonormal eigenvectors [8]. Hence, the covariance matrix Σ_X can be diagonalized by an orthonormal matrix A as

$$A^T \Sigma_X A = D \tag{11.96}$$

where D is a diagonal matrix containing the eigenvalues of Σ_X. From (11.82), it is seen that by defining a transform matrix $T = A^T$ the covariance matrix of $\boldsymbol{Y} = T\boldsymbol{X}$ is

$$\Sigma_Y = T\Sigma_X T^T = A^T \Sigma_X A = D \tag{11.97}$$

In other words, by using a transform matrix where the rows are the eigenvectors of $\boldsymbol{X}$, the transformed vector contains uncorrelated variables. The obtained transform is often referred to as the Karhunen–Loeve transform (KLT) [56, 57, 58].

Since T is an orthogonal transform matrix, its determinant is $|T| = 1$, and thus,

$$|\Sigma_Y| = |T| \cdot |\Sigma_X| \cdot |T^T| = |\Sigma_X| \tag{11.98}$$

If the KLT is used, Σ_Y is a diagonal matrix and the determinant is the product of the diagonal entries,

$$|\Sigma_Y| = \prod_i \sigma_{Y_i}^2 \tag{11.99}$$

From the Hadamard inequality, it is known that the determinant of a positive semidefinite matrix is upper bounded by the product of the diagonal elements [8]. Thus, the determinant of Σ_X can be bounded as

$$|\Sigma_X| \leq \prod_i \sigma_{X_i}^2 \tag{11.100}$$

Summarizing, this means that, however, the variances are distributed over the vector $\boldsymbol{X}$, the geometrical mean will not increase by applying the KLT since

$$\prod_i \sigma_{X_i}^2 \geq |\Sigma_X| = |\Sigma_Y| = \prod_i \sigma_{Y_i}^2 \tag{11.101}$$

Hence, the KLT minimizes the geometrical mean of the variances over all orthonormal transforms, and consequently it maximizes the transform coding gain G_{TC}.

In practice, there are some drawbacks of the KLT since it requires knowledge of the statistics of the vector $\boldsymbol{X}$. In most applications, this is not available and then has to be estimated. In that case, the statistics, or the transform matrix, must be sent together with the compressed data, which will eat much of the compression gain. In practical applications, there are other more suitable transforms, such as the discrete cosine transform introduced in the next section.

11.4.1 Applications to Image and Video Coding

Lossy compression, and especially transform coding, is often applied to image, video, and audio coding, where the quality required at the reconstruction is dependent on the situation. In, for example, a DVD recording, the demands on the quality are much higher than a video call over third-party application, like Skype. That means also there is room for substantial compression.

By experiments, it has been noted that for a typical image the energy content is concentrated in the low-frequency region, and high-frequency regions are not that vital for the perception. From the derivations previously discussed in this section, this is a typical case where transform coding should be very efficient. The most widespread example is the JPEG standard, standardized by the *Joint Photographic Experts Group* as a joint effort from the standardization organizations ITU-T and ISO/IEC JTC1. There are different generations of the standard with JPEG as the first from 1994 [59, 60], followed by JPEG2000. In this text, an overview of the JPEG standard is given. For a more thorough description, refer to, e.g., [61, 12].

Even though the KLT is optimal, it is not a practical transform since it requires knowledge of the covariance matrix of the vector. By comparing different transform, it has been found that the discrete cosine transform (DCT) is the one that works best for images [62]. The DCT can be viewed as a real-valued counterpart of the Fourier transform, and as such it can be defined in some different ways. The dominating version, DCT-II, and the one used in the JPEG standard is given below

Definition 11.5 (DCT) The discrete cosine transform for the vector $\boldsymbol{x} = (x_0, \ldots, x_{n-1})$ is given by

$$y_i = \sum_{j=0}^{n-1} \alpha(i) \sqrt{\frac{2}{n}} \cos\left(\frac{(2j+1)i\pi}{2n}\right) x_j, \quad i = 0, \ldots, n-1 \tag{11.102}$$

where

$$\alpha(i) = \begin{cases} \frac{1}{\sqrt{2}}, & i = 0 \\ 1, & i = 1, \ldots, n-1 \end{cases} \tag{11.103}$$

□

As before, it can be given in the matrix form by the transform matrix

$$T = \left[\alpha(i) \sqrt{\frac{2}{n}} \cos\left(\frac{(2j+1)i\pi}{2n}\right)\right]_{i,j=0,\ldots,n-1} \tag{11.104}$$

For $n = 8$, the matrix becomes

$$T_8 = \left[\frac{\alpha(i)}{2} \cos\left(\frac{(2j+1)i\pi}{16}\right)\right]_{i,j=0,\ldots,7}$$

$$= \begin{pmatrix} 0.354 & 0.354 & 0.354 & 0.354 & 0.354 & 0.354 & 0.354 & 0.354 \\ 0.490 & 0.416 & 0.278 & 0.098 & -0.098 & -0.278 & -0.416 & -0.490 \\ 0.462 & 0.191 & -0.191 & -0.462 & -0.462 & -0.191 & 0.191 & 0.462 \\ 0.416 & -0.098 & -0.490 & -0.278 & 0.278 & 0.490 & 0.098 & -0.416 \\ 0.354 & -0.354 & -0.354 & 0.354 & 0.354 & -0.354 & -0.354 & 0.354 \\ 0.278 & -0.490 & 0.098 & 0.416 & -0.416 & -0.098 & 0.490 & -0.278 \\ 0.191 & -0.462 & 0.462 & -0.191 & -0.191 & 0.462 & -0.462 & 0.191 \\ 0.098 & -0.278 & 0.416 & -0.490 & 0.490 & -0.416 & 0.278 & -0.098 \end{pmatrix} \tag{11.105}$$

The DCT given in Definition 11.5 is for the one-dimensional case, i.e. a column vector $\boldsymbol{x}$ is transformed to $\boldsymbol{y} = T\boldsymbol{x}$. It can be seen that the transform is an orthonormal transform since $T \cdot T^T = I$, and hence the inverse transform is given by $\boldsymbol{x} = T^T\boldsymbol{x}$.

By viewing the pixels as numbers, an image can be represented by an $M \times N$ matrix. In the JPEG standard, to make a transform coding of such matrix, the image is divided into subblocks of 8×8 pixels and encoded separately. The transform used is a two-dimensional version of the above. This is performed by taking the transform of each row and column separately. Letting X be the matrix for the subblock, the transform for each row is given by

$$Y_r = T_8 X \tag{11.106}$$

Similarly, the transform of each column is given by

$$Y_c = X T_8^T \tag{11.107}$$

Thus, the two-dimensional DCT for 8×8 matrix X is

$$Y = T_8 X T_8^T \tag{11.108}$$

Naturally, the transform in the general case for an $n \times n$ matrix is derived from

$$Y = T X T^T \tag{11.109}$$

The following definition summarizes the transform calculation.

Definition 11.6 (2D-DCT) The discrete cosine transform for the matrix $X = \left[x_{ij}\right]_{ij=0,\dots,n-1}$ is given by the matrix $Y = \left[y_{ij}\right]_{ij=0,\dots,n-1}$, where

$$y_{ij} = \sum_{k=0}^{n-1} \sum_{\ell=0}^{n-1} \alpha(i)\alpha(j) \frac{2}{n} \cos\left(\frac{(2k+1)i\pi}{2n}\right) \cos\left(\frac{(2\ell+1)i\pi}{2n}\right) x_{k\ell} \tag{11.110}$$

and

$$\alpha(u) = \begin{cases} \frac{1}{\sqrt{2}}, & u = 0 \\ 1, & u = 1, \dots, n-1 \end{cases} \tag{11.111}$$

□

For the 8×8 case, this gives

$$y_{ij} = \sum_{k=0}^{7} \sum_{\ell=0}^{7} \frac{\alpha(i)\alpha(j)}{4} \cos\left(\frac{(2k+1)i\pi}{16}\right) \cos\left(\frac{(2\ell+1)i\pi}{16}\right) x_{k\ell} \tag{11.112}$$

In Figure 11.12, the left-hand picture contains 512×512 pixels. To compress this according to the JPEG standard, it is divided into 8×8 subpictures, whereas one of them is shown in the figure. The pixels are numbers between 0 and 255, and the subfigure here is represented by the matrix

$$X = \begin{pmatrix} 185 & 191 & 205 & 207 & 208 & 211 & 202 & 169 \\ 164 & 164 & 150 & 164 & 164 & 184 & 201 & 197 \\ 165 & 168 & 139 & 91 & 94 & 87 & 94 & 109 \\ 193 & 208 & 207 & 192 & 201 & 198 & 195 & 165 \\ 214 & 200 & 195 & 189 & 178 & 162 & 165 & 154 \\ 133 & 129 & 134 & 163 & 178 & 171 & 181 & 163 \\ 202 & 196 & 192 & 183 & 187 & 190 & 175 & 153 \\ 170 & 142 & 138 & 126 & 146 & 165 & 148 & 113 \end{pmatrix} \tag{11.113}$$

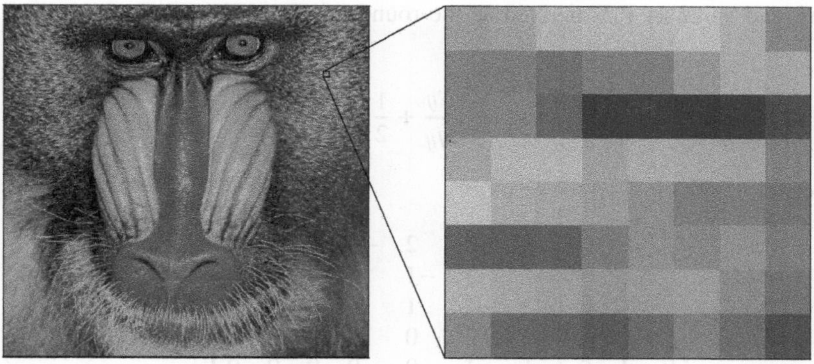

Figure 11.12 The original picture with 512 × 512 pixels, and one of the 8 × 8 subpictures.

To take the transform of this matrix, first the values have to be centered around zero. This is done by subtracting 128 from each value. The transformed matrix is thus given by

$$Y = T_8(X - 128)T_8^T$$

$$= \begin{pmatrix} 326.9 & 38.4 & -1.5 & 34.0 & -20.9 & 7.2 & -6.6 & 9.3 \\ 34.9 & -7.7 & 12.3 & -26.6 & -5.1 & -5.5 & -5.8 & -3.7 \\ -2.2 & -32.7 & -10.2 & 23.6 & -7.7 & 8.3 & 8.5 & -4.7 \\ 109.2 & -48.6 & -45.7 & -2.4 & 14.8 & 14.4 & -7.7 & -3.5 \\ 84.6 & 27.0 & -30.0 & 11.8 & -9.6 & 8.7 & 5.0 & 5.0 \\ 62.3 & 30.1 & -25.7 & -2.8 & -15.4 & -7.6 & 2.6 & 5.0 \\ -128.6 & 4.0 & 4.5 & -0.4 & 2.8 & -5.8 & 0.9 & 7.1 \\ -38.8 & 85.6 & 25.5 & -12.6 & 2.8 & -6.7 & 7.8 & 2.6 \end{pmatrix} \tag{11.114}$$

Even though the subpicture is taken from the fur of the baboon, which is an area with many changes and a typically hard part to compress, it is clearly seen that the high-frequency content is much less than the low-frequency content. The magnitude of the values decreases when moving from the left upper corner to the right lower corner, which is also how the frequencies increase. On top of this, the human perception is more sensitive to low-frequency parts than the high-frequency parts in an image. To meet this, the high-frequency content is not only lower in magnitude than the low-frequency content, it should also be quantized harder. In JPEG, this is solved by having predefined quantization matrices, as the one in (11.115).

$$Q = \begin{pmatrix} 16 & 11 & 10 & 16 & 24 & 40 & 51 & 61 \\ 12 & 12 & 14 & 19 & 26 & 58 & 60 & 55 \\ 14 & 13 & 16 & 24 & 40 & 57 & 69 & 56 \\ 14 & 17 & 22 & 29 & 51 & 87 & 80 & 62 \\ 18 & 22 & 37 & 56 & 68 & 109 & 103 & 77 \\ 24 & 35 & 55 & 64 & 81 & 104 & 113 & 92 \\ 49 & 64 & 78 & 87 & 103 & 121 & 120 & 101 \\ 72 & 92 & 95 & 98 & 112 & 100 & 103 & 99 \end{pmatrix} \tag{11.115}$$

The quantized value of Y is derived as the rounded value of y_{ij} when normalized by q_{ij},

$$y_{Q,ij} = \left\lfloor \frac{y_{ij}}{q_{ij}} + \frac{1}{2} \right\rfloor \tag{11.116}$$

For our example in (11.114), the result is

$$Y_Q = \begin{pmatrix} 20 & 3 & 0 & 2 & -1 & 0 & 0 & 0 \\ 3 & -1 & 1 & -1 & 0 & 0 & 0 & 0 \\ 0 & -3 & -1 & 1 & 0 & 0 & 0 & 0 \\ 8 & -3 & -2 & 0 & 0 & 0 & 0 & 0 \\ 5 & 1 & -1 & 0 & 0 & 0 & 0 & 0 \\ 3 & 1 & 0 & 0 & 0 & 0 & 0 & 0 \\ -3 & 0 & 0 & 0 & 0 & 0 & 0 & 0 \\ -1 & 1 & 0 & 0 & 0 & 0 & 0 & 0 \end{pmatrix} \tag{11.117}$$

To store or transmit the values, they should ideally be compressed using a Huffman code. Since the statistics is not available, it is replaced by a lookup table procedure giving a compression scheme that is close to the optimum. Before performing the compression, however, there are two aspects of the generated data that can be utilized.

The first aspect to notice is that the upper left element of Y, i.e. y_{00}, is a bias term that gives the average tone of this part of the image. For most parts of the full image, this is a slowly varying parameter between subblocks, so to decrease the dynamics of the value, instead the difference to the corresponding value in the subsequent sub-block is encoded. In most cases, this value is more limited and can be more efficiently compressed.

The second observation is again that the magnitude of Y decreases with growing frequency. In Y_Q that typically means the lower right part of the matrix contains only zeros. By reading the values into a vector according to a zigzag pattern over the data as shown in Figure 11.13, this vector should typically end with a sequence of

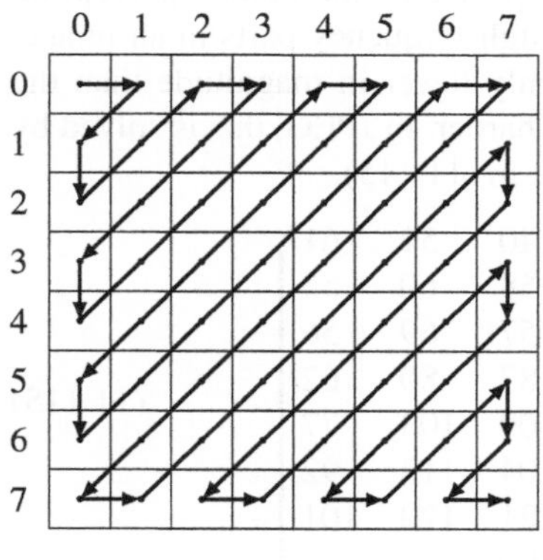

Figure 11.13 Sequence of samples for encoding.

zeros. By applying the zigzag pattern to the matrix in (11.117), the following vector is obtained:

$$y_Q = (20, 3, 3, 0, -1, 0, 2, 1, -3, 8, 5, -3, -1, -1, -1, 0, 0, 1, -2,\\ 1, 3, -3, 1, -1, 0, 0, 0, 0, 0, 0, 0, 0, 0, 0, -1, 1, 0, 0, 0, 0,\\ 0, 0, 0, 0, 0, 0, 0, 0, 0, 0, 0, 0, 0, 0, 0, 0, 0, 0, 0, 0, 0, 0, 0) \tag{11.118}$$

where there are 27 zeros at the end. Instead of coding these trailing zeros individually, a special codeword is introduced called *end of block* (EOB). In this case, the EOB symbol replaces 27 trailing zeros.

At the decoder side, first the binary codewords are decoded so the transform matrix Y_Q is reconstructed. To map the quantized data back to the transform matrix, the following function is used:

$$\hat{y}_{ij} = y_{Q,ij} q_{ij} \tag{11.119}$$

Finally, the reconstructed subblock is derived as

$$\hat{X} = T_8^T \hat{Y} T + 128 \tag{11.120}$$

The additional 128 reinserts the bias term that was removed at the beginning of the encoding process. Hence, the reconstruction of (11.117) gives

$$\hat{X} = \begin{pmatrix} 181 & 189 & 197 & 204 & 213 & 214 & 198 & 177 \\ 164 & 164 & 159 & 158 & 171 & 193 & 204 & 202 \\ 162 & 151 & 124 & 93 & 80 & 84 & 89 & 88 \\ 212 & 215 & 210 & 198 & 193 & 198 & 199 & 194 \\ 192 & 198 & 197 & 187 & 177 & 168 & 153 & 136 \\ 129 & 135 & 141 & 150 & 166 & 182 & 181 & 170 \\ 211 & 202 & 187 & 179 & 185 & 189 & 174 & 153 \\ 153 & 144 & 134 & 135 & 151 & 160 & 142 & 117 \end{pmatrix} \tag{11.121}$$

In Figure 11.14, the matrix $\hat{X}$ is shown as a picture. The left picture in the figure is the resulting distortion after reconstruction introduced by the quantization.

To change the level of compression, the quantization matrix Q is altered. Naturally, the obtained quality in the reconstructed picture is degraded if the compression level is increased. In Figure 11.15, the reconstructed picture for the case with quantization matrix Q is shown on the left. This picture has a good resemblance with the original picture shown in Figure 11.12. In total, there are 141,000 trailing zeros that are removed in the encoding process. Comparing with the total number of pixels in the original picture, which is $512 \times 512 = 262,144$, this is about 54% of the possible values.

In the right-hand picture of Figure 11.15, the quantization matrix is changed to $8Q$, which clearly lower the quality of the picture. The subblocks are clearly visible at several places in the picture and the fur is blurred, especially at the lower left and right parts. On the other hand, 89% of the values in the DCT domain are trailing zeros that can be removed, so the compression is substantially improved. In general, JPEG

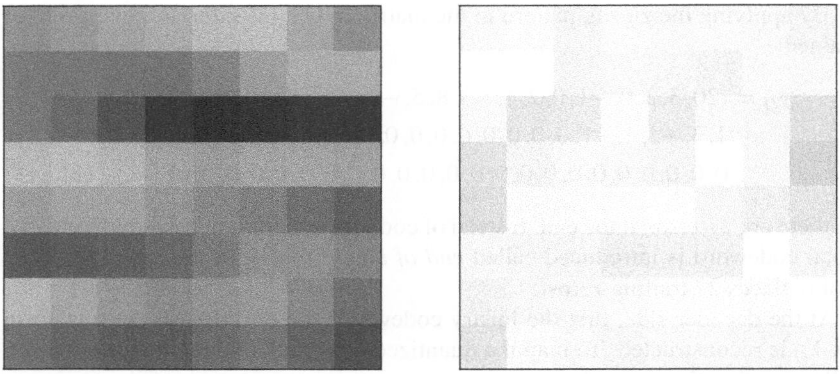

Figure 11.14 The reconstructed subblock and the quantization distortion compared to the original subblock in Figure 11.12.

can often compress pictures to use in average less that 1 bit per pixel without severe distortion in the reconstructed picture.

Image coding is an essential part of video coding, since a movie is composed of a series of images. Moreover, the images in a movie are often strongly correlated which can be utilized. In principle, objects in an image can be followed to neighboring images by motion estimation. By using the data for how objects are moved, the compression rate for the images can be increased substantially. In MPEG coding (Moving Picture Experts Group), the series of images is encoded by three types of frames. A prediction frame, P frame, can be derived from the previous P frame by motion estimation. To allow for a viewer to start the play back in the middle of the movie, and to prevent errors to propagate too long in the series, periodically there are frames containing the full image encoded with JPEG, the so-called I frames. The series of I frames and P frames constitutes the anchor frames of the movie. The

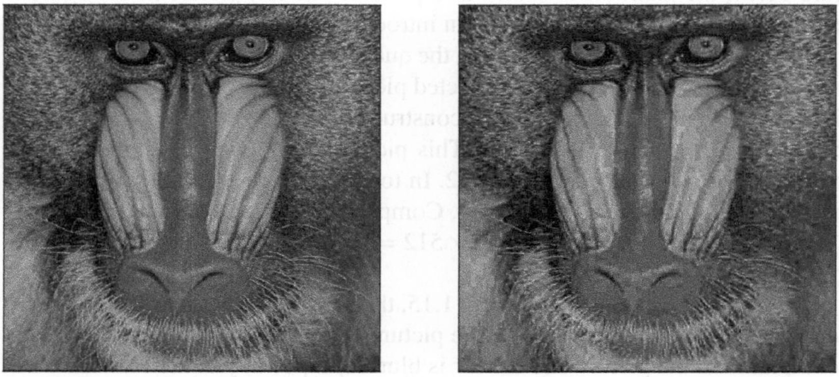

Figure 11.15 The reconstructed image after quantizing using matrix Q, left-hand picture, and $8Q$, right-hand picture.

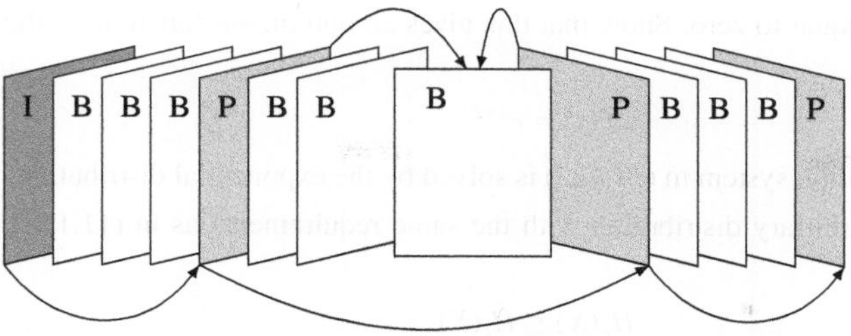

Figure 11.16 The frame structure in MPEG video coding. The arrows show how frames are predicted in the algorithm.

P frame is estimated from the most recent anchor frame. Images in between two consecutive anchor frames are encoded using a bidirectional motion estimation using both the previous and the following anchor frames. These frames are called B frames (see Figure 11.16). The B frames are not used in prediction, so errors in these will only affect this image. An error in a P frame or an I frame will propagate to the preceding frames until the next I frame.

PROBLEMS

11.1 Consider a k-ary source with statistics $P(X = x) = \frac{1}{k}$. Given the Hamming distortion

$$d(x, \hat{x}) = \begin{cases} 0, & x = \hat{x} \\ 1, & x \neq \hat{x} \end{cases}$$

and that the source and destination alphabets are the same, show that the rate-distortion function is (Hint: Fano's lemma can be useful.)

$$R(\delta) = \begin{cases} \log(k) - \delta \log(k-1) - h(\delta), & 0 \leq \delta \leq 1 - \frac{1}{k} \\ 0, & \delta \geq 1 - \frac{1}{k} \end{cases}$$

11.2 In this problem, it is shown that the exponential distribution maximizes the entropy over all one-sided distribution with fixed mean [89]. The idea is to consider a general distribution $f(x)$ such that $f(x) \geq 0$, with equality for $x < 0$, and $E[X] = 1/\lambda$. The result is obtained by maximizing the entropy $H(X) = -\int_0^\infty f(x) \ln f(x) dx$ with respect to the requirements

$$\int_0^\infty f(x)dx = 1$$
$$\int_0^\infty xf(x)dx = 1/\lambda \tag{11.122}$$

For simplicity, the derivation will be performed over the natural base in the logarithm.

(a) Set up a maximization function in f according to the Lagrange multiplier method. (Notice that you will need two side conditions). Differentiate with respect to the function f, and equal to zero. Show that this gives an optimizing function on the form

$$f(x) = e^{\alpha+\beta x}$$

and that the equation system in (11.122) is solved by the exponential distribution.

(b) Let $g(x)$ be an arbitrary distribution with the same requirements as in (11.122). Show that

$$H_g(X) \leq H_f(X)$$

i.e. that $f(x) = \lambda e^{-\lambda x}$ maximizes the entropy.

11.3 To derive the rate distortion function for the exponential distribution, $X \sim \text{Exp}(\lambda)$, i.e. $f(x) = \lambda e^{-\lambda x}$, first introduce a distortion measure and bound the mutual information. Then define a test channel from $\hat{X}$ to X to show that the bound can be obtained with equality. Use the natural logarithm in the derivations.

(a) Consider the distortion measure

$$d(x,\hat{x}) = \begin{cases} x-\hat{x}, & x \geq \hat{x} \\ \infty, & o.w. \end{cases}$$

Use the result that the exponential distribution maximizes the entropy over all one-sided distributions (see Problem 11.2) to show that $I(X;\hat{X}) \geq -\ln(\lambda\delta)$, where $E[d(x,\hat{x})] \leq \delta$.

(b) Define a test channel from $\hat{X}$ to X in order to show equality in the bound for the mutual information. To derive the distribution on X, first show that the Laplace transform of the density function for the exponential distribution, $T \sim \text{Exp}(\lambda)$, where $f_T(t) = \lambda e^{-\lambda t}$, is

$$E\left[e^{sT}\right] = \frac{1}{1+s/\lambda}$$

Derive the distribution on X and argue that the rate distortion function is

$$R(\delta) = \begin{cases} -\ln(\lambda\delta), & 0 \leq \delta \leq 1/\lambda \\ 0, & \delta > 1/\lambda \end{cases}$$

11.4 A source is given with i.i.d. symbols generated according to the Laplacian distribution,

$$f_\alpha(x) = \frac{\alpha}{2}e^{-\alpha|x|}, \quad -\infty \leq x \leq \infty$$

With the distortion measure $d(x,\hat{x}) = |x-\hat{x}|$, show that the rate-distortion function is

$$R(\delta) = \begin{cases} -\log(\alpha\delta), & 0 \leq \delta \leq \frac{1}{\alpha} \\ 0, & \delta \geq \frac{1}{\alpha} \end{cases}$$

11.5 The source variable $X \sim N(0, \sqrt{2})$ is quantized with an 3-bit linear quantizer where the quantization limits are given by the integers

$$\{-3, -2, -1, 0, 1, 2, 3\}$$

(a) If the reconstruction values are located at

$$\{-3.5, -2.5, -1.5, -0.5, 0.5, 1.5, 2.5, 3.5\}$$

derive (numerically) the average distortion.

(b) Instead of the reconstruction levels above, define optimal levels. What is the average distortion for this case?

(c) Assume that the quantizer is followed by an optimal source code, what is the required number of bits per symbol?

11.6 In Example 11.4, it is shown that the optimal reconstruction levels for a 1-bit quantization of a Gaussian variable, $X \sim N(0, \sigma)$ are $\pm\sqrt{2/\pi}\sigma$. Derive the average distortion.

11.7 A random variable X is Gaussian distributed with zero mean and unit variance, $X \sim N(0, 1)$. Let the outcome from this variable be quantized by a uniform quantizer with eight intervals of width Δ. That is, the quantization intervals are given by the limits $\{-3\Delta, -2\Delta, -\Delta, 0, \Delta, 2\Delta, 3\Delta\}$ and the reconstruction values $x_Q \in \{-\frac{7\Delta}{2}, -\frac{5\Delta}{2}, -\frac{3\Delta}{2}, -\frac{\Delta}{2}, \frac{\Delta}{2}, \frac{3\Delta}{2}, \frac{5\Delta}{2}, \frac{7\Delta}{2}\}$. Notice that since the Gaussian distribution has infinite width, i.e., $-\infty \leq x \leq \infty$, the outer most intervals also have infinite width. That means any value exceeding the upper limit 3Δ will be reconstructed to the highest reconstruction value $\frac{7\Delta}{2}$, and vice versa for the lowest values. If Δ is very small, the outer regions, the clipping regions, will have high probabilities and the quantization error will be high. On the other hand, if Δ is large, so the clipping region will have very low probability, the quantization intervals will grow and this will also give high quantization error.

(a) Sketch a plot of the squared quantization error when varying the quantization interval Δ.

(b) Find the Δ that minimizes the quantization error

Note: The solution requires numerical calculation of the integrals.

11.8 Use Jensen's inequality to show that the arithmetic mean exceeds the geometric mean, i.e. that

$$\frac{1}{n}\sum_i a_i \geq \Big(\prod_i a_i\Big)^{1/n} \tag{11.123}$$

for any set of positive real numbers $a_1, a_2, \dots, a_n$.

APPENDIX A

PROBABILITY DISTRIBUTIONS

IN THIS APPENDIX, some of the most common distributions are listed. In Section A.1 the discrete distributions are described, and in Section A.2 the continuous distributions are described. The distributions are given by the probability function, $p_X(x) = P(X = k)$, for the discrete case, and the density function, $f_X(x)$, for the continuous case. Together with the distributions, expressions for the mean and variance are listed together with the entropy function. The entropy is calculated using the natural logarithm,

$$H_e(X) = E\big[-\ln p(x)\big] \tag{A.1}$$

which gives the unit *nats* and not *bits*. To convert to base 2 use

$$H(X) = \frac{H_e(X)}{\ln 2} \tag{A.2}$$

The reason for using nats in the derivations is that with the natural base in the logarithm, integration does not need base conversion.

A.1 DISCRETE DISTRIBUTIONS

In this section, some common discrete distributions are shown. For each of them, the mean, variance, and entropy are listed.

A.1.1 Bernoulli Distribution

If the random variable X has a *Bernoulli distribution*, $X \sim \text{Be}(p)$, it describes a binary outcome, $X = 0$ or $X = 1$. The probability function is

$$p_X(k) = \begin{cases} p, & k = 1 \\ 1-p, & k = 0 \end{cases} \tag{A.3}$$

or, equivalently,

$$p_X(k) = pk + (1-p)(1-k), \qquad k = 0, 1 \tag{A.4}$$

The mean and variance are

$$E\big[X\big] = p \tag{A.5}$$

$$V\big[X\big] = p(1-p) \tag{A.6}$$

Information and Communication Theory, First Edition. Stefan Höst.

The entropy for the Bernoulli distribution is the well-known *binary entropy function*

$$H_e(X) = h_e(p) = -p\ln p - (1-p)\ln(1-p), \quad \text{[nats]} \tag{A.7}$$

A.1.2 Uniform Distribution

If the random variable X has a *uniform distribution*, $X \sim \mathrm{U}(n)$, it describes n equiprobable outcomes,

$$p_X(k) = \begin{cases} \frac{1}{n}, & k = 0, 1, \dots, n-1 \\ 0, & \text{otherwise} \end{cases} \tag{A.8}$$

(see Figure A.1). The mean, variance, and entropy are given by

$$E[X] = \frac{n-1}{2} \tag{A.9}$$

$$V[X] = \frac{n^2-1}{12} \tag{A.10}$$

$$H(X) = \ln n \quad \text{[nats]} \tag{A.11}$$

A.1.3 Geometric Distribution

If the random variable X has a *geometric distribution*, $X \sim \mathrm{Ge}(p)$, it describes the number of binary experiments until the first successful outcome. Let a successful experiment occur with probability p and an unsuccessful experiment occur with probability $1-p$. The probability for the kth experiment being the first successful is

$$p_X(k) = p(1-p)^{k-1}, \qquad k = 1, 2, \dots \tag{A.12}$$

(see Figure A.2). The mean, variance, and entropy are given by

$$E[X] = \frac{1}{p} \tag{A.13}$$

$$V[X] = \frac{1-p}{p^2} \tag{A.14}$$

$$H_e(X) = \frac{h_e(p)}{p} \quad \text{[nats]} \tag{A.15}$$

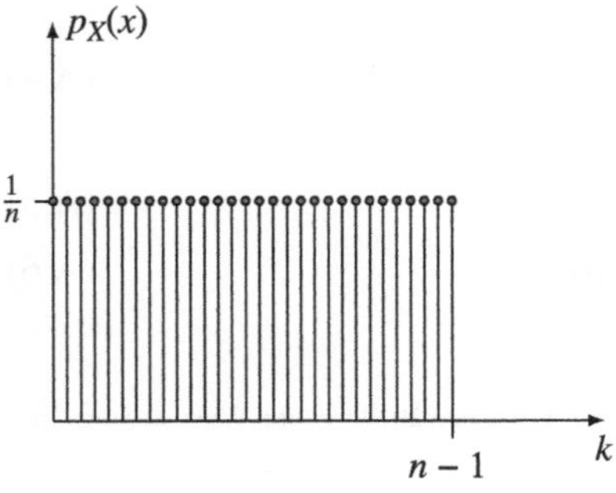

Figure A.1 Probability function of the uniform distribution, $X \sim \mathrm{U}(n)$.

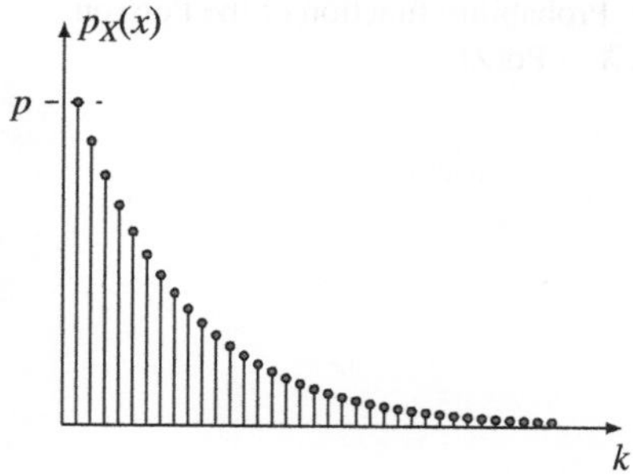

Figure A.2 Probability function of the geometric distribution, $X \sim \text{Ge}(p)$.

A.1.4 Binomial Distribution

If the random variable X has a *binomial distribution*, $X \sim \text{Bin}(n, p)$, it describes the probability of success in exactly k binary experiments out of n possible. Let a successful experiment occur with probability p and an unsuccessful experiment occur with probability $1 - p$. Then the probability function is

$$p_X(k) = \binom{n}{k} p^k (1-p)^{n-k}, \qquad k = 0, 1, \ldots n \tag{A.16}$$

(see Figure A.3). The mean, variance, and entropy are given by

$$E\big[X\big] = np \tag{A.17}$$

$$V\big[X\big] = np(1-p) \tag{A.18}$$

$$H_e(X) = \frac{1}{2}\ln\Big(2\pi enp(1-p)\Big) + O\left(\tfrac{1}{n}\right) \quad \text{[nats]} \tag{A.19}$$

A.1.5 Poisson Distribution

If the random variable X has a *Poisson distribution*, $X \sim \text{Po}(\lambda)$, it describes the probability for a number of events during a time interval, if the intensity of events is λ, and the events occur independent of each other. The probability function is

$$p_X(k) = \frac{\lambda^k e^{-\lambda}}{k!}, \qquad k = 0, 1, 2, \ldots \tag{A.20}$$

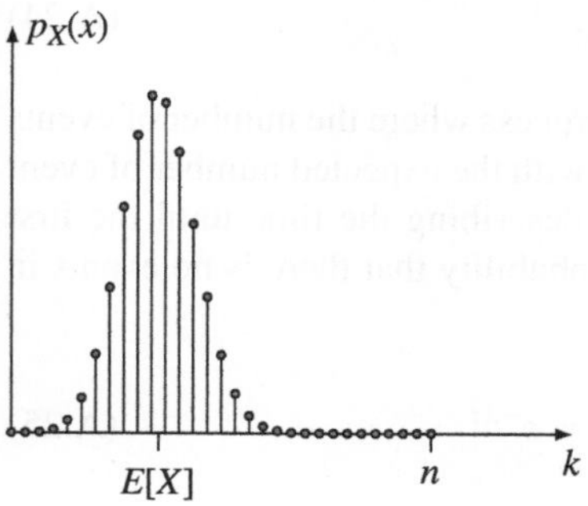

Figure A.3 Probability function of the binomial distribution, $X \sim \text{Bin}(n, p)$.

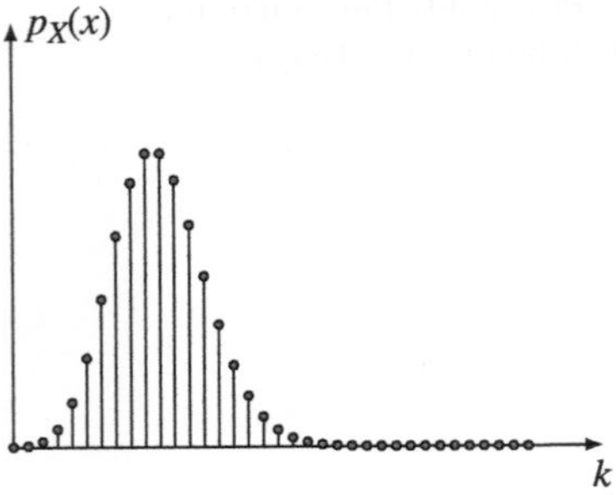

Figure A.4 Probability function of the Poisson distribution, $X \sim \mathrm{Po}(\lambda)$.

(see Figure A.4). The mean, variance, and entropy are given by

$$E[X] = \lambda \tag{A.21}$$

$$V[X] = \lambda \tag{A.22}$$

$$H_e(X) = \lambda \ln \frac{e}{\lambda} + e^{-\lambda} \sum_{k=0}^{\infty} \frac{\lambda^k \ln k!}{k!} \quad \text{[nats]} \tag{A.23}$$

The interpretation of the Poisson distribution described above can be seen from the binomial distribution. View a time interval as n time slots in which an event can occur with probability p. Then the probability of k events during this time follows from the binomial distribution. With $\lambda = np$ being the mean number of events and letting $n \to \infty$, and consequently $p = \frac{\lambda}{n} \to 0$,

$$\begin{aligned}
p_X(k) &= \lim_{n\to\infty} \binom{n}{k} \left(\frac{\lambda}{n}\right)^k \left(1 - \frac{\lambda}{n}\right)^{n-k} \\
&= \lim_{n\to\infty} \frac{(n-k+1)\cdots(n-1)n}{k!} \left(\frac{\frac{\lambda}{n}}{1-\frac{\lambda}{n}}\right)^k \left(1 - \frac{\lambda}{n}\right)^n \\
&= \lim_{n\to\infty} \underbrace{\frac{(n-k+1)\cdots(n-1)n}{n^k}}_{\to 1} \frac{\lambda^k}{k!} \underbrace{\left(\frac{1}{1-\frac{\lambda}{n}}\right)^k}_{\to 1} \underbrace{\left(1 + \frac{-\lambda}{n}\right)^n}_{\to e^{-\lambda}} \\
&= \frac{\lambda^k e^{-\lambda}}{k!}
\end{aligned} \tag{A.24}$$

A Poisson process with intensity λ is a stochastic process where the number of events in a time interval of length τ is Poisson distributed with the expected number of event $\lambda\tau$. For this process, let T be a random variable describing the time until the first event. The probability for T, to exceed τ is the probability that there is no events in the interval $[0, \tau]$,

$$P(T > \tau) = \frac{(\lambda\tau)^0 e^{-\lambda\tau}}{0!} = e^{-\lambda\tau} \tag{A.25}$$

The density function of T can be obtained from the derivative of $P(T < \tau) = 1 - e^{-\lambda\tau}$ as

$$f_T(\tau) = \lambda e^{-\lambda\tau} \tag{A.26}$$

which is recognized as the density function of an exponential distribution, Exp(λ). Hence, the time between two events in a Poisson process is exponential distributed.

A.2 CONTINUOUS DISTRIBUTIONS

In this section, several commonly used continuous distributions are listed. Sometimes the density function is given for a normalized case. The entropy for the nonnormalized case can be found from the relations

$$H_e(aX + c) = H_e(X) + \ln a \tag{A.27}$$

where c is a constant and a a positive constant. There are also a couple of other functions that show up in the description. The *gamma function* can be defined in some different ways, but the most common is

$$\Gamma(z) = \int_0^\infty e^{-t} t^{z-1} dt \tag{A.28}$$

It is a generalization of the factorial function to real values (actually, it can also be defined for complex values). If k is a positive integer then

$$\Gamma(k) = (k-1)! \tag{A.29}$$

The *digamma function* is defined as

$$\psi(x) = \frac{\partial}{\partial x} \ln \Gamma(x) = \frac{\Gamma'(x)}{\Gamma(x)} \tag{A.30}$$

In some cases, the *Euler–Mascheroni constant* shows up in the formulas. It is often defined by the limit value

$$\gamma = \lim_{n\to\infty} \sum_{k=1}^{n} \frac{1}{k} - \ln n \tag{A.31}$$

and a numerical value of is $\gamma = 0.5772156649015$. It is related to the gamma and digamma functions through

$$\psi(x) = -\frac{1}{x} - \gamma - \sum_{n=1}^{\infty}\left(\frac{1}{n+x} - \frac{1}{n}\right) \tag{A.32}$$

If k is an integer,

$$\psi(k) = -\gamma + \sum_{n=1}^{k-1} \frac{1}{n} \tag{A.33}$$

and for the case when $k = 1$,

$$\gamma = -\psi(1) = -\Gamma'(1) \tag{A.34}$$

Finally, the *beta function* is defined as

$$B(p, q) = \frac{\Gamma(p)\Gamma(q)}{\Gamma(p + q)} \tag{A.35}$$

Similar to the gamma function being a generalization of the factorial function, the beta function is a generalization of the binomial function. If n and k are positive integers

$$B(n - k, k) = \frac{1}{(n - k)\binom{n-1}{k-1}} \tag{A.36}$$

In the following, several continuous distributions are given by their density function, together with the corresponding mean, variance, and entropy functions. As in the previous section, the entropy is given for the natural base, i.e. in the unit nats. Many of the expressions for the entropies can be found in [63].

A.2.1 Uniform (Rectangular) Distribution

If the random variable X has a *uniform distribution*, $X \sim \mathrm{U}(a)$, the density function is

$$f_X(x) = \begin{cases} \frac{1}{a}, & 0 \le x \le a, \quad a > 0 \\ 0, & \text{otherwise} \end{cases} \tag{A.37}$$

(see Figure A.5). The mean, variance, and entropy are given by

$$E[X] = \frac{a}{2} \tag{A.38}$$

$$V[X] = \frac{a^2}{12} \tag{A.39}$$

$$H_e(X) = \ln a \quad \text{[nats]} \tag{A.40}$$

Since the entropy is not dependent on a shift in x, the described distribution starts at 0. In a more general case, there can be one starting point and one ending point, $\mathrm{U}(a, b)$.

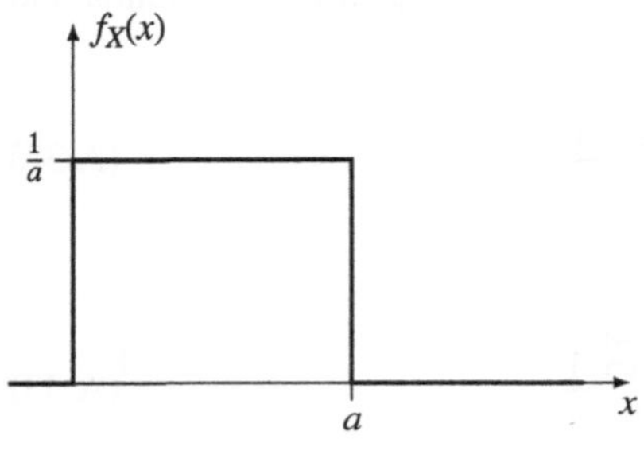

Figure A.5 Density function of the uniform distribution, $X \sim \mathrm{U}(a)$.

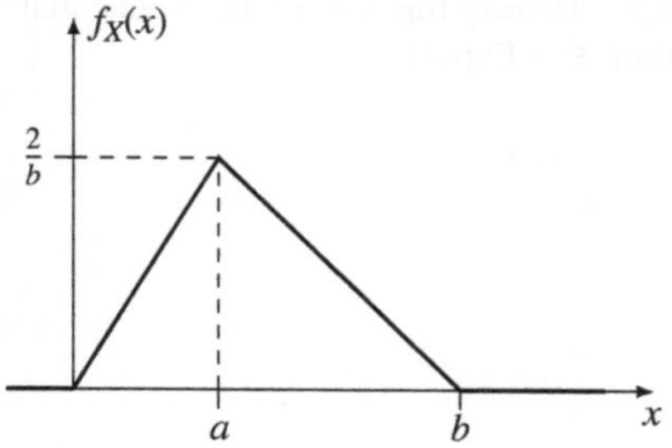

Figure A.6 Density function of the triangular distribution, $X \sim \mathrm{Tri}(a, b)$.

A.2.2 Triangular Distribution

If the random variable X has a *triangular distribution*, $X \sim \mathrm{Tri}(a, b)$, the density function is

$$f_X(x) = \begin{cases} \frac{2}{ab}x, & 0 \leq x \leq a \\ \frac{2}{b(b-a)}(-x+b), & a \leq x \leq b \\ 0, & \text{otherwise} \end{cases} \tag{A.41}$$

(see Figure A.6). The mean, variance, and entropy are given by

$$E[X] = \frac{a+b}{3} \tag{A.42}$$

$$V[X] = \frac{a^2+b^2-ab}{18} \tag{A.43}$$

$$H_e(X) = \frac{1}{2} + \ln\frac{b}{2} \quad \text{[nats]} \tag{A.44}$$

As for the uniform distribution, the start of the density function is chosen to $x = 0$. In a more general description, there is a starting point, highest point, and end point.

A.2.3 Exponential Distribution

If the random variable X has an *exponential distribution*, $X \sim \mathrm{Exp}(\lambda)$, the density function is

$$f_X(x) = \lambda e^{-\lambda x}, \quad x \geq 0 \tag{A.45}$$

(see Figure A.7). The mean, variance, and entropy are given by

$$E[X] = \frac{1}{\lambda} \tag{A.46}$$

$$V[X] = \frac{1}{\lambda^2} \tag{A.47}$$

$$H_e(X) = 1 - \ln\lambda \quad \text{[nats]} \tag{A.48}$$

The exponential distribution shows the remaining lifetime in a system where there is no memory, i.e. the remaining lifetime is independent of the past lifetime.

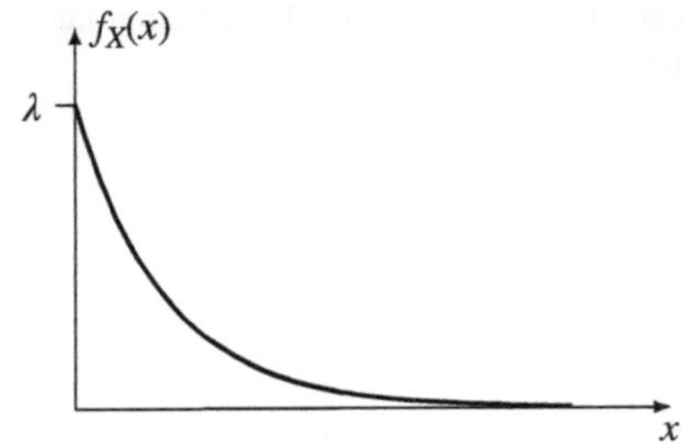

Figure A.7 Density function of the exponential distribution, $X \sim \text{Exp}(\lambda)$.

A.2.4 Normal Distribution

If the random variable X has a *normal distribution*, or *Gaussian distribution*, $X \sim N(\mu, \sigma)$, the density function is

$$f_X(x) = \frac{1}{\sqrt{2\pi\sigma^2}} e^{-(x-\mu)^2/2\sigma^2} \qquad -\infty \leq x \leq \infty \tag{A.49}$$

(see Figure A.8). The mean, variance, and entropy are given by

$$E[X] = \mu \tag{A.50}$$

$$V[X] = \sigma^2 \tag{A.51}$$

$$H_e(X) = \frac{1}{2}\ln\left(2\pi e\sigma^2\right) \quad \text{[nats]} \tag{A.52}$$

The normal distribution is very important in both probability theory and its applications. The sum of two normal distributed variables is again normal distributed,

$$X_1 \sim N(\mu_1, \sigma_1),\ X_2 \sim N(\mu_2, \sigma_2) \Rightarrow X_1 + X_2 \sim N\left(\mu_1 + \mu_2, \sqrt{\sigma_1 + \sigma_2}\right) \tag{A.53}$$

One of its most well-known use is from the *central limit theorem* stating that for n identical and independently distributed (i.i.d.) random variables, X_i, $i = 1, 2, \dots, n$, with mean $E[X_i] = \mu$ and variance $V[X_i] = \sigma^2$, the arithmetic mean

$$X = \frac{1}{n}\sum_{i=1}^{n} X_i \tag{A.54}$$

will be normal distributed with $N(\mu, \frac{\sigma}{\sqrt{n}})$ as n goes to infinity.

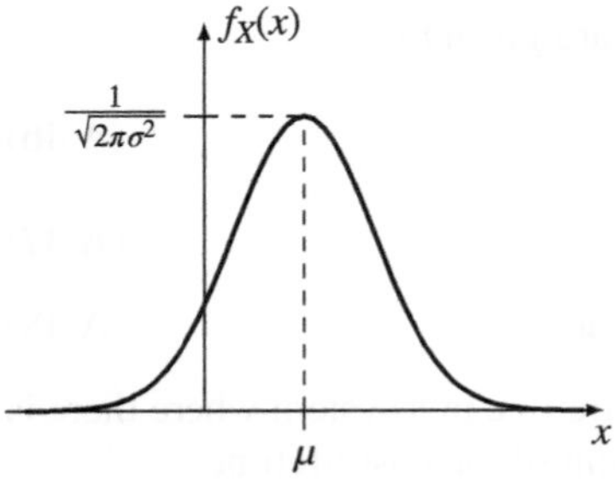

Figure A.8 Density function of the normal distribution, $X \sim N(\mu, \sigma)$.

Often, a normalized Gaussian distribution is used, $Z \sim N(0,1)$. Then the density function, mean, variance, and entropy are given by

$$f_Z(z) = \frac{1}{\sqrt{2\pi}} e^{-z^2/2} \tag{A.55}$$

$$E\big[Z\big] = 0 \tag{A.56}$$

$$V\big[Z\big] = 1 \tag{A.57}$$

$$H_e(Z) = \frac{1}{2}\ln(2\pi e) \quad \text{[nats]} \tag{A.58}$$

In the n-dimensional case, a vector $\boldsymbol{X} = (X_1, \dots, X_n)$ is n-dimensional Gaussian distributed, $\boldsymbol{X} \sim N(\boldsymbol{\mu}, \Sigma)$, then the density function is

$$f_{\boldsymbol{X}}(\boldsymbol{x}) = \frac{1}{\sqrt{(2\pi)^k |\Sigma|}} e^{-\frac{1}{2}(\boldsymbol{x}-\boldsymbol{\mu})\Sigma^{-1}(\boldsymbol{x}-\boldsymbol{\mu})^T} \tag{A.59}$$

where the mean vector is

$$E\big[\boldsymbol{X}\big] = \boldsymbol{\mu} \tag{A.60}$$

and covariance matrix

$$\Sigma = E\big[(\boldsymbol{X}-\boldsymbol{\mu})^T(\boldsymbol{X}-\boldsymbol{\mu})\big] = \Big(E\big[(X_i-\mu_i)(X_j-\mu_j)\big]\Big)_{i,j=1,\dots,n} \tag{A.61}$$

The entropy is

$$H_e(\boldsymbol{X}) = \frac{1}{2}\ln\big((2\pi e)^n|\Sigma|\big) \quad \text{[nats]} \tag{A.62}$$

The above vector representation is assumed to be real valued. It can also be generalized to the complex vector, but that lies outside the scope of this text.

A.2.5 Truncated Normal Distribution

If the random variable X has a *truncated normal distribution*, $X \sim TN(a, \sigma)$, the density function is a truncated version of the normal distribution $N(0, \sigma)$. The interval support of the random variable is $-a \leq X \leq a$, and the corresponding density function is given by

$$f_X(x) = \frac{f_G(x)}{1 - 2Q(\frac{a}{\sigma})}, \quad -a \leq x \leq a \tag{A.63}$$

where

$$f_G(x) = \frac{1}{\sqrt{2\pi\sigma^2}} e^{-x^2/2\sigma^2} \tag{A.64}$$

is the density function of the original normal distribution with zero mean and variance σ^2 (see Figure A.9). The function $Q(z)$ is the one-sided normalized error function

$$Q(z) = \int_z^\infty \frac{1}{\sqrt{2\pi}} e^{-t^2/2} dt \tag{A.65}$$

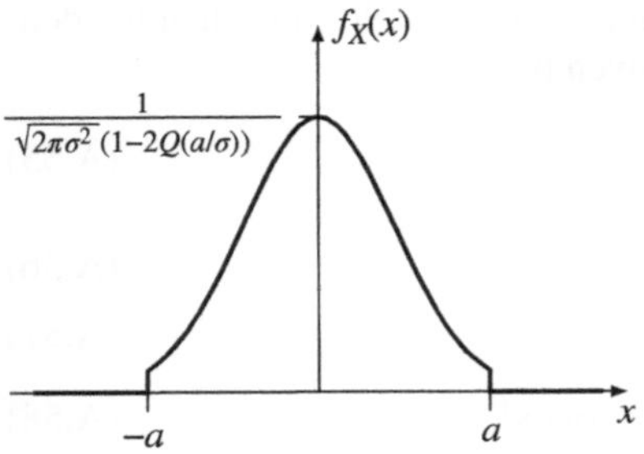

Figure A.9 Density function of the truncated normal distribution, $X \sim \text{TN}(a, \sigma)$.

Sometimes, the two-sided error function is used instead, i.e.,

$$\text{erfc}(\zeta) = \int_\zeta^\infty \frac{2}{\sqrt{\pi}} e^{-t^2} dt = 2Q(\sqrt{2}\zeta)$$

or, alternatively, the erf-function, defined as $\text{erf}(\zeta) = 1 - \text{erfc}(\zeta)$. Then the denominator can be expressed as

$$1 - 2Q(\tfrac{a}{\sigma}) = 1 - \text{erfc}(\tfrac{a}{\sqrt{2}\sigma}) = \text{erf}(\tfrac{a}{\sqrt{2}\sigma}) \tag{A.66}$$

The mean, variance, and entropy for the truncated normal distribution are given by

$$E\big[X\big] = 0 \tag{A.67}$$

$$V\big[X\big] = \sigma^2 - \frac{a\sigma\sqrt{2}}{\sqrt{\pi}\big(1 - 2Q(\frac{a}{\sigma})\big)} e^{-a^2/2\sigma^2} \tag{A.68}$$

$$H_e(X) = \frac{1}{2}\ln\Big(2\pi e\sigma^2\big(1 - Q(\tfrac{a}{\sigma})\big)^2\Big) - \frac{a}{\sqrt{2\pi}\sigma\big(1 - Q(\frac{a}{\sigma})\big)} e^{-a^2/2\sigma^2} \quad \text{[nats]} \tag{A.69}$$

A.2.6 log-Normal Distribution

If the random variable X has an *log-normal distribution*, $X \sim \text{logN}(\mu, \sigma)$, the density function is

$$f_X(x) = \frac{1}{x\sqrt{2\pi\sigma^2}} e^{-(\ln x - \mu)^2/2\sigma^2} \quad 0 \le x \le \infty \tag{A.70}$$

(see Figure A.10). The mean, variance, and entropy are given by

$$E\big[X\big] = e^{\mu + \frac{\sigma^2}{2}} \tag{A.71}$$

$$V\big[X\big] = e^{2\mu+\sigma^2}\big(e^{\sigma^2} - 1\big) \tag{A.72}$$

$$H_e(X) = \frac{1}{2}\ln\big(2\pi e\sigma^2\big) + \mu \quad \text{[nats]} \tag{A.73}$$

If $X \sim \text{logN}(\mu, \sigma)$, then its logarithm is normal, $Y = \ln X \sim \text{N}(\mu, \sigma)$.

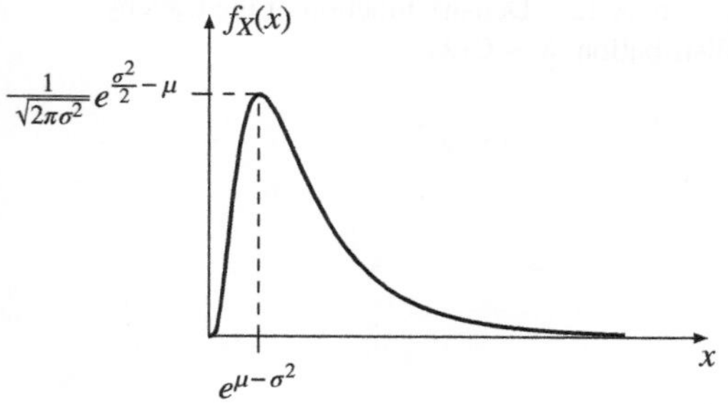

Figure A.10 Density function of the log-normal distribution, $X \sim \log N(\mu, \sigma)$.

A.2.7 Rayleigh Distribution

If the random variable X has a *Rayleigh distribution*, $X \sim R(\sigma)$, the density function is

$$f_X(x) = \frac{x}{\sigma^2} e^{-x^2/2\sigma^2}, \quad x \geq 0 \tag{A.74}$$

(see Figure A.11). The mean, variance, and entropyare given by

$$E[X] = \sigma\sqrt{\frac{\pi}{2}} \tag{A.75}$$

$$V[X] = 2\sigma^2\left(1 - \frac{\pi}{4}\right) \tag{A.76}$$

$$H_e(X) = 1 + \ln\frac{\sigma}{\sqrt{2}} + \frac{\gamma}{2} \quad \text{[nats]} \tag{A.77}$$

If X_1 and X_2 are two independent random variables distributed according to $N(0, \sigma)$, then $\sqrt{X_1^2 + X_2^2}$ is Rayleigh distributed, $R(\sigma)$. This distribution is often used in channel modeling for wireless channels.

A.2.8 Cauchy Distribution

If the random variable X has a *Cauchy distribution*, $X \sim C(\lambda)$, the density function is

$$f_X(x) = \frac{\lambda}{\pi(\lambda^2 + x^2)}, \quad -\infty \leq x \leq \infty \tag{A.78}$$

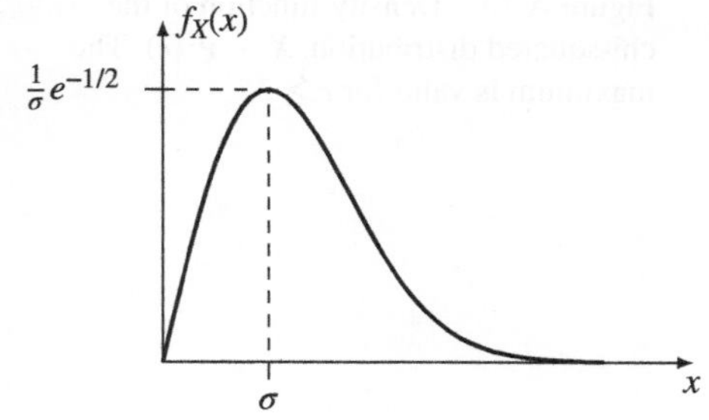

Figure A.11 Density function of the Rayleigh distribution, $X \sim R(\sigma)$.

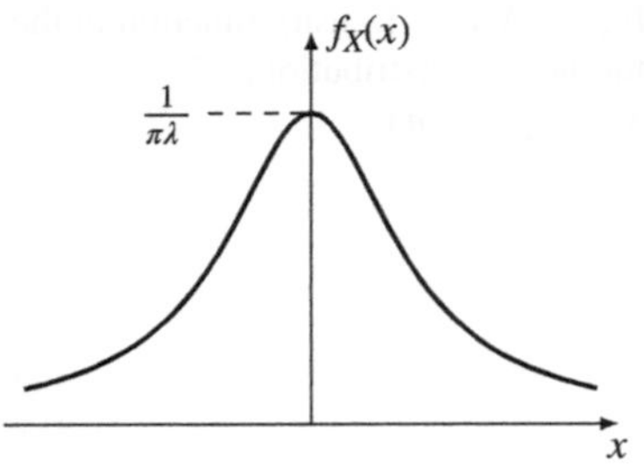

Figure A.12 Density function of the Cauchy distribution, $X \sim \mathrm{C}(\lambda)$.

(see Figure A.12). The mean variance does not exist for this distribution, but the entropy is given by

$$H_e(X) = \ln(4\pi\lambda) \quad \text{[nats]} \tag{A.79}$$

If the angle ϕ is a random variable that is uniformly distributed in the interval $-\frac{\pi}{2} \leq \phi \leq \frac{\pi}{2}$, then $X = \lambda \tan\phi$ is Cauchy distributed, $\mathrm{C}(\lambda)$.

A.2.9 Chi-Squared Distribution

If the random variable X has a *chi-squared distribution*, $X \sim \chi^2(r)$, the density function is

$$f_X(x) = \frac{1}{2^{r/2}\Gamma\left(\frac{r}{2}\right)} x^{\frac{r}{2}-1} e^{-\frac{x}{2}}, \quad x \geq 0 \tag{A.80}$$

(see Figure A.13). The mean, variance, and entropy are given by

$$E\big[X\big] = r \tag{A.81}$$

$$V\big[X\big] = 2r \tag{A.82}$$

$$H_e(X) = \ln\Big(2\Gamma\left(\tfrac{r}{2}\right)\Big) + \Big(1 - \frac{r}{2}\Big)\psi\left(\tfrac{r}{2}\right) + \frac{r}{2} \quad \text{[nats]} \tag{A.83}$$

Let $X_1, \dots, X_r$ be independent random variables, each distributed according to $\mathrm{N}(0, 1)$, then $X = X_1^2 + \cdots + X_r^2$ is chi-squared distributed, $\chi^2(r)$. This also gives that if $Y \sim \chi^2(2)$, then $\sqrt{Y} \sim \mathrm{R}(1)$.

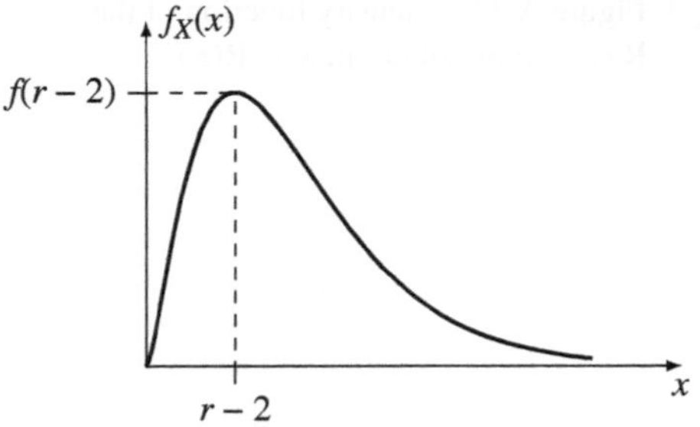

Figure A.13 Density function of the chi-squared distribution, $X \sim \chi^2(r)$. The maximum is valid for $r \geq 3$.

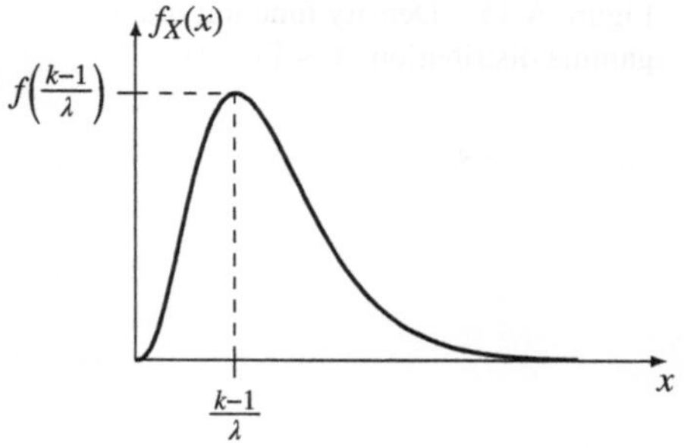

Figure A.14 Density function of the Erlang distribution, $X \sim \mathrm{Er}(k, \lambda)$.

A.2.10 Erlang Distribution

If the random variable X has an *Erlang distribution*, $X \sim \mathrm{Er}(k, \lambda)$, the density function is

$$f_X(x) = \frac{\lambda^k}{(k-1)!} x^{k-1} e^{-\lambda x} \quad x \geq 0 \tag{A.84}$$

(see Figure A.14). The mean, variance, and entropy are given by

$$E[X] = \frac{k}{\lambda} \tag{A.85}$$

$$V[X] = \frac{k}{\lambda^2} \tag{A.86}$$

$$H_e(X) = \ln \frac{(k-1)!}{\lambda} - (k-1)\psi(k) + k \quad \text{[nats]} \tag{A.87}$$

The Erlang distribution is often used to model the waiting time in queuing theory. If $X_1, \dots, X_k$ are independent random variables that are exponentially distributed, $\mathrm{Exp}(\lambda)$, then the sum $X_1 + \cdots + X_k$ is Erlang-distributed $\mathrm{Er}(k, \lambda)$.

A.2.11 Gamma Distribution

If the random variable X has a *gamma distribution*, $X \sim \Gamma(n, \lambda)$, the density function is

$$f_X(x) = \frac{\lambda^n}{\Gamma(n)} x^{n-1} e^{-\lambda x} \quad x \geq 0 \tag{A.88}$$

(see Figure A.15). The mean, variance, and entropy are given by

$$E[X] = \frac{n}{\lambda} \tag{A.89}$$

$$V[X] = \frac{n}{\lambda^2} \tag{A.90}$$

$$H_e(X) = \ln \frac{\Gamma(n)}{\lambda} - (n-1)\psi(n) + n \quad \text{[nats]} \tag{A.91}$$

In the Erlang distribution, k must be a positive integer. The gamma distribution is a generalisation of positive real numbers.

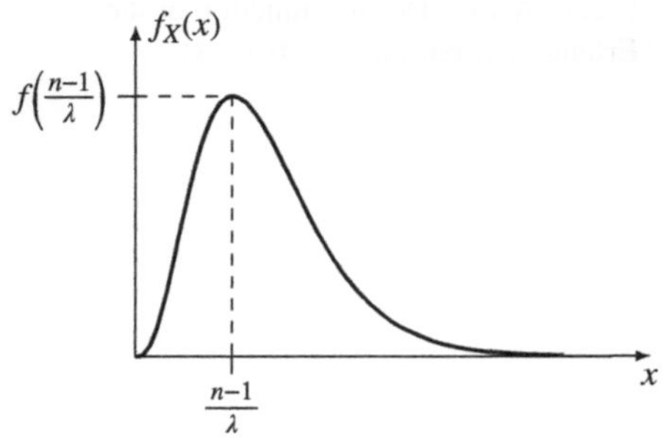

Figure A.15 Density function of the gamma distribution, $X \sim \Gamma(n, \lambda)$.

A.2.12 Beta Distribution

If the random variable X has a *beta distribution*, $X \sim \beta(p, q)$, the density function is

$$f_X(x) = \frac{\Gamma(p+q)}{\Gamma(p)\Gamma(q)} x^{p-1}(1-x)^{q-1}, \quad 0 \le x \le 1,\ p > 0,\ q > 0 \tag{A.92}$$

(see Figure A.16). The mean, variance, and entropy are given by

$$E\big[X\big] = \frac{p}{p+q} \tag{A.93}$$

$$V\big[X\big] = \frac{pq}{(p+q)^2(p+q+1)} \tag{A.94}$$

$$\begin{aligned} H_e(X) = \ln \frac{\Gamma(p)\Gamma(q)}{\Gamma(p+q)} \\ -(p-1)\big(\psi(p) - \psi(p+q)\big) \\ -(q-1)\big(\psi(q) - \psi(p+q)\big) \quad \text{[nats]} \end{aligned} \tag{A.95}$$

The beta distribution is a generalization of the binomial distribution for noninteger values.

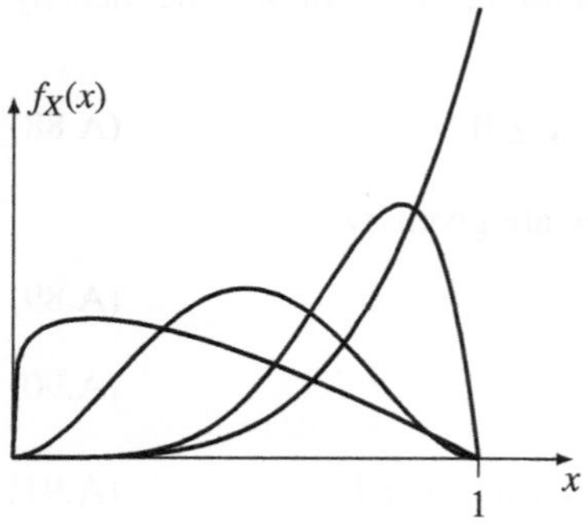

Figure A.16 Density function of the beta distribution, $X \sim \beta(p, q)$. In the figure, four different settings of (p, q) are shown; $(1.2, 2)$, $(3, 3)$, $(6, 2)$, and $(5, 1)$.

APPENDIX **B**

SAMPLING THEOREM

THE SAMPLING THEOREM is a vital part of the derivation of the Gaussian channel, so for completeness a proof of it is included as appendix. It is a vital part when understanding the relation between the analog and digital domains in an electronic system, and is often described in a first course in signal processing, like [85]. Normally, Nyquist gets the credit of the sampling theorem, but in Nyquist's 1928 paper [3] sampling is not mentioned. Instead, he studied the maximum distinguishable pulses in a signal of a certain bandwidth and time duration, which can be viewed as the dual problem to sampling. There is no single source to refer to as the origin of the sampling theorem as normally stated today. Often Shannon is mentioned as one of them [64], but others are also mentioned, such as Kotelnikov [65] and Whittaker [66]. Even so, the theorem is often referred to as the *Nyquist sampling theorem*, or sometimes *Nyquist–Shannon sampling theorem.*

B.1 THE SAMPLING THEOREM

In general, a signal is analog, meaning it is both continuous in time and in amplitude. In order to process the signal in, e.g., a computer, it must first go through an analog-to-digital converter, where both the time axis and the amplitude are discretized. The process of discretizing the amplitude is called quantization and is further described in Chapter 11. Sampling is then the process of discretizing the time, so it can be represented as a vector.

In Figure B.1, a schematic view of sampling is shown. The signal $x(t)$ enters the sampler, which outputs the sampled sequence x_n. Then the samples are the amplitude values sorted as a vector, where

$$x_n = x(nT_s), \quad \forall n \in \mathbb{Z} \tag{B.1}$$

Them T_s is the time between samples, and correspondingly the sampling frequency is $F_s = 1/T_s$. This process is of course only useful if there is a way to get back to the analog signal (digital-to-analog conversion). This reverse process is the reconstruction process, and the sampling theorem sets a limit on the sampling frequency to be able to do perfect reconstruction by using the vector x_n and F_s.

Theorem B.1 (Shannon–Nyquist sampling theorem) Let $x(t)$ be a band-limited signal with the maximum frequency content at $f_{\max} \leq W$. If the signal is sampled

Information and Communication Theory, First Edition. Stefan Höst.

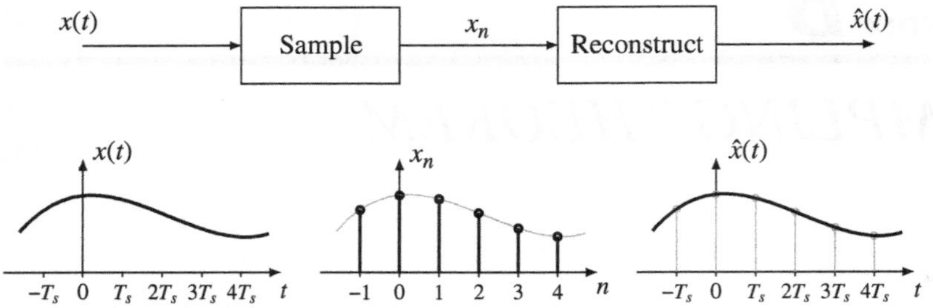

Figure B.1 The process of sampling and reconstruction.

with $F_s \geq 2W$ samples per second to form the sequence $x_n = x(nT_s)$, $T_s = \frac{1}{F_s}$, it can be reconstructed with

$$x(t) = \sum_{n=-\infty}^{\infty} x_n \operatorname{sinc}(t - nT_s) \tag{B.2}$$

where

$$\operatorname{sinc}(t) = \frac{\sin(\pi F_s t)}{\pi F_s t} \tag{B.3}$$

□

The sampling frequency $F_s = 2W$ is often called the *Nyquist rate*.

The proof of this theorem can be derived in some different ways, but most often it involves a detour from time domain to the frequency domain and back. In Shannon's work, it was shown using the Fourier series expansion, and although short and concise, the version here follows the more traditional via Fourier transforms. In this way, the steps in the sampling process becomes more clear. To start, the analog signal is assumed to be band-limited, meaning its frequency contents are zero outside $-W \leq f \leq W$. In other words, the Fourier transform of $x(t)$ has the property $X(f) = 0$, if $|f| > W$.

From a practical aspects, taking a sample value at a certain time instant as in (B.1) is not straightforward. Instead the signal is averaged over a short-time duration, Δ, i.e. an estimate of the sample value is given by

$$x_{n,\Delta} = \frac{1}{\Delta} \int_{nT_s}^{nT_s+\Delta} x(t)dt \tag{B.4}$$

In Figure B.2, the gray areas correspond to the sample intervals. In the sample interval, the argument can be viewed as the multiplication between the function $x(t)$ and a rectangular pulse of width Δ and unit area,

$$\Pi_\Delta(t) = \begin{cases} 1/\Delta, & 0 \leq t \leq \Delta \\ 0, & \text{otherwise} \end{cases} \tag{B.5}$$

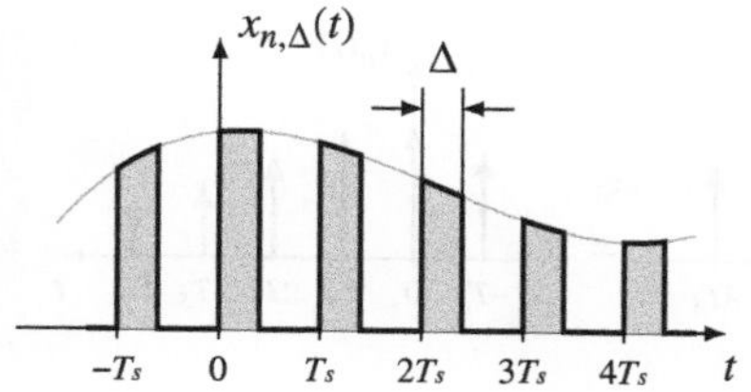

Figure B.2 Sampling as average over a time interval Δ.

Putting the pulses together in a sequence gives the pulse train in

$$P_\Delta(t) = \sum_n \Pi_\Delta(t - nT_s) \tag{B.6}$$

which is shown in Figure B.3. Then the gray-shaded function in Figure B.2 is given by $x_{n,\Delta}(t) = x(t)P_\Delta(t)$.

Naturally, the smaller interval Δ, the better the accuracy of the sample value. From a mathematical point, letting the interval go to zero means

$$\lim_{\Delta\to 0} \Pi_\Delta(t) = \delta(t) \tag{B.7}$$

and, thus, the sample function becomes a series of Dirac pulses,

$$\lim_{\Delta\to 0} P_\Delta(t) = \lim_{\Delta\to 0} \sum_n \Pi_\Delta(t - nTs) = \sum_n \delta(t - nTs) \tag{B.8}$$

Such infinite pulse train of Dirac pulses is a well-studied function, and sometimes it goes under the name the *Shah function*, from the Cyrillic letter *Shah*, Ш.

Definition B.1 The Shah function, $Ш(t)$, is an infinite series of Dirac functions at unit distance,

$$Ш(t) = \sum_k \delta(t - k) \tag{B.9}$$

□

To get the same function as in (B.8), the function must be scaled in time with the sampling time T_s. To preserve the unit area of the pulses, their amplitude must also be scaled. The corresponding function becomes

$$\frac{1}{T_s}Ш\left(\frac{t}{T_s}\right) = \sum_n \delta(t - nT_s) \tag{B.10}$$

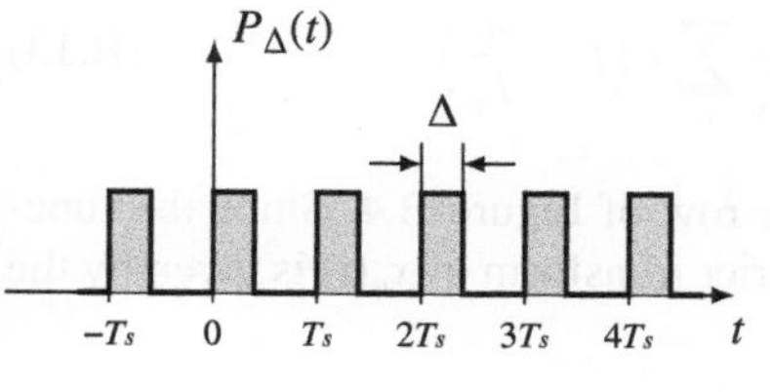

Figure B.3 The sampling function as block pulses.

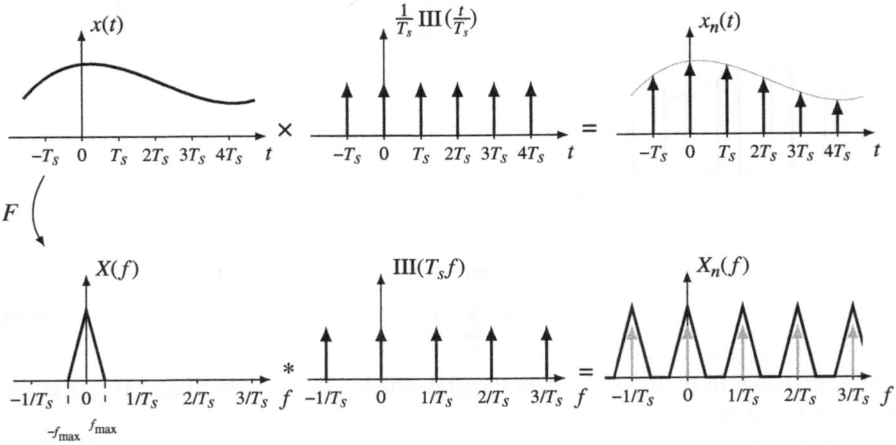

Figure B.4 Sampling viewed in time domain and frequency domain.

Thus, from a mathematical point of view, the sampled function can be seen as the sequence of scaled Dirac pulses in

$$x_n(t) = x(t) \cdot \frac{1}{T_s}\text{Ш}\left(\frac{t}{T_s}\right) = \sum_n x_n \delta(t - nT_s) \tag{B.11}$$

As seen in the upper row of Figure B.4, multiplication with the Shah function corresponds to the sampling process.

It is further known that the sampled function $x(t)$ is band-limited, and its Fourier transform can schematically be viewed as the left function in the lower row of Figure B.4. From, e.g., [7], the Fourier transform of the Shah function can be found.

Theorem B.2 The Shah function is its own Fourier transform,

$$\text{Ш}(t) \xrightarrow{\mathcal{F}} \text{Ш}(f) \tag{B.12}$$

□

With the scaling of the Shah function that means

$$\frac{1}{T_s}\text{Ш}\left(\frac{t}{T_s}\right) \xrightarrow{\mathcal{F}} \text{Ш}(T_s f) = \frac{1}{T_s}\sum_n \delta\left(f - \frac{n}{T_s}\right) \tag{B.13}$$

which is shown in the middle plot of the lower row of Figure B.4. Since the functions are multiplied in the time domain, the Fourier transform of $x_n(t)$ is given by the following convolution:

$$X_n(f) = X(f) * \text{Ш}(T_s f)$$
$$= \frac{1}{T_s} \sum_n X_n(f) * \delta\left(f - \frac{n}{T_s}\right)$$
$$= \frac{1}{T_s} \sum_n X_n\left(f - \frac{n}{T_s}\right) \quad \text{(B.14)}$$

which is shown as the right-hand plot in the lower row of Figure B.4. It is seen that sampling in the time domain corresponds to periodic repetition of the signal in the frequency domain, with period $F_s = 1/T_s$.

Since $F_s \geq f_{\max}$, the different pulses in the frequency domain do not overlap. This fact facilitates the reconstruction substantially. By filtering the pulse centered around $f = 0$ and deleting the others, the original signal will be reconstructed. As shown in Figure B.5, the frequency function for the sampled function, $X_n(f)$, can be multiplied by an ideal low-pass filter $H(f)$ to reconstruct the frequency contents for the original signal, here denoted by $\hat{X}(f)$. The low-pass filter is given by

$$H(f) = \begin{cases} 1, & |f| \leq \frac{F_s}{2} \\ 0, & \text{otherwise} \end{cases} \quad \text{(B.15)}$$

Effectively, this means that the reconstruction takes whatever is in the interval $-\frac{F_s}{2} \leq f \leq \frac{F_s}{2}$ and use as an estimate of $X(f)$. As long as the pulses are separated in the frequency domain, this method will give perfect reconstruction. However, if the sampling frequency is chosen too low, i.e. $F_s < f_{\max}$, the pulses will overlap and sum together. Then the reconstructed pulse is not the same as the original. This distortion is called *aliasing*.

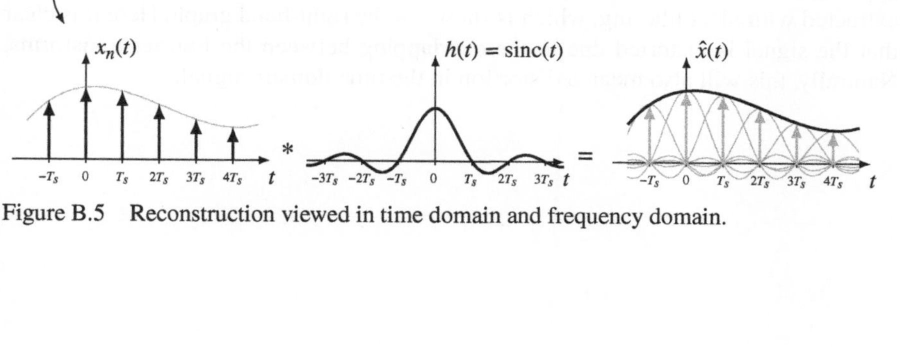

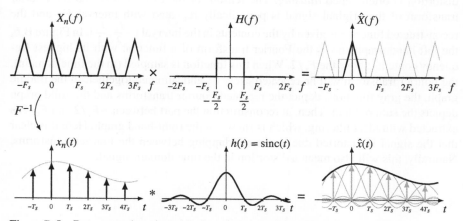

Figure B.5 Reconstruction viewed in time domain and frequency domain.

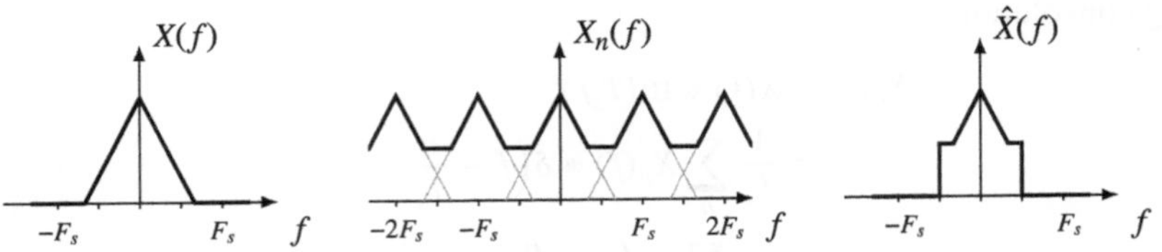

Figure B.6 Sampling below the Nyquist rate, resulting in aliasing.

To see what the reconstruction corresponds in the time domain, the lower row in Figure B.5 shows the inverse Fourier transforms. The periodic repetition of the frequency pulses corresponds to the sampled function $x_n(t)$, and the multiplication a convolution. The ideal low-pass filter is a box in the frequency domain thus a sinc in the time domain,

$$h(t) = \text{sinc}(t) = \frac{\sin(2\pi Wt)}{2\pi Wt} \tag{B.16}$$

Since the Dirac pulse is the identity for convolution, a pulse in the $x_n(t)$ will be replaced by a sinc function with the same amplitude,

$$\hat{x}(t) = x_n(t) * h(t) = \text{sinc}(t) * \sum_n \delta(t - nT_S) \tag{B.17}$$

$$= \sum_n x_n\big(\text{sinc}(t) * \delta(t - nT_s)\big) \tag{B.18}$$

$$= \sum_n x_n \,\text{sinc}(t - nT_s) \tag{B.19}$$

This completes the proof.

When the sampling frequency is chosen below the Nyquist rate, i.e. $F - s < 2W$, the reconstructed signal will be a distorted version of the original signal. This distortion is often called *aliasing*. The reason for the distortion is that the Fourier transform of the original signal is periodically repeated with interval F_s, and the reconstructed function is given by the contents in the interval $[-\frac{F_s}{2}, \frac{F_s}{2}]$. In Figure B.6, the left-hand graph shows the Fourier transform of a function with its highest frequency contents above the $F_s/2$. When this function is sampled (in the time domain), the Fourier transform will be repeated every F_s and summed together. In the middle graph, the gray functions depict the repeated Fourier transforms and the solid graph depicts the sum of them. Then, at reconstruction the part between $-F_s/2$ and $F_s/2$ is extracted with ideal filtering, which is shown in the right-hand graph. Here it is clear that the signal is distorted due to the overlapping between the Fourier transforms. Naturally, this will also mean a distortion in the time domain signal.

BIBLIOGRAPHY

[1] C. E. Shannon. A mathematical theory of communication. *Bell Syst. Tech. J.*, vol. 27, pp. 379–423, 623–656, 1948.

[2] H. Nyquist. Certain factors affecting telegraph speed. *Bell Syst. Tech. J.*, vol. 3, pp. 324–346, 1924.

[3] H. Nyquist. Certain topics in telegraph transmission theory. *Trans. AIEE*, vol. 47, pp. 617–644, Apr. 1928.

[4] R. V. L. Hartley. Transmission of information. *Bell Syst. Tech. J.*, vol. 7, pp. 535–563, July 1928.

[5] A. Gut. *An Intermediate Course in Probability*. Springer-Verlag, Berlin, Germany, 1995.

[6] A. Papoulis. *Probability, Random Variables, and Stochastic Processes*, 3rd ed. McGraw-Hill, New York, 1991.

[7] R. Bracewell. *The Fourier Transform and Its Applications*. McGraw-Hill, New York, 1999.

[8] R. A. Horn and C. R. Johnson. *Matrix Analysis*. Cambridge University Press, Cambridge, UK, 1985.

[9] S. Kullback and R. A. Leibler. On information and sufficiency. *Ann. Math. Stat.*, vol. 22, no. 1, pp. 79–86, Mar. 1951.

[10] J. Lin. Divergence measures based on the Shannon entropy. *IEEE Trans. Inform. Theory*, vol. 37, pp. 145–151, 1991.

[11] T. M. Cover and J. A. Thomas. *Elements of Information Theory*, 2nd ed. John Wiley & Sons, Inc., Hoboken, NJ, 2006.

[12] K. Sayood. *Introduction to Data Compression*. Elsevier, Amsterdam, the Netherlands, 2006.

[13] R. G. Gallager. *Information Theory and Reliable Communication*. John Wiley & Sons, Inc., New York, 1968.

[14] J. L. Massey. *Lecture Notes: Applied Digital Information Theory I and II*. Available at http://www.isiweb.ee.ethz.ch/archive/massey_scr/, 1980–1998.

[15] R. B. Ash. *Information Theory*. Dover Publication, Mineola, NY, 1990 (first published by John Wiley and Sons, 1965).

[16] L. Kraft. *A device for quantizing, grouping, and coding amplitude modulated pulses* [master's thesis]. Massachusetts Institute of Technology, Cambridge, MA, 1949.

[17] B. McMillan. Two inequalities implied by unique decipherability. *IEEE Trans. Inform. Theory*, vol. 2, pp. 115–116, 1956.

[18] R. K. Sundaram. *A First Course in Optimization Theory*. Cambridge University Press, Cambridge, UK, 1996.

[19] B. McMillan. The basic theorems of information theory. *Ann. Math. Stat.*, vol. 24, pp. 196–219, 1953.

[20] D. A. Huffman. A method for the construction of minimum redundancy codes. *Proc. Inst. Radio Eng.*, vol. 40, pp. 1098–1101, 1952.

[21] N. Faller. An adaptive system for data compression. In *Record of 7th Asilomar Conference on Circuits, Systems, and Computers*, 1973, pp. 593–597.

[22] R. G. Gallager. Variations on a theme by Huffman. *IEEE Trans. Inform.Theory*, vol. IT-24, pp. 668–674, 1978.

[23] D. E. Knuth. Dynamic Huffman coding. *J. Algorithms*, vol. 6, pp. 163–180, 1985.

[24] J. S. Vitter. Design and analysis of dynamic Huffman codes. *J. ACM*, vol. 34, no. 4, pp. 825–845, Oct. 1987.

Information and Communication Theory, First Edition. Stefan Höst.

[25] J. Ziv and A. Lempel. A universal algorithm for sequential data compression. *IEEE Trans. Inform. Theory*, vol. 23, pp. 337–343, 1977.
[26] J. Ziv and A. Lempel. Compression of individual sequences via variable-rate coding. *IEEE Trans. Inform. Theory*, vol. 24, pp. 530–536, 1978.
[27] J. A. Storer and T. G. Szymanski. Data compression via textual substitution. *J. ACM*, vol. 29, pp. 928–951, 1982.
[28] T. Welch. A technique for high-performance data compression. *IEEE Comput.*, vol. 17, pp. 8–19, 1984.
[29] P. Deutsch. RFC 1951 DEFLATE compressed data format specification version 1.3, May 1996.
[30] W3C Recommendation, 10 November, 2003, Portable network graphics (png) specification (second edition) [online]. Available at https://www.w3.org/TR/PNG/.
[31] G. Roelofs. *PNG The Definite Guide*. O'Reilly & Associates, Sebastopol, CA, 2003.
[32] R. Johannesson. *Informationsteori – grundvalen för (tele-) kommunikation*. Studentlitteratur, Lund, Sweden, 1988 (in Swedish).
[33] W. C. Huffman and V. Pless. *Fundamentals of Error-Correcting Codes*. Cambridge University Press, Cambridge, UK, 2003.
[34] R. Johannesson and K. Zigangirov. *Fundamentals of Convolutional Codes*. IEEE Press, Piscataway, NJ, 1999.
[35] S. Lin and D. J. Costello. *Error Control coding*, 2nd ed. Prentice Hall, Upper Saddle River, NJ, 2004.
[36] F. J. MacWilliams and N. J. A. Sloane. *The Theory of Error-Correcting Codes*. North Holland, Amsterdam, the Netherlands, 1977.
[37] ITU-T Recommendation G.992.5, Asymmetric digital subscriber line 2 transceivers (ADSL2) Extended bandwidth ADSL2 (ADSL2plus), 2009.
[38] Recommendation G.993.2, Very high speed digital subscriber line transceivers 2 (VDSL2), 2015.
[39] H. Robbins. A remark on Stirling's formula. *Am. Math. Monthly*, vol. 62, pp. 26–29, Jan. 1955.
[40] P. Elias. Coding for noisy channels. *IRE Conv. Rec.*, Pt. 4, pp. 37–46, Mar. 1955.
[41] A. J. Viterbi. Error bounds for convolutional codes and an asymptotically optimum decoding algorithm. *IEEE Trans. Inform. Theory*, vol. IT-13, pp. 260–269, Apr. 1967.
[42] G. D. Forney Jr. The Viterbi algorithm. *Proc. IEEE*, vol. 61, pp. 268–278, 1973.
[43] J. A. Heller. Short constraint length convolutional codes. *Jet Propulsion Laboratory, Space Program Summary 37–54*, vol. III, pp. 171–177, Oct./Nov. 1968.
[44] L. Bahl, J. Cocke, F. Jelinek, and J. Raviv. Optimal decoding of linear codes for minimizing symbol error rate. *IEEE Trans. Inform. Theory*, vol. IT-20, pp. 284–287, Mar. 1974.
[45] I. Reed and G. Solomon. Polynomial codes over certain finite fields. *J. Soc. Ind. Appl. Math.*, vol. 8, no. 2, pp. 300–304, 1960.
[46] G. D. Forney Jr. *Concatenated Codes*. MIT Press, Cambridge, MA, 1966.
[47] C. Berrou, A. Glavieux, and P. Thitimajshima. Near Shannon limit error-correction. In *ICC'93*, Geneva, Switzerland, May 1993, pp. 1064–1070.
[48] R. G Gallager. *Low-Density Parity-Check Codes*. MIT Press, Cambridge, MA, 1963.
[49] D. J. Costello. Free distance bounds for convolutional codes. *IEEE Trans. Inform. Theory*, vol. 20, pp. 356–365, 1974.
[50] D. Bertsekas and R. Gallager. *Data Networks*, 2nd ed. Prentice Hall, Englewood Cliffs, NJ, 1992. Available at web.mit.edu/dimitrib/www/datanets.html.
[51] J. G. Proakis. *Digital Communications*, 4th ed. McGraw-Hill, New York, 2001.
[52] I. S. Gradshteyn and I. M. Ryzhik. *Table of Intergrals, Series and Products*, 7th ed. Academic Press, New York, 2007.
[53] C. E. Shannon. Coding theorems for a discrete source with a fidelity criterion. *IRE National Convention Record*, Part 4, pp. 142–163, 1959.
[54] R. McEliece. *The Theory of Information and Coding*. Cambridge University Press, Cambridge, UK, 2004.
[55] V. K. Goyal. Theoretical foundation of transform coding. *IEEE Signal Process. Mag.*, vol. 18, pp. 9–21, 2001.

[56] K. Karhunen. Über lineare methoden in der Wahrscheinlichkeitsrechnung. *Ann. Acad. Sci. Fennicae. Ser. A. I. Math.-Phys.*, vol. 37, pp. 1–79, 1947.

[57] M. Loeve. Fonctions aleatoires de second ordre. *Processus Stochastiques et Mouvement Brownien*, 1948.

[58] H. Hetelling. Analysis of a complex of statistical variables into principal components. *J. Educational Psychol.*, vol. 24, pp. 417–441 and 498–520, 1933.

[59] ITU-T Rec. T.81 (09/92) and ISO/IEC 10918-1:1994. Information technology – Digital compression and coding of continuous-tone still images – Requirements and guidelines.

[60] JPEG Group. http://www.jpeg.org/.

[61] W. B. Pennebaker and J. L. Mitchell. *JPEG Still Image Data Compression Standard*. Kluwer Academic Publishers, Dordrecht, the Netherlands, 1992.

[62] N. Ahmed, T. Natarajan, and K. R. Rao. Discrete cosine transform. *IEEE Trans. Comput.*, vol. 23, pp. 90–93, 1974.

[63] A. C. G. Verdugo and P. N. Rathie. On the entropy of continuous probability distributions. *IEEE Trans. Inform. Theory*, vol. IT-24, no. 1, pp. 120–122, Jan. 1978.

[64] C. E. Shannon. Communication in the presence of noise. *Proc. IRE*, vol. 37, pp. 10–21, 1949.

[65] V. A. Kotelnikov. On the transmission capacity of the ether and of cables in electrical communications. In *Proceedings of the first All-Union Conference on the Technological Reconstruction of the Communications Sector and the Development of Low-Current Engineering*, Moscow, 1933 (translated from Russian).

[66] E. T. Whittaker. On the functions which are represented by the expansions of the interpolation theory. *Proc. Royal Soc. Edinburgh.*, vol. 35, pp. 181–194, 1915.

[67] N. Abramson. *Information Theory and Coding*. McGraw-Hill, New York, 1963.

[68] T. Berger and J. D. Gibson. Lossy source coding. *IEEE Trans. Inform. Theory*, vol. 44, pp. 2693–2723, 1998.

[69] I. Bocharova. *Compression for Multimedia*. Cambridge University Press, Cambridge, UK, 2009.

[70] M. Bossert. *Channel Coding for Telecommunications*. John Wiley & Sons, Inc., New York, 1999.

[71] A. el Gamal and T. Cover. Multiple user information theory. *Proc. IEEE*, vol. 68, pp. 1466–1483, 1980.

[72] G. D. Forney Jr. Review of random tree codes. Technical report, Appendix A, Final Report, Contract NAS2-3637, NASA CR73176, NASA Ames Research Center, December 1967.

[73] G. D. Forney Jr. and G. Ungerboeck. Modulation and coding for linear Gaussian channels. *IEEE Trans. Inform. Theory*, vol. 44, pp. 2384–2415, 1998.

[74] G. D. Forney Jr. and L.-F. Wei. Multidimensional constellations—Part I: Introduction, figures of merit, and generalized cross constellations. *IEEE J. Sel. Areas Commun.*, vol. 7, pp. 877–892, 1989.

[75] R. G. Gallager. Circularly-symmetric Gaussian random vectors [online]. Available at www.rle.mit.edu/rgallager/documents/CircSymGauss.pdf, January 1, 2008.

[76] R. G. Gallager. *Principles of Digital Communication*. Cambridge University Press, Cambridge, UK, 2008.

[77] Gnu.org. https://www.gnu.org/philosophy/gif.html. Last visited: 2015-02-25.

[78] R. M. Gray. *Source Coding Theory*. Kluwer Academic Publishers, Dordrecht, the Netherlands, 1990.

[79] R. Hamming. Error detecting and error correcting codes. *Bell Syst. Tech. J.*, vol. 29, pp. 147–160, 1950.

[80] B. Kudryashov. *Theori Informatii*. Piter, St. Petersburg, Russia, 2009 (in Russian).

[81] S. Kullback. *Information Theory and Statistics*. Dover Publications, Mineola, NY, 1997 (first published by John Wiley and Sons in 1959).

[82] E. A. Lee and D. G. Messerschmitt. *Digital Communication*. Kluwer Academic Publishers, Dordrecht, the Netherlands, 1994.

[83] R. McEliece. *Finite Fields for Computer Scientists and Engineers*. Springer, Berlin, Germany, 1986.

[84] J. R. Pierce. *An Introduction to Information Theory*. Dover Publications, Mineola, NY, 1980 (first published by Harper and Brothers in 1961).

[85] J. G. Proakis and D. G. Manolakis. *Digital Signal Processing*. Pearson Prentice Hall, Upper Saddle River, NJ, 2007.

[86] F. M. Reza. *An Introduction to Information Theory*. McGraw-Hill, New York, 1961.
[87] S. Ross. *Introduction to Probability Models*, 10th ed. Elsevier, Amsterdam, the Netherlands, 2010.
[88] K. Sayood. *Lossless Compression Handbook*. Academic Press, New York, 2003.
[89] S. Verdu. The exponential distribution in information theory. *Problems Inform. Transmission*, vol. 32, pp. 86–95, 1996.
[90] A. J. Viterbi and J. K. Omura. *Principles of Digital Communication and Coding*. McGraw Hill, New York, 1979.
[91] J. M. Wozencraft and I. M. Jacobs. *Principles of Digital Communication Engineering*. John Wiley and Sons, Inc., New York, 1965.
[92] J. F. Young. *Information Theory*. Butterworth & Co., London, 1971.

INDEX

Information and Communication Theory, First Edition. Stefan Höst.

IEEE PRESS SERIES ON DIGITAL AND MOBILE COMMUNICATION

John B. Anderson, *Series Editor*
University of Lund

---

1. *Future Talk: The Changing Wireless Game*
   Ron Schneiderman
2. *Digital Transmission Engineering*
   John B. Anderson
3. *Fundamentals of Convolutional Coding*
   Rolf Johannesson and Kamil Sh. Zigangirov
4. *Mobile and Personal Communication Services and Systems*
   Raj Pandya
5. *Wireless Video Communications: Second to Third Generation and Beyond*
   Lajos Hanzo, Peter J. Cherriman, and Jürgen Streit
6. Wireless Communications in the 21st Century
   Mansoor Shafi, Shigeaki Ogose, and Takeshi Hattori
7. *Introduction to WLLs: Application and Deployment for Fixed and Broadband Services*
   Raj Pandya
8. *Trellis and Turbo Coding*
   Christian B. Schlegel and Lance C. Perez
9. *Theory of Code Division Multiple Access Communication*
   Kamil Sh. Zigangirov
10. *Digital Transmission Engineering,* Second Edition
    John B. Anderson
11. *Wireless LAN Radios: System Definition to Transistor Design*
    Arya Behzad
12. *Wireless Broadband: Conflict and Convergence*
    Vern Fotheringham and Chetan Sharma
13. *Millimeter Wave Communication Systems*
    Kao-Cheng Huang and Zhaocheng Wang
14. *Channel Equalization for Wireless Communications: From Concepts to Detailed Mathematics*
    Gregory E. Bottomley
15. *Handbook of Position Location: Theory, Practice and Advances*
    Seyed A. Reza Zekavat and R. Michael Buehrer
16. *Digital Filters: Principles and Applications with MATLAB*
    Fred J. Taylor
17. *Resource Allocation in Uplink OFDMA Wireless Systems: Optimal Solutions and Practical Implementations*
    Elias Yaacoub and Zaher Dawy

18. *Non-Gaussian Statistical Communication Theory*
David Middleton
19. *Frequency Stability: Introduction & Applications*
Venceslav F. Kroupa
20. *Mobile Ad Hoc Networking: Cutting Edge Directions*, Second Edition
Stefano Basagni, Marco Conti, Silvia Giordano, and Ivan Stojmenovic
21. *Surviving Mobile Data Explosion*
Dinesh C. Verma and Paridhi Verma
22. *Cellular Communications: A Comprehensive and Practical Guide*
Nishith Tripathi and Jeffrey H. Reed
23. *Fundamentals of Convolutional Coding*, Second Edition
Rolf Johannesson and Kamil Sh. Zigangirov
24. *Trellis and Turbo Coding*, Second Edition
Christian B. Schlegel and Lance C. Perez
25. *Problem-Based Learning in Communication Systems Using MATLAB and Simulink*
Kwonhue Choi and Huaping Liu
26. *Bandwidth Efficient Coding*
John B. Anderson
27. *Visible Light Communications: Modulation and Signal Processing*
Zhaocheng Wang, Qi Wang, Wei Huang, and Zhengyuan Xu
28. *Switch/Router Architectures: Shared-Bus and Shared-Memory Based Systems*
James Aweya
29. *Handbook of Position Location: Theory, Practice, and Advances*, Second Edition
Edited by S. A. (Reza) Zekavat and R. Michael Buehrer
30. *Information and Communication Theory*
Stefan Höst